IMPEDANCE SPECTROSCOPY

IMPEDANCE SPECTROSCOPY
Applications to Electrochemical and Dielectric Phenomena

Vadim F. Lvovich

A JOHN WILEY & SONS, INC., PUBLICATION

Published by John Wiley & Sons, Inc., Hoboken, New Jersey
Published simultaneously in Canada

For general information on our other products and services or for technical support, please contact our Customer Care Department within the United States at (800) 762-2974, outside the United States at (317) 572-3993 or fax (317) 572-4002.

Wiley also publishes its books in a variety of electronic formats. Some content that appears in print may not be available in electronic formats. For more information about Wiley products, visit our web site at www.wiley.com.

Library of Congress Cataloging-in-Publication Data:

Lvovich, Vadim F., 1967–
  Impedance spectroscopy with application to electrochemical and dielectric phenomena / Vadim F. Lvovich.
    p. cm.
  Includes index.
   ISBN 978-0-470-62778-5 (hardback)
  1. Impedance spectroscopy. 2. Electrochemistry. I. Title.
  QD116.I57L86 2012
  543'.4—dc23   2011028940

Printed in the United States of America.

10 9 8 7 6 5 4

Contents

Preface

Since its conceptual introduction in the late 19th century, the impedance spectroscopy has undergone a tremendous evolution into a rich and vibrant multidisciplinary science. Over the last decade Electrochemical Impedance Spectroscopy (EIS) has become established as one of the most popular analytical tools in materials research. The technique is being widely and effectively applied to a large number of important areas of materials research and analysis, such as corrosion studies and corrosion control; monitoring of properties of electronic and ionic conducting polymers, colloids and coatings; measurements in energy storage, batteries, and fuel cells-related systems; biological analysis and biomedical sensors; measurements in semiconductors and solid electrolytes; studies of electrochemical kinetics, reactions and processes. Impedance spectroscopy is a powerful technique for investigating electrochemical systems and processes. EIS allows to study, among others, such processes as adsorption, charge- and mass-transport, and kinetics of coupled sequential and parallel reactions.

In a broader sense, EIS is an extraordinarily versatile, sensitive, and informative technique broadly applicable to studies of electrochemical kinetics at electrode-media interfaces and determination of conduction mechanisms in various materials through bound or mobile electronic, ionic, semiconductor, and mixed charges. Impedance analysis is fundamentally based on a relatively simple electrical measurement that can be automated and remotely controlled. Its main strength lies in its ability to interrogate relaxation phenomena whose time constants ranging over several orders of magnitude from minutes down to microseconds. In contrast to other analytical techniques, EIS is noninvasive technique that can be used for on-line analysis and diagnostics. The method offers the most powerful on-line and off-line analysis of the status of electrodes, monitors and probes in many different complex time- and space-resolved processes that occur during electrochemical experiments. For instance, the EIS technique has been broadly practiced in the development of sensors for monitoring rates of materials' degradation, such as metal corrosion and biofouling of implantable medical devices.

EIS is useful as an empirical quality-control procedure that can also be employed to interpret fundamental electrochemical and electronic processes. Experimental impedance results can be correlated with many practically useful chemical, physical, mechanical, and electrical variables. With the current availability of ever evolving automated impedance equipment covering broad frequency and potential ranges, the EIS studies have become increasingly popular as more and more electrochemists, material scientists, and engineers understand the theoretical basis for impedance spectroscopy and gain skill in the impedance data interpretation.

The impedance technique appears destined to play an increasingly important role in fundamental and applied electrochemistry and material science in the coming years. However, broader practical utilization of EIS has been hindered by the lack of comprehensive and cohesive explanation of the theory, measurements, analysis techniques, and types of acquired data for different investigated systems. These factors may be connected with the fact that existing literature reviews of EIS are very often difficult to understand by non-specialists. As will be shown later, the ambiguity of impedance data interpretation and the establishment of direct relationships with practical physical, chemical, electrical, and mechanical parameters constitute the main disadvantages of the technique. These general weaknesses are amplified especially when considering a great variety of practical impedance applications, where a practical investigator or researcher often has to decide if any of the previously known impedance response models and their interpretations are even remotely applicable to the problem in hand. EIS data demonstrate the investigated system's response to applied alternating or direct electrical fields. It becomes the investigators' responsibility to convert the electrical data into parameters of interest, whether it is a concentration of bioanalyte, corrosion rate of metal surfaces, performance characteristics of various components of a fuel cell, or rate of oxidative decomposition of polymer films.

As industrial scientist and engineer with career encompassing multiple senior technology and product development positions in leading R&D divisions in Specialty Chemicals, Electronics, BioMedical, and Aerospace industrial corporations, the author has learned over the years to greatly appreciate the investigative power and flexibility of EIS and impedance-based devices in commercial product development. This book was born out of acute need to catalog and explain multiple variations of the EIS data characteristics encountered in many different practical applications. Although the principle behind the method remains the same, the impedance phenomena investigation in different systems presents a widely different data pattern and requires significant variability in the experimental methodology and interpretation strategy to make sense of the results. The EIS experimental data interpretation for both unknown experimental systems, and well-known systems investigated by other (non–electrochemical) means is widely acknowledged to be the main source of the method's application challenge, often listed at the main impediment to the method's broader penetration into scientific and technological markets. This book attempts to at least partially standardize the catalog of EIS responses across many practically encountered fields of use and to present a coherent approach to the analysis of experimental results.

This book is intended to serve as a reference on the topic of practical applications of impedance spectroscopy, while also addressing some of the most basic aspects of EIS theory. The theory of the impedance spectroscopy has been presented in great details and with remarkable skill in well-received monographs by J. R. MacDonald, and recently by M. Orazem and B. Tribollet, as well as in many excellent review chapters referenced in this book. There are a number of short courses, several monographs and many independent publications on the impedance spectroscopy. However, the formal courses on the topic are rarely offered in the university settings. At the same time, there is a significant worldwide need to offer independent, direct and comprehensive training on practical applications of the impedance analysis to many industrial scientists and engineers relatively unfamiliar with the EIS theory but eager to apply impedance analysis to address their everyday product development technological challenges. This manuscript emphasizes practical applications of the impedance spectroscopy. This book in based around a catalogue representing a typical impedance data for large variety of established, emerging, and non-conventional experimental systems; relevant mathematical expressions; and physical and chemical interpretation of the experimental results. Many of these events are encountered in the field by industrial scientists and engineers in electrochemistry, physical and analytical chemistry, and chemical engineering.

This book attempts to present a balance of theoretical considerations and practical applications for problem solving in several of the most widely used fields where electrochemical impedance spectroscopy analysis is being employed. The goal was to produce a text that would be useful to both the novice and the expert in EIS. It is primarily intended for industrial researchers (material scientists, analytical and physical chemists, chemical engineers, material researchers), and applications scientists, wishing to understand how to correctly make impedance measurements, interpret the results, compare results with expected previously published results form similar chemical systems, and use correct mathematical formulas to verify the accuracy of the data. A majority of these individuals reside in the specialty chemicals, polymers, colloids, electrochemical renewable energy and power sources, material science, electronics, biomedical, pharmaceutical, personal care and other smaller industries. The book intends to provide a working background for the practical scientist or engineer who wishes to apply EIS as a method of analysis without needing to become an expert electrochemist. With that in mind, both somewhat oversimplified electrochemical models and in-depth analysis of specific topics of common interest are presented. The manuscript covers many of the topics needed to help readers identify whether EIS may be an appropriate method for their particular practical application or research problem. A number of practical examples and graphical representations of the typical data in the most common practical experimental systems are presented. In that respect the book may also be addressed to students and researchers who may found the presented catalog of impedance phenomenological data and the relevant discussions to be of assistance in their introduction to theoretical and practical aspects of electrochemical research.

Starting with general principles, the book emphasizes practical applications of the electrochemical impedance spectroscopy to separate studies of bulk solution and interfacial processes, using of different electrochemical cells and equipment for experimental characterization of different systems. The monograph provides relevant examples of characterization of large variety of materials in electrochemistry, such as polymers, colloids, coatings, biomedical species, metal oxides, corroded metals, solid-state devices, and electrochemical power sources. The book covers many of the topics needed to help readers identify whether impedance spectroscopy may be an appropriate method for their particular research problem.

This book incorporates the results of the last two decades of research on the theories and applications of impedance spectroscopy, including more detailed reviews of the impedance methods applications in industrial colloids, biomedical sensors and devices, and supercapacitive polymeric films. The book is organized so each chapter stands on its own. The book should assist readers to quickly grasp how to apply their new knowledge of impedance spectroscopy methods to their own research problems through the use of features such as:

- Equations and circuit diagrams for the most widely used impedance models and applications
- Figures depicting impedance spectra of typical materials and devices
- Theoretical considerations for dealing with modeling, equivalent circuits, and equations in the complex domain
- Best measurement methods for particular systems and alerts to potential sources of errors
- Review of impedance instrumentation
- Review of related techniques and impedance spectroscopy modifications
- Extensive references to the scientific literature for more information on particular topics and current research

It is hoped that the more advanced reader will also find this book valuable as a review and summary of the literature on the subject. Of necessity, compromises have been made between depth, breadth of coverage, and reasonable size. Many of the subjects such as mathematical fundamentals, statistical and error analysis, and a number of topics on electrochemical kinetics and the method theory have been exceptionally well covered in the previous manuscripts dedicated to the impedance spectroscopy. Similarly the book has not been able to accommodate discussions on many techniques that are useful but not widely practiced. While certainly not nearly covering the whole breadth of the impedance analysis universe, the manuscript attempts to provide both a convenient source of EIS theory and applications, as well as illustrations of applications in areas possibly unfamiliar to the reader. The approach is first to review the fundamentals of electrochemical and material transport processes as they are related to the material properties analysis by impedance / modulus / dielectric spectroscopy (Chapter 1), discuss the data representation (Chapter 2) and modeling (Chapter 3) with relevant examples (Chapter 4). Chapter 5 discusses separate components of the impedance circuit, and Chapters 6 and 7 present several typical examples of combining these components into practically encountered complex distributed systems. Chapter 8 is dedicated to the EIS equipment and experimental design. Chapters 9 through 12 are dedicated to detailed discussions of impedance analysis applications to specific experimental systems, representing both well-studied and emerging fields. Chapter 13 offers a brief review of EIS modifications and closely related analytical methods.

I owe thanks to many others who have helped with this project. I am especially grateful to John Wiley & Sons, Inc. and Lone Wolf Enterprises, Ltd. for their conscientious assistance with many details of preparation and production. Over the years many valuable comments and encouragement have been provided by colleagues through the electrochemical community who assured that there would be a demand for this book. I also would like to thank my wife Laura and my son William for affording me the time and freedom required to undertake such a project.

CHAPTER 1

Fundamentals of Electrochemical Impedance Spectroscopy

1.1. Concept of complex impedance

The concept of electrical impedance was first introduced by Oliver Heaviside in the 1880s and was soon afterward developed in terms of vector diagrams and complex numbers representation by A. E. Kennelly and C. P. Steinmetz [1, p. 5]. Since then the technique has gained in exposure and popularity, propelled by a series of scientific advancements in the field of electrochemistry, improvements in instrumentation performance and availability, and increased exposure to an ever-widening range of practical applications.

For example, the development of the double-layer theory by Frumkin and Grahame led to the development of the equivalent circuit (EC) modeling approach to the representation of impedance data by Randles and Warburg. Extended studies of electrochemical reactions coupled with diffusion (Gerisher) and adsorption (Eppelboin) phenomena, effects of porous surfaces on electrochemical kinetics (de Levie), and nonuniform current and potential distribution dispersions (Newman) all resulted in a tremendous expansion of impedance-based investigations addressing these and other similar problems [1]. Along with the development of electrochemical impedance theory, more elaborate mathematical methods for data analysis came into existence, such as Kramers-Kronig relationships and nonlinear complex regression [1, 2]. Transformational advancements in electrochemical equipment and computer technology that have occurred over the last 30 years allowed for digital automated

1

impedance measurements to be performed with significantly higher quality, better control, and more versatility than what was available during the early years of EIS. One can argue that these advancements completely revolutionized the field of impedance spectroscopy (and in a broader sense the field of electrochemistry), allowing the technique to be applicable to an exploding universe of practical applications. Some of these applications, such as dielectric spectroscopy analysis of electrical conduction mechanisms in bulk polymers and biological cell suspensions, have been actively practiced since the 1950s [3, 4]. Others, such as localized studies of surface corrosion kinetics and analysis of the state of biomedical implants, have come into prominence only relatively recently [5, 6, 7, 8].

In spite of the ever-expanding use of EIS in the analysis of practical and experimental systems, impedance (or complex electrical resistance, for a lack of a better term) fundamentally remains a simple concept. Electrical resistance R is related to the ability of a circuit element to resist the flow of electrical current. Ohm's Law (Eq. 1-1) defines resistance in terms of the ratio between input voltage V and output current I:

$$R = \frac{V}{I} \qquad (1\text{-}1)$$

While this is a well-known relationship, its use is limited to only one circuit element—the ideal resistor. An ideal resistor follows Ohm's Law at all current, voltage, and AC frequency levels. The resistor's characteristic resistance value R [ohm] is independent of AC frequency, and AC current and voltage signals though the ideal resistor are "in phase" with each other. Let us assume that the analyzed sample material is ideally homogeneous and completely fills the volume bounded by two external current conductors ("electrodes") with a visible area A that are placed apart at uniform distance d, as shown in Figure 1-1. When external voltage V is applied, a uniform current I passes through the sample, and the resistance is defined as:

$$R = \rho \frac{d}{A} \qquad (1\text{-}2)$$

where ρ [ohm cm] is the characteristic electrical resistivity of a material, representing its ability to resist the passage of the current. The inverse of resistivity is conductivity σ [1 / (ohm cm)] or [Sm/cm], reflecting the material's ability to conduct electrical current between two bounding electrodes.

An ideal resistor can be replaced in the circuit by another ideal element that completely rejects any flow of current. This element is referred as an "ideal" capacitor (or "inductor"), which stores magnetic energy created by an applied electric field, formed when two bounding electrodes are separated by a non-conducting (or "dielectric") medium. The AC current and voltage signals though the ideal capacitor are completely "out of phase" with each other, with current following voltage. The value of the capacitance presented in Farads [F] depends on the area of the electrodes A, the distance between the electrodes d, and the properties of the dielectric medium reflected in a "relative permittivity" parameter ε as:

$$C = \frac{\varepsilon_0 \varepsilon A}{d}$$
(1-3)

where ε_0 = constant electrical permittivity of a vacuum $(8.85 \cdot 10^{-14}\,\text{F/cm})$. The relative permittivity value represents a characteristic ability of the analyzed material to store electrical energy. This parameter (often referred to as simply "permittivity" or "dielectric constant") is essentially a convenient multiplier of the vacuum permittivity constant ε_0 that is equal to a ratio of the material's permittivity to that of the vacuum. The permittivity values are different for various media: 80.1 (at 20°C) for water, between 2 through 8 for many polymers, and 1 for an ideal vacuum. A typical EIS experiment, where analyzed material characteristics such as conductivity, resistivity, and permittivity are determined, is presented in Figure 1-1.

Impedance is a more general concept than either pure resistance or capacitance, as it takes the phase differences between the input voltage and output current into account. Like resistance, impedance is the ratio between voltage and current, demonstrating the ability of a circuit to resist the flow of electrical current, represented by the "real impedance" term, but it also reflects the ability of a circuit to store electrical energy, reflected in the "imaginary impedance" term. Impedance can be defined as a complex resistance encountered when current flows through a circuit composed of various resistors, capacitors, and inductors. This definition is applied to both direct current (DC) and alternating current (AC).

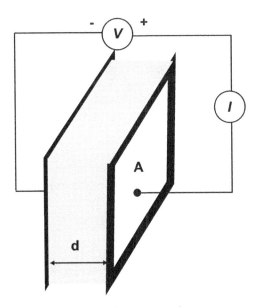

FIGURE 1-1   Fundamental impedance experiment

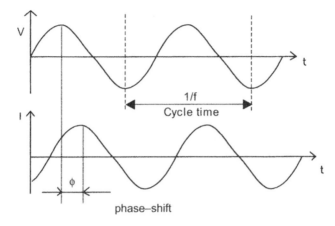

FIGURE 1-2 Impedance experiment: sinusoidal voltage input V at a single frequency f and current response I

In experimental situations the electrochemical impedance is normally measured using excitation AC voltage signal V with small amplitude V_A (expressed in volts) applied at frequency f (expressed in Hz or 1/sec). The voltage signal $V(t)$, expressed as a function of time t, has the form:

$$V(t) = V_A \sin(2\pi f t) = V_A \sin(\omega t) \tag{1-4}$$

In this notation a "radial frequency" ω of the applied voltage signal (expressed in radians/second) parameter is introduced, which is related to the applied AC frequency f as $\omega = 2\pi f$.

In a linear or pseudolinear system, the current response to a sinusoidal voltage input will be a sinusoid at the same frequency but "shifted in phase" (either forward or backward depending on the system's characteristics)—that is, determined by the ratio of capacitive and resistive components of the output current (Figure 1-2). In a linear system, the response current signal $I(t)$ is shifted in phase (ϕ) and has a different amplitude, I_A:

$$I(t) = I_A \sin(\omega t + \phi) \tag{1-5}$$

An expression analogous to Ohm's Law allows us to calculate the complex impedance of the system as the ratio of input voltage $V(t)$ and output measured current $I(t)$:

$$Z^* = \frac{V(t)}{I(t)} = \frac{V_A \sin(\omega t)}{I_A \sin(\omega t + \phi)} = Z_A \frac{\sin(\omega t)}{\sin(\omega t + \phi)} \tag{1-6}$$

The impedance is therefore expressed in terms of a magnitude (absolute value), $Z_A = |Z|$, and a phase shift, ϕ. If we plot the applied sinusoidal voltage signal on the x-axis of a graph and the sinusoidal response signal $I(t)$ on the y-axis, an oval known as a "Lissajous figure" will appear (Figure 1-3A).

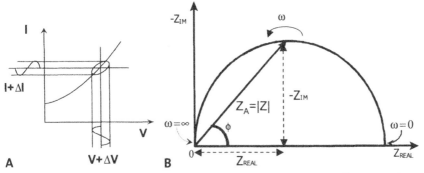

FIGURE 1-3 Impedance data representations: A. Lissajous figure; B. Complex impedance plot

Analysis of Lissajous figures on oscilloscope screens was the accepted method of impedance measurement prior to the availability of lock-in amplifiers and frequency response analyzers. Modern equipment allows automation in applying the voltage input with variable frequencies and collecting the output impedance (and current) responses as the frequency is scanned from very high (MHz-GHz) values where timescale of the signal is in micro- and nanoseconds to very low frequencies (µHz) with timescales of the order of hours.

Using Euler's relationship:

$$\exp(j\phi) = \cos\phi + j\sin\phi \tag{1-7}$$

it is possible to express the impedance as a complex function. The potential $V(t)$ is described as:

$$V(t) = V_A \, e^{j\omega t} \tag{1-8}$$

and the current response as:

$$I(t) = I_A \, e^{j\omega t - j\phi} \tag{1-9}$$

The impedance is then represented as a complex number that can also be expressed in complex mathematics as a combination of "real," or in-phase (Z_{REAL}), and "imaginary," or out-of-phase (Z_{IM}), parts (Figure 1-3B):

$$Z^* = \frac{V}{I} = Z_A e^{j\phi} = Z_A \, (\cos\phi + j\sin\phi) = Z_{REAL} + jZ_{IM} \tag{1-10}$$

and the phase angle ϕ at a chosen radial frequency ω is a ratio of the imaginary and real impedance components:

$$\tan\phi = \frac{Z_{IM}}{Z_{REAL}} \quad \text{or} \quad \phi = \arctan(\frac{Z_{IM}}{Z_{REAL}}) \tag{1-11}$$

1.2.   Complex dielectric, modulus, and impedance data representations

In addition to the AC inputs such as voltage amplitude V_A and radial frequency ω, impedance spectroscopy also actively employs DC voltage modulation (which is sometimes referred to as "offset voltage" or "offset electrochemical potential") as an important tool to study electrochemical processes. Alternative terms, such as "dielectric spectroscopy" or "modulus spectroscopy," are often used to describe impedance analysis that is effectively conducted only with AC modulation in the absence of a DC offset voltage (Figure 1-4).

Dielectric analysis measures two fundamental characteristics of a material—permittivity ε and conductivity σ (or resistivity ρ)—as functions of time, temperature, and AC radial frequency ω. As was discussed above, permittivity and conductivity are two parameters characteristic of respective abilities of analyzed material to store electrical energy and transfer electric charge. Both of these parameters are related to molecular activity. For example, a "dielectric" is a material whose capacitive current (out of phase) exceeds its resistive (in phase) current. An "ideal dielectric" is an insulator with no free charges that is capable of storing electrical energy. The Debye Equation (Eq. 1-12) relates the relative permittivity ε to a concept of material polarization density P [C/m²], or electrical dipole moment [C/m] per unit volume [m³], and the applied electric field V:

$$P = (\varepsilon - 1)\varepsilon_0 V \qquad (1\text{-}12)$$

Depending on the investigated material and the frequency of the applied electric field, determined polarization can be electronic and atomic (very small translational displacement of the electronic cloud in THz frequency range), orientational or dipolar (rotational moment experienced by permanently polar molecules in kHz-MHz frequency range), and ionic (displacement of ions with respect to each other in Hz-kHz frequency region).

The dielectric analysis typically presents the permittivity and conductivity material properties as a combined "complex permittivity" ε* parameter, which is analogous to the concept of complex impedance Z^* (Figure 1-4A). Just as complex impedance can be represented by its real and imaginary components, complex permittivity is a function of two parameters—"real" permittivity (often referred to as "permittivity" or "dielectric constant") ε' and imaginary permittivity (or "loss factor") ε" as:

$$\varepsilon^* = \varepsilon' - j\varepsilon'' \qquad (1\text{-}13)$$

In dielectric material ε' represents the alignment of dipoles, which is the energy storage component that is an inverse equivalent of Z_{IM}. ε" represents the ionic conduction component that is an inverse equivalent of Z_{REAL}. Both real permittivity and loss factor can be calculated from sample resistance R, conductivity σ, resistivity ρ, and capacitance C measured in a fundamental experimental setup (Figure 1-1) as:

$$\varepsilon' = \frac{Cd}{\varepsilon_0 A} \text{ and } \varepsilon'' = \frac{d}{RA\omega\varepsilon_0} = \frac{\sigma}{\omega\varepsilon_0} = \frac{1}{\rho\omega\varepsilon_0} \qquad (1\text{-}14)$$

Permittivity and conductivity values and their relative contributions to the measured voltage to current ratio (impedance) are often dependent on the material's temperature, external AC frequency, and magnitude of the applied voltage. In fact, real permittivity is often not quite appropriately referred to as the "dielectric constant," a parameter that should always be specified at a standard AC frequency (usually about 100 kHz) and temperature conditions (typically 25°C) and therefore is not exactly "constant." The concepts of "conductivity" and "resistivity" for a chosen material are also vague. These parameters have to be specified at standard temperature conditions and be carefully measured with full consideration of the impedance dependence on the applied electric field. In practice these rules are often not followed. For instance, conductivity of many solutions is often measured by hand-held meters operating at an arbitrary frequency around ~ 1kHz, which, as will be shown in Section 6-3, may or may not be appropriate conditions for many materials even when they belong to the same family (such as aqueous solutions). Alternatively, operating at much higher frequencies may result in the measurement being dominated by the out-of-phase capacitive impedance, which is a function of the sample's dielectric constant and not of its conductivity. For instance, saline (ρ = 100 ohm cm, ε = 80) is a conductor below 250MHz and a capacitor above 250MHz.

Dielectric spectroscopy, although using the same type of electrical information as impedance spectroscopy, is logically different in its analysis and approach to data representation. Dielectric response is based on a concept of "energy storage" and resulting "relaxation" per release of this energy by the system's individual components. Initially the concept of dielectric relaxation was introduced by Maxwell and expanded by Debye, who used it to describe the time required for dipolar molecules to reversibly orient themselves in the external AC electric field. In the experiments of Debye a step function excitation was applied to the system, and the system was allowed to "relax" to equilibrium after the excitation was removed. The time required for that process to take place was called "relaxation time" $\tau = 1/2\pi f_c$ that is inversely related to "critical relaxation frequency" f_c. Dielectric spectroscopy measures relaxation times by detecting frequency dependence of complex permittivity ε^* and determining f_c values from positions of the peaks in the $\varepsilon^* = f(f)$ plot as the input voltage signal is scanned over the experimental AC frequency range.

Dispersion, or frequency dependence, according to the laws of relaxation, is the corresponding frequency domain expression of complex permittivity ε^* as a function of radial (or cycling) frequency $\omega = 2\pi f$. For example, as the applied frequency ω is increased, a steplike decrease in complex permittivity is observed due to the fact that polarized molecules that are fully aligned with each change in direction of the AC field at lower frequencies cannot follow the higher frequency field at each direction reversal (Figure 1-4B). As the high-frequency AC field changes direction faster, these molecules "relax" to nonaligned positions where they cannot store energy. Large nonpolar molecules typically lose their orientation with the field at low Hz frequencies and have relaxation times on the order of seconds. Smaller and more polar ionic species "relax" at kHz-MHz frequencies and show millisecond to microsecond relaxation times. The components of a sample typically have high permittivity (capacitance) values at low frequencies where more different types of molecules can completely align with the field and store the maximum possible energy and are being effectively charged as dielectric dipoles. At high frequencies fewer dipoles store energy, and the total measured capacitance and permittivity of the system are low. Comparative analysis of dielectric dependencies of pure components presents an opportunity to identify and separate these species in complex mixed systems based on the AC frequency dependence of their dielectric response $\varepsilon^* = f(\omega)$.

The above frequency dependence of dielectric material properties, such as capacitance $C(\omega)$, permittivity $\varepsilon'(\omega)$, and conductivity $\sigma(\omega)$, can be expressed by Debye's "single relaxation" model [3, p. 65]. The Debye model is a popular representation used to illustrate bulk relaxation processes in ideal dielectrics, such as highly resistive polymers, where it is assumed that there is no conduction (or "loss") through the bulk material as the sample resistance R is infinitely high and conductivity $\sigma \rightarrow 0$ [3, 4]. This model is a classical representation of a simple dielectric or fully capacitive experimental system, where transition occurs from high-frequency permittivity ε_∞ (or capacitance C_2) to low-frequency permittivity ε_{LF} (or capacitance C_1) where $C_1 > C_2$ (Section 4-4). In the Debye model the response is ideally capacitive at both high and low frequency extremes, with the transition between the two regimes characterized by permittivity increment $\Delta\varepsilon = \varepsilon_{LF} - \varepsilon_\infty$. It is usually expressed for a medium permittivity ε as a function of the AC field's radial frequency ω, where τ is the characteristic "relaxation time" of the system. Expansion of Equation 1-13 leads to:

$$\varepsilon^*(\omega) = \varepsilon' - j\varepsilon'' = \varepsilon_\infty + \frac{\Delta\varepsilon}{1 + j\omega\tau} \text{ where } \varepsilon'(\omega) = \varepsilon_\infty + \frac{\Delta\varepsilon}{1 + \omega^2\tau^2} \text{ and } \varepsilon''(\omega) = \frac{\Delta\varepsilon\omega\tau}{1 + \omega^2\tau^2} \quad (1\text{-}15)$$

Complex permittivity of a more realistic "lossy" dielectric where non-zero parallel DC conductivity $\sigma(\omega)$ exists can be represented on the basis of a more complex Havriliak-Negami model. This model also accounts for non-idealities of both capacitive and resistive components accounting for the asymmetry and broadness of the dielectric dispersion curve and resulting frequency-dependent conductivity $\sigma(\omega)$ and permittivity $\varepsilon(\omega)$ contributions:

$$\varepsilon^*(\omega) = \left(\varepsilon_\infty + \frac{\Delta\varepsilon}{(1 + (j\omega\tau)^\alpha)^\beta} \right) - j \left(\frac{\sigma(\omega)}{\varepsilon_0\omega} \right)^N \quad (1\text{-}16)$$

Where: $\tau = 1/2\pi f_c$ = characteristic relaxation time, ω = radial frequency, N = parameter that defines the frequency dependence of the conductivity term (typically $N \rightarrow 1$ and equals the slope of the low-frequency increase in $\varepsilon^* = f(\omega)$ or $\varepsilon'' = f(\omega)$ plot due to the low-frequency conduction through the system, as shown in Figure 1-4B), α and β = shape parameters accounting for symmetric and asymmetric broadening of the relaxation peak.

In addition to the Debye model for dielectric bulk materials, other dielectric relaxations expressed according to Maxwell-Wagner or Schwartz "interfacial" mechanisms exist. For example, the Maxwell-Wagner "interfacial" polarization concept deals with processes at the interfaces between different components of an experimental system. Maxwell-Wagner polarization occurs

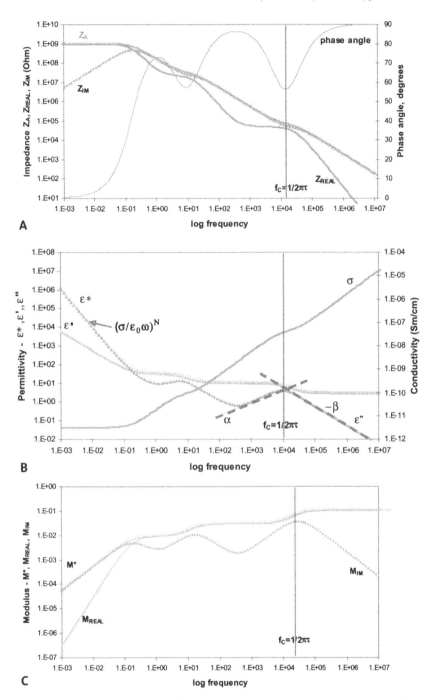

FIGURE 1-4   Representations of complex impedance data as function of AC frequency: A. impedance and phase angle; B. permittivity and conductivity; C. modulus

either at inner boundary layers separating two dielectric components of a sample or more often at an interface between the sample and an external ideal electrical conductor (electrode). In both cases this leads to a significant separation of charges over a considerable distance. This contribution of Maxwell-Wagner polarization to dielectric loss can be orders of magnitude larger than the molecular fluctuation's dielectric response described by the Debye mechanism. Maxwell-Wagner interfacial effects are prominent in electrochemical studies dealing with heterogeneous interfacial kinetics.

The Debye model is primarily describing a bulk material "dielectric" response. Traditional dielectric spectroscopy has found significant use in characterization of multicomponent resistive materials with mixed or particle-based conduction mechanisms, such as polymers, nonpolar organics (lubricants), and moderately resistive aqueous (cellular) colloids. However, there are relatively few practical cases of nearly ideal dielectric materials where more than one well-resolved dielectric relaxations can be identified at a constant temperature. The Maxwell-Wagner electrode-sample interface phenomenon and other interfacial effects (such as double layer charging and Faradaic kinetics) that result in apparent high interfacial capacitance masking settled capacitance changes in bulk material at frequencies below ~10 kHz are viewed as severe restricting factors in studies of dielectric materials. With the exception of extreme cases, such as ion-free insulating media with polarized particle conduction (such as electrorheological fluids) or subfreezing sample temperatures, AC frequency range relatively free of the effects of interfacial polarization is often limited to high kHz-GHz. In such a limited frequency range at a constant temperature the appearance of several types of conducting species showing significant frequency dependence, not interfering with one another and present in a significant and balanced range of concentrations, is a rare occurrence. Hence, dielectric analysis often relies not on the AC frequency but on wide temperature modulation at a few selected AC frequencies as the primary interrogation mode to extract details of sample analytical information.

Another representation of the dielectric properties of analyzed media is complex modulus M^* (Figure 1-4C). The modulus is the inverse of complex permittivity ε^* and can also be expressed as a derivative of complex impedance Z^*:

$$M^* = \frac{1}{\varepsilon^*} = M' - jM'' = j\omega\varepsilon_0 Z^* = -\omega\varepsilon_0 Z_{IM} + j\omega\varepsilon_0 Z_{REAL} \qquad (1\text{-}17)$$

Fundamentally, complex electrochemical impedance (Z^*), modulus (M^*), and permittivity (ε^*) parameters are all determined by applying an AC potential at a variable frequency and measuring output current through the sample (Figure 1-1). In a broader sense dielectric, modulus, and impedance analysis represent the same operational principles and can be referred to as subsets of a universal broadband electrochemical impedance spectroscopy (UBEIS). This technique analyzes both resistive and capacitive components of the AC current signal response, containing the excitation frequency and its harmonics. This current signal output can be analyzed as a sum of sinusoidal functions (a Fourier series). Depending on applied AC frequency and voltage, the output current can be supported by various conductive mechanisms through the analyzed system. These conductive mechanisms can be related to a single process, or be supported by a combination of various ionic, electronic, and particle conductors and their relative concentrations. The conduction process occurs both in bulk sample and at the interface between the sample and the electrodes, where a series of electron-exchanging reactions may take place. Critical relaxation frequencies can be determined from peak positions in complex impedance, modulus and permittivity plots.

The same data can be presented in modulus, dielectric, and impedance domains (Figure 1-4) and can be converted directly between the domains using expressions based on Maxwell equations. Although the data representing the electrochemical relaxation phenomenon fundamentally contain the same information and are independent of the chosen representation method, presenting the data separately as dielectric, impedance, and modulus plots often allows extracting additional useful information about the analyzed system. For instance, localized relaxations result in the peaks appearing at different frequencies in complex impedance or complex modulus vs. frequency plots, whereas long-range fundamental conductivity results in exact overlapping of the modulus and impedance peaks. Also, as will be shown later, these data representations have different resolving capabilities to present the results as a function of the applied experimental conditions.

Nevertheless, historically a differentiation exists between "dielectric" and "impedance" spectroscopies. Traditional dielectric analysis has been applied primarily to the analysis of bulk "dielectric" properties of polymers, plastics, composites, and nonaqueous fluids with very high bulk material resistance. The dielectric method is characterized by using higher AC voltage amplitudes, temperature modulation as an independent variable, lack of DC voltage perturbation, and often operating frequencies above 1 kHz or measurements at several selected discrete frequencies [2, p. 33].

Traditional impedance spectroscopy is preoccupied with investigating charge and material-exchange ("electrochemical") kinetic processes that occur at electrode-sample interfaces. This technique actively employs DC modulation just as most electrochemical techniques do, typically uses low AC voltage amplitudes to maintain linearity of signal response, and employs wide frequency ranges from MHz down to μHz. Unlike dielectric spectroscopy, the analytical aspect of traditional impedance analysis is based not on bulk material investigations but rather on quantification of sample species through relevant interfacial impedance parameters. The majority of traditional EIS studies emphasize involved analysis of Faradaic and double-layer interfacial kinetics and the effects of DC potential modulation. As opposed to the dielectric spectroscopy experimental approach, traditional impedance analysis often attempts to minimize or keep constant the effects of bulk material impedance and related dielectric relaxations due to capacitance effects of energy-storage components in the system. That is typically achieved by operating in highly conductive samples (such as aqueous solutions with supporting electrolytes) with a total bulk solution resistance of just several ohms and negligible capacitive effects. The impedance response of the electrode-solution interface, which is located in series with the bulk-solution impedance, can be easily determined at relatively low, typically Hertzian, AC frequencies. As demonstrated in Figure 1-4A, the impedance response of the conducting species and medium properties are qualitatively reflected in an integrative manner. Therefore the appearance of additional interfacial impedance at low frequencies results in an increase in the measured impedance to reflect a combined effect of both bulk and interfacial impedances. However, if the bulk impedance is kept very low by operating in highly conductive media, the total measured low-frequency impedance effectively becomes equal to the interfacial impedance.

The impedance output is fundamentally determined by the characteristics of an electrical current conducted through the system and is based on the concept that the obtained data represents the sample's "least impedance to the current." A significant portion of this book is dedicated to analysis of various possible parallel and sequential conducting mechanisms through the analyzed systems of interest. In any given experimental system there are always several competing paths for the current to travel through a sample. The current, however, chooses one or several closely matched predominant "paths of least resistance" between two electrical conductors (electrodes) under applied conditions such as AC frequency and voltage amplitude, DC voltage, electrode geometry and configuration, sample composition and concentration of main conducting species, temperature, pressure, convection, and external magnetic fields. Only the conduction through this predominant path of least resistance, or, to be more exact, the path of the least impedance, is measured by the EIS.

For example, for a sample represented by a parallel combination of a capacitor C and a resistor R (defined there as R/C), at high frequencies ω the impedance to current is the lowest through the capacitive component where impedance is inversely proportional to the frequency, as $Z^* \sim (\omega C)^{-1}$, and therefore is smaller than the impedance of the finite resistor R. At lower frequencies the opposite becomes true—the capacitive impedance component becomes large, and the current predominantly flows through the resistor; the total measured impedance reflects the resistance value as $Z^* \sim R$. The detected impedance output is determined by measuring the current passing through the least impeding segment of the circuit. The characteristic parameters of this ideal system can therefore be determined from the total impedance response $Z^* = f(\omega)$ as pure capacitance C at higher frequencies and pure resistance R at lower frequencies. Characteristic relaxation frequency f_C corresponds to a value where switching between the two conduction mechanisms occurs. The inverse of this frequency is characteristic "relaxation time" $\tau = 2\pi f_c$ for this circuit. For example, in aqueous conductive solution with permittivity of ~80 and bulk resistance ~ 1 ohm, it is easy to determine that at all frequencies under $f_C \sim 1$ GHz the current is conducted through a very small bulk solution resistor, and the capacitive characteristics are not contributing to the measured impedance signal. Similar results were shown for the above example illustrating current conduction through a saline solution.

These examples represent a circuit composed of ideal capacitive and resistive electrical components. The path of least impedance through a real-life sample placed between two conducting metal electrodes can be represented by a combination of chemical and mechanical elements that can only to some degree be approximated by these ideal electrical elements. These conducting venues through the sample may be the most plentiful and mobile electrons, ions, and particles; double-layer capacitive charging effects, specific adsorption, charge transfer "resistance" to tunneling electrons crossing the interface between the sample and the electrodes, and transport of discharging species to the surface of the electrode through diffusion layer concentration gradients at the electrode/sample interfaces. To interpret the EIS results, scientists and engineers have to intelligently devise the experiment to extract the needed analytical information from the experimentally obtained impedance. This interpretation includes initial development of the relationships between the electrical parameters representing the path of least impedance to the current and the "real life" investigated chemical and mechanical components that result in the measured impedance response. Through these relationships it may be possible to analyze the actual chemical, physical, and mechanical processes inside the analyzed system.

To the great benefit of analytical chemists concerned with the accurate analysis of concentrations of analytes and material scientists preoccupied with sample properties, the mechanisms of transporting current are largely determined by limited types of species that are present in the analyzed sample at significant levels. In that respect UBEIS is indeed a "spectroscopy," where a combination of carefully conducted experiments, some degree of initial preconceived knowledge of the analyzed sample, selection of experimental conditions, and careful interpretation of results are required to determine the sample composition and the mechanisms responsible for the impedance response, quantify them, and develop a mechanical model of the entire system's physical and chemical response to applied external electric field.

With that in mind, one cannot always determine "all" species present in the experimental system, as not "all" species may participate in the conduction through the path of least impedance, even if their response is dependent on the controlled experimental parameters, such as AC frequency. As with many other spectroscopy methods, UBEIS has the advantage of often being able to separate and identify several types of species inside an experimental sample, as various species may create a path of least impedance through the sample at different (and variable) experimental conditions. However, many potential analytes may still not be detectable, as they do not conduct current or are present in such low concentrations that their contribution to the overall conduction process may be negligible. However, they may have a disproportionably strong activity at the electrode interface. The same general limitations are present in optical and other "spectroscopies" based on "detection of the most prominent contributors" principle. It is the task of good experimental scientists to develop experimental designs that would provide the necessary information about the task at hand.

As will be further discussed, current traveling through a sample has to overcome successive impedances of two fundamentally different regions. One of them is the "bulk" of the sample, and the other is the "interface" where the sample meets the current conducting electrodes. Electrochemistry in general and the UBEIS in particular are often used to analyze both bulk sample conduction mechanisms and interfacial processes, where electron transfer, mass transport, and adsorption are often present. EIS analysis has often treated the bulk and interfacial processes separately [2, p. 28]. The analysis is achieved on the basis of selective responses of bulk and interfacial processes to sampling AC frequencies. The peaks appearing in the impedance spectrum can be described according to the theory of dielectric/impedance relaxations. Again, as in the case of any other spectroscopy method, the subject of the UBEIS analysis is the detection and interpretation of these peaks.

1.3. Electrochemical experiment: charge and material transport

In the basic electrochemical experiment there are at least two electrodes bounding a sample (such as electrolyte solution) with external potential (voltage) V difference applied between the two (Figure 1-5). Under the influence of the electric field the current I passes through a complete circuit, transported by electrons in the metal conductors, the electrodes, and ionic migration in the electrolyte. The electrons are free to move in the conduction bands inside the metal. A current that seems to flow quickly is not due to a "very fast" speed of the electrons in metal conductors (which is, in fact, rather slow at ~ 0.3 mm/sec), but because there are different electrons entering and leaving a conductor. In principle, when an electron is supplied to a wire end, "another" electron is coming out of the other end.

Ionic liquid (electrolyte), such as water with added ionic salts, is one of the common samples investigated by EIS. An average ion electrophoretic migration velocity in aqueous solution is ~ 10 mm/sec. As was shown above, the resistance to ionic migration current in the aqueous bulk solution within the frequency range of a typical impedance measurement can be simplified by a

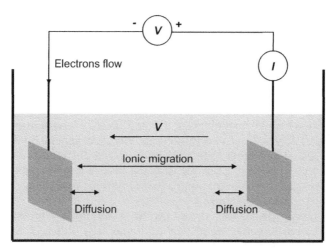

FIGURE 1-5   The basic electrochemical experiment with material transport directions

small "bulk solution resistance" component R_{SOL} (Figure 1-6). Current in the bulk solution is transported predominantly by migration, which is dependent on the applied external voltage V (often referred to in the electrochemical literature as "electrochemical potential difference $\Delta\phi$"). For migration processes involving electroactive species of type i, the Kohlrausch equation applies as does the familiar Ohm's Law equation, and for flux of charged species J_i [mol / sec] under the influence of electrochemical potential gradient $\Delta\phi/\Delta x$:

$$J_i = \frac{\Delta V(x)}{\Delta x} AF \sum z_i u_i C^*_i = \frac{\Delta\phi(x)}{\Delta x} A \sum \sigma_i \qquad (1\text{-}18)$$

Where: A = surface area of electrode, z_i = full charge of electroactive species of type i, F = Faradaic constant (96500 C/mol) representing free charge of 1 mol of elementary charges, C^*_i = concentration (mol/cm^3) of ions in solution participating in migration, u_i = the mobility constant, and $\sigma_i = F \sum z_i u_i C^*$ = bulk solution conductivity of this type of ions [Sm/cm].

Other types of samples can be subjected to electrochemical analysis, such as aqueous and non-aqueous solutions of ~ 1 μm-sized colloidal particles, where a charged double layer of "counterions" surround each particle and the particles can be regarded as "macro-ions." The colloidal particles will migrate and contribute to the solution conductivity. Solid ionic electrolytes and composite materials are also a possibility. There are also mixed conductors with both ionic and electronic conductance, such as many polymers. Lastly, semiconductors possess "forbidden gaps" that prevent electrons from entering conduction bands, resulting in low conductivity. The conduction can be significantly and selectively enhanced by adding impurities and creating local energy centers.

At the electrode-electrolyte interface separating the electrodes from the sample, there is a charge carrier shift—a transition between electronic and ionic conduction. The transfer of electric charge across the solution/electrode interface is accompanied by an electrochemical reaction at each electrode, a phenomenon known as electrolysis. Electrochemical discharge of the species at the interface is determined by their electrochemical properties, such as their ability to release or accept an electron at a voltage (or electrochemical potential) of this "electrochemically polarized" electrode. The electrode potential is essentially a measure of an excess of electrons at the electrode, which is charged negatively, or a lack of electrons on a positively charged electrode. In electrical terms, the impedance of the system to combined current generated by the discharge processes occurring in a nanometer thick "Helmholtz" interfacial layer of the sample immediately adjacent to the electrode can be represented by the so-called "charge transfer" resistance R_{CT} (Figure 1-6).

As a result of electrolysis reactions, depletion or accumulation of matter and charge may occur next to the electrodes. This reaction consumes or releases additional ions or neutral species from or into the bulk solution, resulting in a concentration gradient ΔC^*_i when concentration of the species in the bulk solution is different from that in the vicinity of the electrodes. However, all species always attempt to maintain equal concentration distribution inside any sample volume. For instance, when the local concentration gradient is created in the vicinity of the electrodes as a result of the species' consumption over the course of the electrochemical reaction, the more abundant solution species always attempt to replenish continuously depleting species at the interface by moving or "diffusing" to the electrode surface. Alternatively, the release of new species as a result of completion of the electrochemical reaction leads to their diffusion away from the electrode into the bulk solution where their concentration is lower than that at the electrode-solution interface. This concentration gradient results in a diffusion mass transport process occurring in a thick "diffusion layer" that can be measured by electrochemical impedance and is represented by a complex diffusion impedance element Z_{DIFF}.

Many species that are also deposited at the interface do not participate in electrolysis at the electrode potential. The charges of these species are countered by electronic charges of the opposite sign on the electrode's interface, producing a "double layer capacitance" with a value C_{DL}. In conductive solution an electric double layer is formed at the electrode as soon as the electrode is wetted. It may be useful to remember that at all interfaces (such as transition areas between the electrode and solution, tissue, or gel, between the solution and the surface of a colloidal particle, or inside the sample) there will be a non-uniform distribution of charges and a resulting electrochemical potential gradient. The capacitive element is created by species that are different from the species that are discharging and diffusing to the electrode, producing a parallel ("energy storage") electrical pathway. Therefore the double layer capacitance is placed in parallel with charge transfer resistance and diffusion pathway

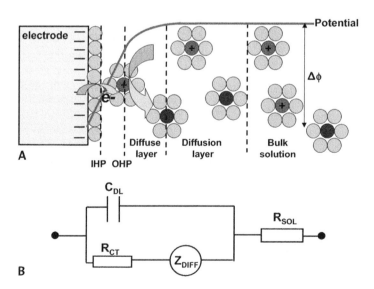

FIGURE 1-6   A. Interfacial electrochemical reaction with diffusion and double layer components; B. Representative electrical circuit

(Figure 1-6). Maxwell-Wagner, Helmholtz, Gouy-Chapman, Stern, and Grahame theories have been used to describe the interfacial and double layer dynamics [9].

In addition to the double layer electrochemical reactions and diffusion effects, specific adsorption (or chemisorption) effects can be present in some systems. Adsorption is a process where species are chemically bound to the metal surface of an electrode due to their chemical affinity and not to coulombic forces based on charge difference or polarity. Adsorption and electrochemical reactions take place on the electrode surface as a function of applied DC electrochemical potential and are determined by specific electrochemical properties of participating species. For the overall analyzed system, the voltage difference between two electrodes and the resulting current can be measured with an ordinary voltmeter and amperemeter (Figure 1-5).

Unlike heterogeneous electrochemical reactions that typically occur at the interface, the overall mass transport in the bulk of a sample is a homogeneous phase phenomenon that has to be carefully controlled. Ohm's Law is completely valid under the assumption of a homogeneous and isotropic bulk sample medium when the current direction and the field coincide and other effects, primarily related to electrochemical interfacial reactions (such as diffusion, electrolysis, etc.) are absent. In a bulk solution with free ions the electroneutrality condition $\sum_i z_i C^*_i = 0$ applies, and the sum of charges is zero. Electroneutrality does not prevail, however, at the interfacial boundaries where space charge regions. The hypothesis often made is that of a dilute solution for which the flux of a species i can be separated into a flux due to diffusion and a flux due to migration in an electric field. At the electrochemical interfacial region the current is transported by both migration and diffusion; the latter is driven by the concentration gradient at applied DC potential. In some situations an external mixing, or "convection," is added, moving the sample with a flow rate of v. Overall current equation, which includes diffusion (first term), migration (second term), and convection (third term), can be expressed as the Nernst-Planck equation [2, p. 45, 6]:

$$I = -R_G TF \sum_i (z_i \Delta C^*_i u_i) - F^2 \sum_i (z_i^2 C^*_i u_i \Delta\phi) + Fv \sum_i (z_i C^*_i)$$
$$= -F \sum_i (z_i \Delta C^*_i D_i) - F \sum_i (z_i \sigma_i \Delta\phi) + Fv \sum_i (z_i C^*_i) \qquad (1-19)$$

Where: the mobility constant u_i is related to the diffusion coefficient D_i and gas constant R_G through the Nernst-Einstein equation $u_i = D_i / R_G T$ and conductivity is $\sigma_i = F \sum_i z_i C^*_i u_i$.

In principle, the mobility can be viewed as a balance between a drag force $F_{DRAG} = 6\pi\eta v_i a_p$ on a particle of size a_p in a sample of viscosity η, which is moving under the influence of an electric field $F_{EL} = z_i e_0 \Delta\phi$. Assuming that the transport properties (D_i, u_i) are uniform in the solution bulk and hence are independent of C^*_i, the concentration change of species type i in the absence of a chemical reaction becomes:

$$\frac{\partial C^*_i}{\partial t} = D_i \nabla^2 C^*_i + z_i F u_i \nabla (C^*_i \Delta\phi) - v \nabla C^*_i \qquad (1-20)$$

Practical electrochemistry always attempts to analyze a complicated system by minimizing the effects of some of its components until a limited number of unknowns can be solved. The aim of the electrochemist is to be able to study each elementary phenomenon in isolation from the others. Hence, a technique capable of extracting the data and allowing these phenomena to be separated should be employed. Mobility dominates the conduction mechanism when the electrolyte has mobile ionic species on reversible electrodes or at high frequencies where only the bulk solution resistance is visible in the impedance spectrum and polarization at the electrodes does not develop. If the electroneutrality of the solution is assured through the presence of major ionic species (or so called "supporting electrolytes") that are not taking part in or consumed by the electrochemical reaction and therefore are not participating in concentration-driven diffusion, a small migration impedance term essentially does not change with applied electrochemical potential. The constant migration term therefore can be relatively easily identified and subtracted. Concentration calculations in the presence of convection are difficult unless very particular hydrodynamic conditions are fulfilled. The well-known example of the rotating disc electrode introduces a steady convection and a constant concentration gradient at the electrode-solution interface. However, in the majority of situations, analyzed media are stagnant or move at constant speeds, and the contribution of convection becomes negligible or easily identifiable. This happens when Schmidt's number, η/D_i, is sufficiently high (several thousands). If the effects of constant mobility and convection can be identified and subtracted from the overall impedance response, the remaining diffusion process results in a concentration gradient located in a so-called "Nernst" or "diffusion" layer of thickness L_D, within which the liquid is nearly motionless ($v = 0$). For such conditions the transport Equation 1-20 can then be reduced with good accuracy to the Fick's diffusion equation:

$$\frac{\partial C^*_i}{\partial t} = D_i \nabla^2 C^*_i$$ (1-21)

Diffusion coefficient in aqueous solution can be expressed as [3]:

$$D = \frac{7.4 \times 10^{-8} \sqrt{yM}}{\eta V_M^{0.6}} T$$ (1-22)

Where: y = the association parameter accounting for hydrogen bonds (for water it is 2.6), M = solvent's molecular weight, T = temperature (K), η = solvent's viscosity (cP), and V_M = molar volume (mL/mol), typically ~ 200 cm²/ mol. Typical values of diffusion coefficients in aqueous solutions are on the order of 10^{-7} cm²/sec (large molecules) to 10^{-5} cm²/sec (small molecules) at room temperatures and atmospheric pressure.

When the above conditions of convection and migration are realized, the resulting current is limited only by the diffusion-driven transport of the electroactive species to the electrode-solution interface. At the interface charge transfer reactions take place that can be studied as a function of the electrochemical potential (Figure 1-6). That experimental situation contains the fundamental premise of the traditional impedance analysis—make mass-transport impedance in the vicinity of the electrode interface exclusively diffusion-limited and investigate the interfacial kinetic phenomena composed of the diffusion (Z_{DIFF}) and electrochemical (R_{CT}) reaction (such as discharge or electrolysis) impedances that can be combined in so-called "Faradaic impedance." The premise of the traditional dielectric analysis is largely different. It attempts to avoid the influence of all interfacial effects by operating at higher AC frequencies where there is no time to develop double layers and electrolysis reactions and to study exclusively migration-driven conduction and bulk sample dielectric relaxation effects under the influence of voltage difference between two bounding electrodes. The combined data are often presented as "equivalent circuits" composed of a combination of series and parallel resistances, capacitances, and inductors, representing combined impedance to the current passage though the bulk solution (at high frequencies) and through the interfacial region (at low frequencies) as shown in Figure 1-6.

1.4.   Fundamental ambiguity of impedance spectroscopy analysis

EIS involves a significant body of research that has been the subject of many controversies. Impedance analysis is based on detection and interpretation of the meaning of processes that occur in response to external voltage perturbation applied at some predetermined AC frequencies. As a response to this external perturbation, the current response indicative of the physical/mechanical structure and chemical composition of the analyzed system or sample is recorded.

EIS was historically applied to two very large groups of systems—those with a significant interfacial electrochemical electron and mass-transport phenomena (such as electrically conductive ionic fluids) and those with predominantly bulk media relaxation phenomena (highly resistive materials, such as polymers, ceramics, and organic colloids). For traditional impedance analysis focused on an understanding of Faradaic processes, mass transport can be made diffusion-related while the migration dependence in the bulk solution can be controlled by adding supporting electrolyte and convection effects can be made constant or absent. For traditional dielectric spectroscopy, material relaxations (or energy storage and release effects) and migration-driven conduction are often studied using temperature modulation, while the low-frequency electrochemical interfacial polarization processes become of secondary importance. This "interfacial polarization" effects are often not studied by dielectric scientists, just as solution effects are often not studied in the traditional impedance literature. The above discussion of experimental strategies offers seemingly uncomplicated concepts of the UBEIS implementation, at least in the laboratory environment.

The experimental approaches and interpretation strategies for investigators working in the traditional impedance and dielectric analysis areas have drifted apart significantly over the years, often leading to misconceptions about assignments of frequency relaxation ranges and broader data interpretation. A combination of these two techniques in a universal broadband EIS

(UBEIS) measurement presents interpretation challenges, as both solution and interfacial effects come to play. For example, the same critical frequency of 100Hz can be assigned to a double layer charging in an aqueous system and to a bulk solution capacitive dispersion when a highly resistive material with a bulk resistance of 1 Mohm is analyzed. One of the intentions of this book is to bring these two types of analysis together and to attempt to reconcile possible interpretation contradictions in examples of several practical common systems. Therefore, it is important to first consider the analyzed sample and to look at its fundamental bulk material properties and geometrical dimensions, as the current passes through the bulk first before reaching the interface. The second step includes electrochemical analysis of possible interfacial kinetic processes (Section 8-7).

An even larger problem in impedance analysis is related to the fact that narrowly defined experimental conditions, such as limiting the measured impedance response by diffusion mass transport or rejection of the interfacial effects, cannot be imposed or controlled in many real-life applications. Very frequently an unknown system or phenomenon has to be characterized, and the ensuing complexity of analysis with many potential unknowns has to be faced. Therefore, a preliminary concept for the investigated system often has to be developed even before the system is analyzed.

While UBEIS is a powerful and versatile method applicable to many areas of science and technology, it is completely inapplicable to blind analysis of systems about which the investigator has no preliminary knowledge. At least in very general terms one almost has to have a significant amount of advance knowledge about the analyzed subject. A series of basic questions about the system has to be answered. Is the system a solution, solid, or gas? Is it aqueous or nonaqueous? Is the analyzed medium composed of ions or conductive or polarizable particles, or is it a semiconductor? Are there any electrochemically active components in the system that may be able to discharge and/or adsorb on the metal electrodes? Interpretation of the data depends greatly on initial understanding of basic components of the system and their projected response to the applied AC and DC electric fields. In response to these questions, it is highly recommended that the scientist develop a series of preliminary expectations and assumptions about the chemical, the physical and mechanical characteristics of the analyzed system, and the anticipated general type of the experimental setup, expected impedance data, and possible interpretation strategy. Intelligent characterization of a studied system is typically based on iterative comparison of the obtained experimental impedance data with these expectations. The expectations are typically based on previously published examples of electrochemical, physical, mechanical, and chemical analysis for similar types of systems, both in application-driven and laboratory experimentation environments. In this manuscript multiple examples of impedance data and analysis in the systems that investigators meet in practical applications or research laboratories are presented as a starting reference point.

References

1. M. E. Orazem, B. Tribollet, *Electrochemical impedance spectroscopy*, J. Wiley & Sons, Hoboken, New Jersey, 2008.
2. E. Barsukov, J. R. MacDonald, *Impedance spectroscopy*, J. Wiley & Sons, Hoboken, New Jersey, 2005.
3. S. Grimnes, O.G. Martinsen, *Bioimpedance and bioelectricity basics*, Academic Press, Oxford, UK, 2000.
4. K. Asami, *Evaluation of colloids by dielectric spectroscopy*, HP Application Note 380-3, 1995, pp. 1–20.
5. S. Krause, *Impedance methods*, in Encyclopedia of Electrochemistry, A. J. Bard (ed.), Wiley-VCH, Vol. 3, 2001.
6. C. Gabrielli, *Identification of electrochemical processes by frequency response analysis*, Solartron Analytical Technical Report 004/83,1998, pp. 1–119.
7. A. Lasia, *Electrochemical impedance spectroscopy and its applications, in modern aspects of electrochemistry*, B.E. Conway, J. Bockris, R. White (Eds.), vol. 32, Kluwer Academic/ Plenum Publishers, New York, 1999, pp. 143–248.
8. B. E. Conway, *Electrochemical supercapacitors*, Kluwer Academic, New York, 1999.
9. A. J. Bard, L. R. Faulkner, *Electrochemical methods, fundamentals and applications*, J. Wiley & Sons, New York, 2001.

CHAPTER 2

Graphical Representation of Impedance Spectroscopy Data

2.1. Nyquist and Bode representation of complex impedance data for ideal electrical circuits

It is convenient to start the discussion of the fundamentals of impedance data representation with an analysis of very simple systems. If a sinusoidal voltage is applied to a pure resistor of value R, then the measured complex impedance is entirely resistive at all frequencies as $Z^* = R$ and the impedance magnitude $|Z| = R$. If a sinusoidal voltage is applied across a pure capacitor, the measured impedance can be calculated according to the relationship $Z^* = -j(\omega C)^{-1}$ where C is the capacitance. The magnitude of the impedance for a pure capacitor is $|Z| = (\omega C)^{-1}$. This impedance depends on the frequency and is entirely capacitive (see Chapter 3).

Many real-life systems cannot be represented by an ideal resistor or capacitor but rather by a circuit that combines resistive and capacitive elements. One typical circuit is the $R\,|\,C$ circuit, represented by a parallel combination of an ideal resistor R=10000 ohm and an ideal capacitor $C = 1\ \mu F$ (Figure 2-1A). The impedances of the two parallel branches of this circuit are equal to a constant resistance R and $-j(\omega C)^{-1}$, respectively. According to Kirchhoff's Law for a parallel circuit, the potentials across both circuit elements are equal, while the total current can be calculated from the sum of the currents flowing through the resistor and capacitor branches:

$$I(t) = \frac{V(t)}{Z^*} = \frac{V(t)}{R} - \frac{\omega C V(t)}{j} = V(t)\left[\frac{1}{R} - \frac{\omega C}{j}\right] = V(t)\frac{1 + (\omega R C)^2}{R - j\omega R^2 C} \qquad (2\text{-}1)$$

23

The expression for the resulting impedance $Z^* = \dfrac{R}{1+(\omega RC)^2} - j\dfrac{\omega R^2 C}{1+(\omega RC)^2}$ is composed of "real" and "imaginary" parts. If the real part $Z_{REAL}(\omega) = \dfrac{R}{1+(\omega RC)^2}$ is plotted on the x-axis and the imaginary part $Z_{IM}(\omega) = -\dfrac{\omega R^2 C}{1+(\omega RC)^2}$ on the y-axis of a chart, a so called "Nyquist plot," or complex plane impedance diagram, is revealed. As shown in Figure 2-1B, this plot has the shape of a semicircle. Notice that in this plot the y-axis was chosen

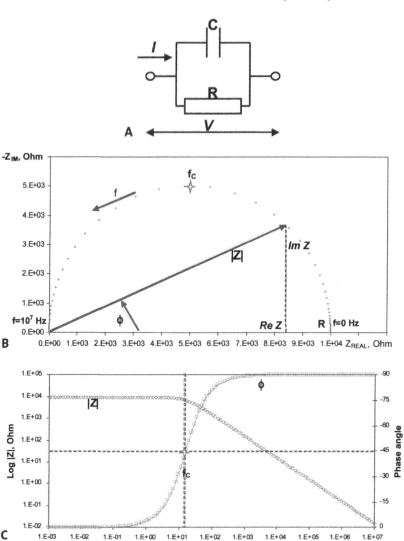

FIGURE 2-1 A. Equivalent circuit; B. Nyquist; C. Bode plots with impedance vector for the R|C circuit

as negative notation and that each point on the Nyquist plot is the impedance at one frequency [1]. Figure 2-1B has been specifically annotated to show that low-frequency data are on the right side of the plot and higher frequencies are on the left. This is often true for EIS data, where impedance usually falls as frequency rises, but is not always true for all circuits and at all AC frequencies. On the Nyquist plot the impedance can be represented as a vector of length $|Z|$. The angle between this vector and the x-axis is ϕ, or "phase angle" which also has a negative notation, as (from Eq. 1-11):

$$\phi = \arctan \frac{Z_{IM}}{Z_{REAL}} \tag{2-2}$$

There is a parameter $\tau = RC$ called "time constant," which is associated with this circuit, and a corresponding "characteristic circular" frequency $\omega_c = 1/\tau$ and "characteristic" or "critical relaxation" frequency $f_c = 1/2\pi\tau = 1/2\pi RC = 15.9$ Hz. At very high frequencies the impedance is completely capacitive, while at low frequencies it becomes completely resistive and approaches the value of R, which equals the diameter of the Nyquist plot semicircle. The phase angle ϕ tends towards $-90°$ at high frequency and towards $0°$ at low frequency, and critical frequency f_c corresponds to a midpoint transition where the phase angle is $-45°$ and $Z_{IM} = Z_{REAL} = R/2$.

As was already discussed in Chapter 1, for current passing through a circuit composed of a parallel combination of capacitance and resistance, the impedance magnitude of the capacitive segment of the circuit is dependent on the inverse of the AC frequency ω, while the impedance of the resistor is frequency-independent. Therefore, if within the frequency range of the analysis the contribution of the impedances in the resistive and capacitive segments of the circuit are comparable, than at frequencies higher than f_c the capacitive portion of the circuit with an impedance that is inversely proportional to the large AC frequency value presents a "path of least impedance" to the current. The measured impedance becomes largely capacitive with larger absolute values of the phase angle. At AC frequencies lower than f_c the reverse becomes true—the capacitive portion of the circuit will have high impedance values as they become inversely proportional to low values of the AC frequency. The current starts flowing predominantly through the constant resistor, and the measured total impedance becomes resistive with the phase angle approaches $0°$.

An interesting aspect of the impedance analysis is revealed in this simple example. Certainly the above statement that "the contribution of the resistive and the capacitive impedances are comparable" is somewhat vague. However, if within the analyzed frequency range either the capacitive or resistive impedance component is consistently significantly smaller that the other (let's say in the ratio of 1:100), then the other (larger) impedance component of the circuit consistently presents a relatively very high impedance to the passing current. Therefore, as the current always bypasses the high impedance component, its contribution to the total measured impedance of the system is not measured, and the high impedance component becomes undetectable by the impedance analysis.

Nyquist plots have proven to be very useful in estimating impedance parameters. The impedance spectra often appear as single or multiple arcs in the complex plane. Nyquist plots are also very popular because the shape of the plot yields insight into possible conduction mechanisms or kinetic governing phenomena. For example, if the plot displays a perfect semicircle (Figure 2-1B), the impedance response corresponds to a single activation-energy-controlled (or charge-transfer) process. A depressed semicircle, as will be shown later, indicates that a more detailed model is required, and multiple loops indicate that more than one time constant is required to describe a process. Often only a portion of one or more semicircles is seen within the frequency range of analysis.

However, the information the Nyquist plot presents is not complete, since there is typically no detailed indication of the frequency at which the impedance is measured for every point. For a detailed presentation of the experimental data, the points in the Nyquist plot must be labeled with the corresponding frequency values. The Nyquist plot also makes determination of low impedance values (typically observed at very high frequencies) very difficult.

Another form of data presentation method is the "Bode" plot. More commonly, it shows a plot of the phase angle and the logarithm of the impedance magnitude as a function of the logarithm of frequency. The Bode plot for the electric circuit of Figure 2-1A is shown in Figure 2-1C. Unlike the Nyquist plot, the Bode plot explicitly shows frequency information and reveals an important low impedance behavior seen at high frequencies. The slope of transition between low-frequency and high-frequency asymptotes can provide useful information concerning the nature of the impedance response if characteristic time constants are well separated. The popularity of the Bode plot representation is based on its utility in circuit analysis. Phase-angle plots are sensitive to system parameters and therefore provide a good means of comparing the model to the experimental results.

However, for more complicated electrochemical systems exhibiting additional impedance components, the Bode plot often makes it difficult to estimate characteristic frequencies. Let us examine a slightly more complicated system where the resistance $R_1 = 1000$ ohm is added in series to a parallel combination of the capacitance $C = 1\mu F$ and resistance $R = 10000$ ohm, producing $R_1 - R \,|\, C$ circuit (Figure 2-2 A) with the corresponding Nyquist and Bode plots shown in Figure 2-2B and Figure 2-2C, respectively. Symbol "– "defines a "series" and "|" a "parallel" connection of the relevant elements in the equivalent circuit diagram.

For this type of circuit the presence of even relatively small resistor $R_1 < R$ significantly changes the phase-angle response. The high-frequency phase angle is at $0°$, and the impedance is equal to the added resistance R_1 and is therefore purely resistive. At lower frequencies the phase-angle absolute value increases due to the appearance of the capacitive effect, reaching $-56°$ at the highest point and decreasing to $0°$ value as the resistor R dominates again at low frequencies where the total impedance becomes equal to $R_1 + R$. The previously described rule to identify the critical frequency based on the phase angle being $-45°$ is not applicable any longer, as the phase-angle ex-

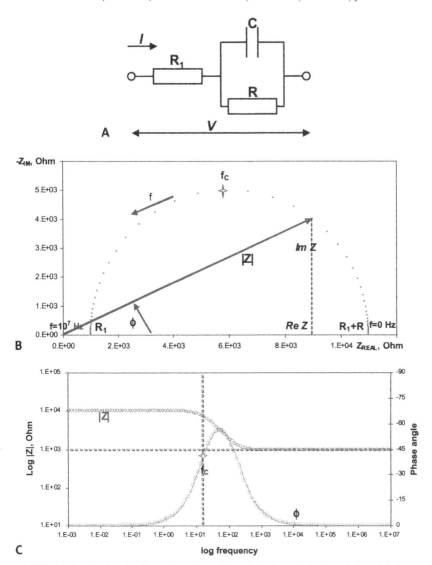

FIGURE 2-2   A. Equivalent circuit; B. Nyquist plot; C. Bode plots with impedance vector for the $R_1 - R|C$ circuit

pression $\phi = \arctan(Z_{IM} / Z_{REAL})$ depends on the Z_{REAL} parameter always having a constant additional contribution equaling R_1. At the critical frequency the phase-angle expression for an ideal semicircle with radius $R/2$ becomes:
$\phi = \arctan\dfrac{R/2}{R/2 + R_1}$. The peak frequency for the circuit corresponds to the phase angle of $-45°$ on the downward slope of the phase-angle plot only in the case of $R_1 \ll R$, as would be in the case of $R_1 \le 100$ ohm. For the case of

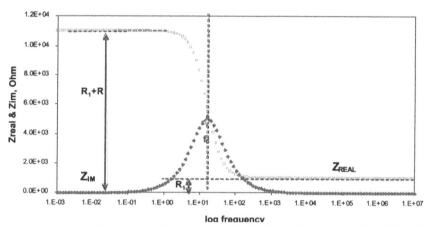

FIGURE 2-3 Real and imaginary parts of the impedance as a function of frequency for a $R_1 - R|C$ circuit

$R_1 = 1000$ ohm the phase-angle value at the critical frequency is $\sim\!-40°$. The critical frequency $f_C = 15.9$ Hz corresponds to the same value of time constant $\tau = RC$ as in the simple $R|C$ circuit.

The impedance magnitude on the Bode plot is much less sensitive to system parameters, and the asymptotic values at low and high frequencies provide values for the "near DC" $(R_1 + R)$ and R_1 resistances, respectively. The difficulties of the Bode plot representation can be somewhat alleviated by plotting the real and imaginary components of the impedance as a function of frequency (Figure 2-3) [2]. The real part of the impedance provides essentially the same information as is available from the $|Z|$ vs. frequency plot, with the high-frequency asymptote revealing the serial resistance R_1, and the low-frequency asymptote corresponding to a sum of resistances R and R_1. The imaginary impedance is independent of the serial resistance R_1. The imaginary part of the impedance has a significant advantage of being able to easily identify the characteristic frequency $f_C = 15.9$ Hz corresponding to the Z_{IM} peak value. For an ideal system at peak frequency measured real impedance becomes $Z_{REAL} = R_1 + 0.5R$, and Z_{IM} is at peak value.

Multiple processes with well-separated time constants can be distinguished by the presence of several peaks in the imaginary impedance plot. This representation may still have a difficulty in resolving two $R|C$ processes with very large differences in the amplitudes of the resistive and capacitive components. In a situation where impedance of one of the processes is significantly larger than the other, logarithmic scales become useful. Real and imaginary impedance can be separately plotted as log-log functions of the AC frequency (Figure 2-4). Plotting log Z_{IM} vs. log frequency presents an easy opportunity to identify the "distributed processes" where the capacitance in the circuit is not ideal due to a presence of several dispersions with similar order of magnitude critical relaxation time constants (as well as several other processes that will be discussed in the following chapters) and the Nyquist plot shows depressed or deformed semicircular features. Plotting log Z_{IM} vs.

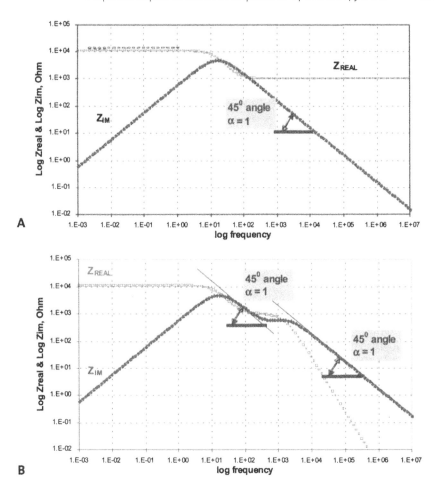

FIGURE 2-4   A. Logarithm of Z_{REAL} and Z_{IM} as a function of log frequency for $R_1 - R|C$ circuit; B. Logarithm of Z_{REAL} and Z_{IM} as a function of log frequency for $R_1|C_1 - R|C$ circuit

log frequency results in a straight line with a constant slope within the frequency range $\Delta \log f$ above the critical relaxation frequency f_c. The slope value can be estimated as $\alpha = \arctan \dfrac{\Delta Z_{IM}}{\Delta \log f}$ and the value of the α–parameter becomes indicative of a degree of nonideality and capacitive dispersion of a single relaxation. The $-45°$ angle corresponds to $\alpha = 1$, indicating that the imaginary impedance component Z_{IM} is described by an ideal capacitor (Figure 2-4A). Deviation from the $-45°$ angle value indicates a nonideal or distributed process with $\alpha \neq 1$ (Section 3-2). That representation can also be used to differentiate a depressed semicircle due to two different $R|C$ processes with $\alpha = 1$ and a depressed semicircle due to a single non-ideal $R|C$ process with $\alpha \neq 1$ (Section 3-2). Visible transition between two linear dependencies with $\alpha = 1$ in $\log Z_{IM}$ vs. $\log f$ plot indicates the presence of at least two distributed processes with separate time constants, such as two $R|C$ circuits in series (Figure 2-4B).

The example of the $R_1 - R | C$ circuit is also interesting practically, as it is broadly used as the initial impedance modeling approximation of a simple electrochemical reaction mechanism. This type of circuit representation is very common in practical and experimental electrochemistry—for example, in many highly conductive ionic solutions, where bulk solution impedance is low and can be adequately represented by a purely resistive element $R_{SOL} = R_1$. After passing though the solution resistance element, the current has to cross the interface where the heterogeneous charge transfer exchange between the solution ions and electrons in the electrodes takes place. This process is controlled by the discharge kinetics process, which can be represented by a "charge transfer" resistance $R_{CT} = R$. Deposition of both discharging species and other solution species not participating in the charge transfer at the interface is countered by electronic charges of the opposite sign on the electrode interface, producing a capacitor with a "double layer capacitance" value $C_{DL} = C$. The parameters of the circuit will be adjusted to better represent a "typical" conductive solution electrochemical process, with $R_{SOL} = 10$ ohm, $C_{DL} = 10$ μF, and $R_{CT} = 100$ ohm.

For the realistic $R_{SOL} - R_{CT} | C_{DL}$ circuit the presence of an even relatively small R_{SOL} element significantly changes the phase-angle response. The high-frequency phase angle equals 0° as the system impedance response is practically all resistive and dominated by R_{SOL}. At lower frequencies the interfacial kinetics process represented by $C_{DL} | R_{CT}$ dominates, and the phase angle absolute value initially increases as the impedance in the medium frequencies is largely affected by the double layer capacitance C_{DL} element; at the lowest frequencies the phase angle decreases to 0° value as charge-transfer resistance becomes the predominant contributor to the total impedance. The critical frequency corresponds to a value of $f_C = 1/2\pi\tau = 1/2\pi R_{CT} C_{DL} = 159$ Hz. The peak frequency for the circuit approaches −45° only in the case of $R_{SOL}/R_{CT} = 0$. The situation becomes even more complex in practical applications, where modification of the sample bulk resistance to achieve the $R_{SOL}/R_{CT} = 0$ condition is not possible. Provision should always be made for a possibly significant large contribution of the high-frequency bulk sample effects, including sometimes substantial capacitive contributions to the high-frequency impedance so $Z_{SOL} \neq R_{SOL}$. That may introduce additional complications into an attempt to utilize the Bode plot notation in the circuit analysis.

The Nyquist plot is a very convenient representation of the $R_{SOL} - R_{CT} | C_{DL}$ circuit process, as it shows an ideal semicircle as an indication of the activation-energy-controlled charge-transfer process. A depressed semicircle in the Nyquist plot is an indication of multiple processes with similar relaxation time constants, or distributed non-ideal kinetics. These ambiguities can be resolved in the "original" and modified Bode plots, as shown in Figures 2-3 and 2-4. The departure of the slope of log Z_{IM} vs. log frequency dependency from the unity indicates a distributed process, with a characteristic frequency that may not even correspond to the highest peak value of Z_{IM}.

In laboratory experimental conditions a well-known liquid sample can often be intelligently modified through the addition of a large excess of highly conducting ions ("supporting electrolyte") that do not participate in the electrochemical discharge on the electrodes at the experimentally applied AC and

DC electrochemical potentials (Section 1-3). This modification results in the practical elimination of the capacitive contributions to the bulk solution impedance for AC frequencies under ~100 MHz, as the bulk solution resistance R_{SOL} is decreased to an extent that all current in the bulk solution passes through R_{SOL}. However, even this condition does not always fulfill the criterion $R_{SOL}/R_{CT} = 0$.

In [2, 3, p. 317] an approach of simplifying the Bode plot analysis by subtracting the estimated value of the high-frequency solution resistance R_{1_EST} from the total impedance was proposed, leaving the system represented by an "adjusted" $R \mid C$ circuit. The adjusted values of the total impedance and phase angle become:

$$Z_{ADJ} = \sqrt{(Z_{REAL} - R_{1\_EST})^2 + Z_{IM}^2} = \sqrt{\left[(R_1 - R_{1\_EST}) + \frac{R}{1 + (\omega RC)^2}\right]^2 + \left(\frac{\omega CR^2}{1 + (\omega RC)^2}\right)^2}$$

(2-3)

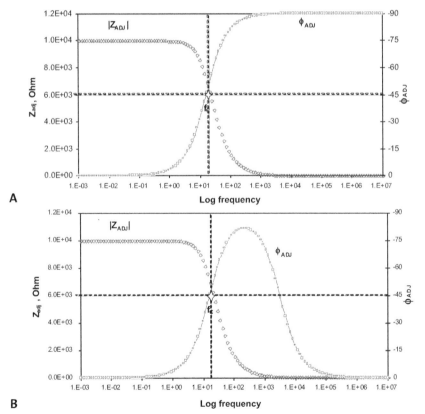

FIGURE 2-5 Adjusted Bode plots for a $R_1 - R|C$ circuit, $R_1 = 1000$ ohm, $R = 10000$ ohm, $C = 1\mu F$, with subtracted A. $R_{1\_EST} = 1000$ ohm; B. $R_{1\_EST} = 950$ ohm

$$\phi_{ADJ} = \tan^{-1}\frac{Z_{IM}}{Z_{REAL} - R_{1\_EST}} = \tan^{-1}\left(\frac{\omega CR^2}{R + (R_1 - R_{1\_EST})(1 + (\omega RC)^2)}\right) \quad (2\text{-}4)$$

The impedance data become easier to interpret, as the critical frequency can be determined from the Bode plot when the condition of phase angle $\phi = -45°$ is fulfilled (Figure 2-5A). At high frequencies the phase angle becomes $-90°$, just as in the case of a simple $R \mid C$ system without additional resistance. In an adjusted plot where R_1 is subtracted, it is also often easier to determine the α-parameter from the slope of log Z_{IM} vs. log frequency plot. A downside of this approach is the less than ideal estimation of R_1, where even ~ 5% error results in significant distortion of the plot features (Figure 2-5B) and the appearance of artificial impedance "features" that can be interpreted as additional high-frequency relaxation processes. The subtraction may also lead to additional noise in the adjusted phase-angle values at high frequencies.

2.2.   Dielectric data representation

In many practical cases increasing the analyzed sample conductivity or the mathematical subtraction of the high-frequency impedance components is not possible. In such situations the high-frequency impedance response must be carefully analyzed and separated from the total measured impedance to identify the interfacial low-frequency impedance components. The experimental systems represented by the $R_{SOL} - R_{CT} \mid C_{DL}$ circuit allow for a relatively simple and trouble-free analysis of the interfacial electrochemistry, while the bulk phenomena are greatly simplified by the analysis in highly conductive solutions with minimal bulk resistance R_{SOL}. Many more complex practical systems include significant bulk material high-frequency relaxation phenomena. This is the case with the analysis of many dielectrics such as organic colloids and resistive composites. As the first approximation, Figure 2-6A illustrates a simplified circuit model [4, 5] to describe the electrochemical system of highly resistive colloidal dispersion where both bulk solution resistance (R_{BULK}) and capacitance (C_{BULK}) are present instead of a simple R_{SOL}. R_{INT} indicated the interfacial resistance (which combines various mass-transport, adsorption and charge-transfer contributions), and C_{DL} is the double layer capacitance.

Practical UBEIS is preoccupied with understanding both bulk and interfacial phenomena. One must assume for a moment the following simplified logical pattern connecting the sample geometry with the applied AC frequency range. As the AC voltage is applied to the sample, the signal is typically "scanned" from the highest to the lowest available experimental frequencies. At a high frequency only "fast timescale" electrochemical processes that have sufficient time to respond to the rapidly changing direction of the AC electric field can be detected by UBEIS. Typically, these processes are representative of the charge's (ions or electrons) migration or dipole relaxations in the bulk sample. At lower AC frequencies an additional contribution of the "slower" processes is observed. The electrochemical kinetics processes that are taking place at the electrode-sample interface are typically slow, with diffusion and charge

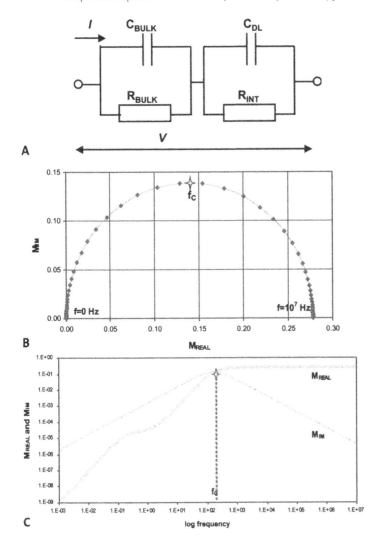

FIGURE 2-6   Impedance analysis of dielectric phenomenon: R_{BULK} = 10^6ohm, C_{BULK} = 10^{-10}F, R_{INT} = 10^6ohm, C_{BULK} = 10^{-6}F A. Equivalent circuit; B. Modulus "Nyquist"; C. Modulus "Bode" plots

transfer becoming limiting steps. Therefore, in a typical experimental situation UBEIS analyzes "bulk sample" processes at high frequencies and then "interfacial kinetics" phenomena at lower frequencies. In the geometrical realm that can be represented by the "first step" in the bulk and the "second step" at the electrochemical interface. Total impedance response measured at a low frequency includes both contributions. As was discussed above, traditional dielectric analysis is intentionally performed at sufficiently high frequencies to include only the bulk material response. The bulk media characteristic resistive and capacitive parameters can theoretically be resolved in both modulus and impedance plots. In practice, good resolution is often difficult to obtain, especially if one relies exclusively on the complex impedance plane (Nyquist or Bode) representations.

There is a significant value in alternative graphical expressions of the impedance data to the traditional EIS Nyquist and Bode impedance plot representations. Appearance of kHz and MHz relaxations on the impedance plots may indicate the presence of bulk sample "dielectric" effects. The impedance representation articulates the low-frequency interfacial response where impedance values are typically high and masks the high-frequency response where impedance is low. The dielectric analysis becomes practical when the bulk sample information at higher frequencies has to be analyzed—for example, to determine its effect on the low-frequency impedance data. Dielectric analysis uses so-called "Cole-Cole" plots of imaginary (loss factor ε'') vs. real (ε') portions of permittivity, which is similar to the Nyquist plot for the impedance or to the Bode plot for the function of frequency (Figure 1-4B). Even more practical are the M* modulus plots expressed both in "Nyquist-type" form of imaginary vs. real modulus (Figure 2-6B) and in the "Bode-type" plots of total real and imaginary modulus vs. logarithm of frequency (Figure 2-6C). This observation is largely related to the different mathematical expressions for impedance (Eq. 1-10), permittivity (Eq. 1-13 and 1-14), and modulus (Eq. 1-17). The real and imaginary portions of the modulus function M* can be expressed from Eq. 1-17 as:

$$M_{REAL} = -2\pi f \varepsilon_o Z_{IM} \tag{2-5}$$

$$M_{IM} = 2\pi f \varepsilon_o Z_{REAL} \tag{2-6}$$

For the data presented in Figure 1-4, it can be noted that measured permittivity is low at high frequencies and often becomes very high at low frequencies (due to interfacial polarization). The modulus representation shows very small absolute modulus values at low frequencies (and the low frequency semicircle is invisible in Figure 2-6B) where the impedance notation is particularly powerful, but will resolve high-frequency dielectric responses well. Figure 2-6 illustrates this example for the $R_{BULK} \mid C_{BULK} - R_{INT} \mid C_{DL}$ circuit.

The usefulness of complex modulus representations in addition to impedance plots is related partially to the composition of the analyzed samples, especially in cases of multicomponent dispersions where the migration-type ionic or particle-based conduction in the bulk sample can be realized by two or more competing processes. As will be shown later, the graphical representation of the modulus often resolves well the resistive differences in the bulk conduction processes, while the impedance representation is preferred to resolve capacitance-related differences [6].

References

1. C. Gabrielli, *Identification of electrochemical processes by frequency response analysis*, Solartron Analytical Technical Report 004/83, 1998, pp. 1–119.

2. M. E. Orazem, N. Pebere, B. Tribollet, *Enhanced graphical representation of electrochemical impedance data*, J. Electrochem. Soc., 2006, 153, 4, pp. B129-B136.

3. M. E. Orazem, B. Tribollet, *ElectrochemicalImpedance spectroscopy*, J. Wiley & Sons, Hoboken, New Jersey, 2008.

4. W. Olthuis, W. Steekstra, P. Bergveld, *Theoretical and experimental determination of cell constants of planar-interdigitated electrolyte conductivity sensors*, Sens. Act. B: Chem.,1995, 24-25, pp. 252–256.

5. V. F. Lvovich, C. C. Liu, M. F. Smiechowski, *Optimization and fabrication of planarInterdigitated impedance sensors for highly resistive non-aqueous industrial fluids*, Sensors and Actuators, 2007, 119, 2, pp. 490–496.

6. D. G. Han, G. M. Choi, *Simulation of impedance for 2-D composite*, Electrochim. Acta,1999)], 44, pp. 4155–4161.

CHAPTER 3

Equivalent Circuits Modeling of the Impedance Phenomenon

3.1.   Ideal circuit elements

Impedance data, represented in impedance, dielectric, or modulus graphical plots, is commonly analyzed by fitting it to an equivalent electrical circuit model. In Chapter 2 several simple electrical circuits containing combinations of ideal resistors and capacitors were used to generate representative impedance data plots. In an experimental laboratory or field application the situation is reversed—initially the impedance of an unknown system is measured, producing impedance data plots. These plots can be compared to those of well-known equivalent circuits, resulting in an initial selection of a circuit that produces the best fit with the experimentally obtained impedance data and the best physical explanation of the evaluated phenomenon. These conditions represent both the great convenience and the great interpretational danger of equivalent circuit analysis.

Equivalent circuits (EC) should always be selected on the basis of an intuitive understanding of the electrochemical system, as long as they are based on the chemical and physical properties of the system and do not contain arbitrarily chosen circuit elements [1]. The condition of "the best fit" between the model and the experimental data still applies, but producing the best fit does not necessarily mean that the developed equivalent circuit model has physical meaning. One needs to apply knowledge of the physical processes involved, compare models with experimental data, and attempt to simplify the representation as much as possible [2, p. 97].

Most of the circuit elements in the model are common electrical elements such as resistors, capacitors, and inductors. As an example, many models contain a resistor that models the cell's solution resistance R_{SOL}. The basic impedance components, such as resistors, capacitors, and inductors, are ideal. Some knowledge of the impedance of the standard circuit components is therefore quite useful. The impedance of a resistor is independent of frequency and has only a real impedance component. Because there is no imaginary impedance, the current through a resistor is always in phase with the voltage. The impedance of an inductor increases as frequency increases. Inductors have only an imaginary impedance component. As a result, an inductor's current is phase-shifted by +90° with respect to the voltage. The impedance vs. frequency behavior of a capacitor is opposite to that of an inductor. A capacitor's impedance decreases as the frequency is raised. Capacitors also have only an imaginary impedance component. The current through a capacitor is phase-shifted −90° with respect to the voltage. Table 3-1 lists the common circuit elements, the equation for their current-voltage relationship, and their impedance.

One must always remember that the impedance EC analysis is an attempt to represent a complex phenomenon combining chemical, physical, electrical, and mechanical components in purely electrical terms. Impedance data are frequently fitted with an equivalent circuit made up of circuit elements related to the physical processes in the investigated system. In many cases, ideal circuit elements such as resistors and capacitors can be applied. Mostly, however,

TABLE 3-1   Ideal Circuit Elements Used in the Models

Component	Equivalent Element	Current vs. Voltage	Impedance
Resistor	R [ohm]	$V = IR$	R
Capacitor	C [F, or ohm^{-1} s]	$I = C\, dV/dt$	$1/j\omega C$
Inductor	L [H, or ohm s]	$V = L\, di/dt$	$j\omega L$
Infinite diffusion	Z_w [ohm]		$R_w / \sqrt{j\omega}$
Finite diffusion	Z_o [ohm]		$R_o \tanh(\sqrt{(j\omega L_D^2)/D})\big/\sqrt{(j\omega L_D^2/D)}$ $R_o \coth(\sqrt{(j\omega L_D^2)/D})\big/\sqrt{(j\omega L_D^2/D)}$
Constant phase element - CPE	Q [ohm^{-1} s$^{\alpha}$]		$\dfrac{1}{Q(j\omega)^{\alpha}}$

distributed circuit elements are required in addition to the ideal circuit elements to describe the impedance response of real systems adequately. As our understanding of impedance analysis compared to real-life phenomena expanded, several additional "distributed" EC elements were introduced to better represent the nonideal nature of the EC-modeled real-life processes. These elements, such as the constant phase element (CPE) and the Warburg diffusion impedance, are listed in Table 3-1 and are discussed in Section 3-2.

3.2. Nonideal circuit elements

In all real systems, some deviation from ideal behavior can be observed. If a potential is applied to a macroscopic system, the total current is the sum of a large number of microscopic current filaments, which originate and end at the electrodes. If the electrode surfaces are rough or if one or more of the dielectric materials in the system is inhomogeneous, many of these microscopic current filaments would be different. In a response to a small-amplitude excitation signal this would lead to frequency-dependent effects, which can often be modeled with simple distributed circuit elements. For example, many capacitors in EIS experiments, most prominently the double-layer capacitor C_{DL}, often do not behave ideally due to the distribution of currents and electroactive species. Instead, these capacitors often act like a constant phase element (CPE), an element that has found widespread use in impedance data modeling.

The term "constant phase element" stems from the fact that the phase angle of the portion of a circuit represented by such an element is AC frequency-independent. The impedance of a CPE has the form of dependency on an effective CPE coefficient Q [ohm^{-1} s$^\alpha$] [3, p. 211] as:

$$Z_{CPE} = \frac{1}{Q(j\omega)^\alpha} = \frac{1}{Q\omega^\alpha}\left[\cos(\alpha\pi/2) - j\sin(\alpha\pi/2)\right] \qquad (3\text{-}1)$$

This equation can describe an impedance of a pure capacitor C for a condition $Q = C$ and the exponent $\alpha = 1$. For the case $\alpha = 0$ the equation describes an impedance response of an ideal resistor with $Q = 1/R$. The CPE behavior and corresponding α value can be determined from a slope of a plot of $\log Z_{IM} = f(\log$ frequency), with ideal capacitive behavior producing a slope of 1 and CPE behavior a slope of less than 1 (Figure 2-4). As will be shown in Chapter 5, a special case exists for a condition of $\alpha = 0.5$, which describes the Warburg impedance for homogeneous semi-infinite diffusion $R_W/\sqrt{j\omega}$, where R_W is diffusion resistance and $Q = 1/R_W$. Another special case is "finite" Warburg diffusion, which is often defined for two electrode boundary conditions. The diffusion impedance for a case of "transmitting" boundary, when the diffusion species are instantly consumed at the electrode, becomes $R_D \tanh(\sqrt{(j\omega L_D^2)/D}/\sqrt{(j\omega L_D^2/D}$. In the case of "reflecting" boundary, when the discharge does not occur and the electrode is totally "blocking," the expression is modified as $R_D \coth(\sqrt{(j\omega L_D^2)/D}/\sqrt{(j\omega L_D^2/D}$, where L_D = thickness of the diffusion layer and D = the diffusion coefficient of the diffusing species.

In the Nyquist plot a circuit composed of a single CPE element produces a straight line, which makes an angle of $-\alpha*90°$ with the x-axis (Figure 3-1). A circuit composed of a resistor R in parallel with a CPE produces a plot with a depressed semicircular arc with an angle of $(\alpha-1)*90°$ [3, p. 212]. Both of these situations are common. For example, there is often a "leakage" conduction path present in parallel with a CPE, represented by an ideal DC resistor R, especially when CPE is used to represent the interfacial processes such as double-layer charging C_{DL}. In that case a depressed arc will often appear in the Nyquist diagram instead of a straight line.

Many intermediate cases exist when distributed processes, such as combinations of several capacitive and resistive elements, can be expressed by α values different from 0, 0.5, or 1 (Section 2-1). Figure 3-2 represents several practical examples of a single CPE, such as the case of an ideally polarizable electrode, where no electrochemical charge-transfer reaction takes place and CPE represents a double-layer charging in series with small solution resistance (Figure 3-2A). Figure 3-2B shows an example of the data representing the electrochemical charge-transfer-reaction process occurring on a nonideally polarizable electrode represented by $R_{CT}|CPE_{DL}$ combination.

CPE with a depressed semicircle has been used to describe both the double-layer capacitance and the low-frequency pseudocapacitance (Section 5-3). The "double layer capacitor" in electrochemical experiments often shows a CPE-like distributed behavior instead of that of a pure capacitor. Several theories have been proposed to account for the nonideal behavior of the double layer, but none has been universally accepted. As a first approximation, one can treat α as an empirical constant and not worry about its physical basis. For double-layer analysis, the parameter Q, expressed in [s″ ohm⁻¹], is

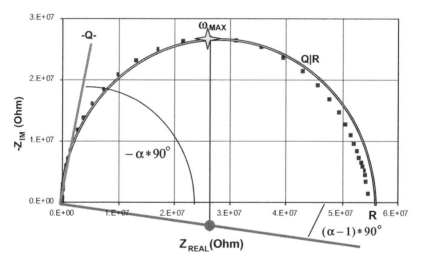

FIGURE 3-1 The Nyquist plot of a single CPE (Q) and a parallel resistor and CPE circuits.

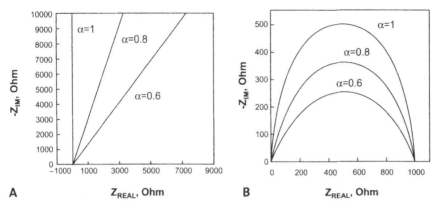

FIGURE 3-2 The Nyquist plots of: A. single CPE (ideally polarizable electrode); B. parallel resistor and CPE circuits

essentially a constant representing an interfacial charging by both surface and electroactive species. CPE element with $0<\alpha<1$ can be further separated into a frequency-dependent "pure" capacitor element with $\alpha = 1$ and "pure" resistor with $\alpha = 0$ that can be parallel or in series [3, p. 210]. The choice between parallel or series representation is arbitrary and not important for CPE [3, p. 210].

When the resistor R and the CPE are in parallel and $0.8 < \alpha < 1$, the circuit impedance can be expressed as $Z = \dfrac{R}{1 + j(2\pi f)^{\alpha} RQ}$. The CPE parameters and "effective equivalent capacitance" C_{EFF} can be easily estimated from the value of Z_{IM}^{MAX} corresponding to the frequency ω_{MAX}, where Z_{IM} is at maximum, following the derivation proposed by Hsu and Mansfeld [4]. From the imaginary portion of Eq. 3-1:

$$Q = \sin(\alpha\pi / 2)\frac{-1}{Z_{IM}(\omega_{MAX})^{\alpha}} \qquad (3\text{-}2)$$

$$C_{EFF} = Q\frac{1}{(\omega_{MAX})^{1-\alpha}\sin(\alpha\pi / 2)} \sim Q(\omega_{MAX})^{\alpha-1} \qquad (3\text{-}3)$$

When $\alpha = 1$, the CPE coefficient Q becomes a pure capacitance C_{EFF}:

$$\omega_{MAX} = \frac{1}{C_{EFF}R} \text{ and } Q = C_{EFF} = \frac{1}{\omega_{MAX}R} = \frac{-1}{Z_{IM}^{MAX}(\omega_{MAX})} \qquad (3\text{-}4)$$

Another series of equations was derived by Brug, allowing one to estimate the "effective equivalent capacitance" C_{EFF} value for several widely encountered circuits [5, 6]. For a series connection of the CPE and R, the value of C_{EFF} is defined by:

$$C_{EFF} = \frac{(RQ)^{1/\alpha}}{R} \qquad (3\text{-}5)$$

And for a $R_1 - CPE | R$ circuit:

$$C_{EFF} = \left[\frac{QR_1R}{R_1+R} \right]^{1/\alpha} \frac{R_1+R}{R_1R} \qquad (3\text{-}6)$$

Both the Hsu and Mansfeld formulas (Eqs. 3-2 through 3-4) and the Brug formulas (Eqs. 3-5 and 3-6) have been widely used to extract effective capacitance values from CPE parameters for studies of various kinetic and double-layer processes, passive films, polymer coatings, and corrosion inhibitors.

One of the main problems associated with CPE is that it is only valid over a limited frequency range. CPE may be distorted at extremely low and high frequencies. In any physically valid model, there should be a shortest and a longest possible relaxation time. This is, however, not true for CPE. Therefore, CPE usually needs to be modified at both ends of the frequency spectrum. In many cases this does not pose a problem, since impedance measurements are not carried out at extreme frequencies and/or the phenomenon represented by the CPE element is prominent only in a narrow range of frequencies, typically 1–3 orders of magnitude.

A more significant practical interpretation problem is related to CPE element flexibility. A CPE element very often can be used to fit into any type of experimental impedance data simply by varying the α-parameter. The choice of CPE representation should always be based on an understanding of the nature of studied distributed process. For the charge-transfer-related low-frequency impedance data, several realistic possibilities grounded in the actual physics of the process exist for a CPE and not for a pure capacitive, representation [2, p. 121]. In the process of heterogeneous electron transfer, the nonuniformity of rough electrode surfaces (inhomogeneity, roughness, edge orientation, and porosity) is frequently responsible for nonideality and observed frequency dispersions and their resulting CPE behavior [7, 8]. One type of distribution can arise along the surface area of the electrode (or parallel to it), also referred to as a two-dimensional (2D) distribution. The other type of distribution can arise along the axis perpendicular to the electrode surface as three-dimensional (3D) distribution. The following processes may be responsible for the deviations from ideal resistor and capacitor response and the resulting CPE representation [3, p. 223]:

- electrode inhomogeneity and surface roughness (3D)
- electrode porosity (3D)
- variability in thickness and conductivity of surface coating (3D)
- slow, uneven adsorption process (2D)
- nonuniform potential and current distribution at the surface (2D)
- grain boundaries and crystal phases on polycrystalline electrode (2D)

Two types of time-constant distributions around the electrode, such as radial current distribution along the electrode's surface and distribution of kinetic rates of adsorption-controlled electrochemical reaction, may lead to the CPE behavior [8, 9, 10, 11]. It has been demonstrated that for the expressions relating the effective capacitance C_{EFF} and the CPE parameter Q values Eqs. 3-3 through 3-6 may become more or less accurate depending on the nature of the current distributions. However, for a given set of CPE parameters the Brug formulas and the Hsu and Mansfeld formula may yield different values for the effective capacitance. In the case of 3D ("normal" or perpendicular to the electrode plane) distribution, the experimental data is better described by the Hsu and Mansfeld formulas (Eqs. 3-3 and 3-4). The 2D ("surface" or parallel to the electrode plane) distribution is better described by Brug's Eqs. 3-5 and 3-6 [12]. Inappropriate use of the formulas—for example, by using an incorrect effective resistances R and R_1 in the C_{EFF} calculations (Eqs. 3-4 through 3-6)—may lead to macroscopic errors, since the values of electrolyte resistance, charge-transfer resistance, and film resistance may be quite different. As in so many impedance data-related practical situations, the selection of the correct formula should rest on the knowledge of the system under investigation.

Local EIS (LEIS) analysis with microelectrodes scanning over a macroelectrode allows distinguishing between 3D and 2D—related interfacial kinetics processes that may appear as the CPE representation. 2D distribution due to current/potential and adsorption nonuniformities show a typical $R \,|\, C$ behavior on small surfaces analyzed by LEIS, while 3D processes demonstrate a $R \,|\, CPE$ type of behavior [13]. For an interfacial kinetics on a macroelectrode, the LEIS scan reveals several different $R \,|\, C$ combinations representing 2D distribution or several $R \,|\, CPE$ combinations in the case of 3D distribution. Just as with a "traditional" macroscale EIS, the LEIS can distinguish between the CPE or pure capacitive double later representation from a slope of a plot of $log\text{-}Z_{IM} = f(log\ frequency)$, with capacitive behavior giving a slope of 1 and CPE behavior a slope of <1.

The analysis of bulk solution processes may also involve CPE representation, with the center of the arc located below the Z_{REAL} axis (Figure 3-1) for two reasons:

- presence of distributed elements in the system with several relaxation times τ distributed around the mean $\tau_{MAX} = (\omega_{MAX})^{-1}$ [2, p. 16]; typically that occurs due to inhomogeneity in bulk solution [3, p. 223] with several conducting species present with time constants within two orders of magnitude from the "distributed" time constant τ_{MAX} of the single semicircle [1, p. 16]

- many-body interactions between clusters in bulk solution [3, p. 223]

These conditions result in modification of the Debye equation (Eq. 1-15) into a Cole-Cole equation, which accounts for a distribution of relaxation states:

$$\varepsilon^*(\omega) = \varepsilon_\infty + \frac{\Delta\varepsilon}{1+(j\omega\tau)^\alpha} \tag{3-7}$$

As was discussed in Section 1-2, the presence of conduction in parallel with capacitance and non-idealities in the capacitive and resistive components of the relaxation process described by $R_{BULK} \,|\, CPE_{BULK}$ structure result in a more complicated Havriliak-Negami model (Eq. 3-8).

$$\varepsilon^*(\omega) = \varepsilon_\infty + \frac{\Delta\varepsilon}{(1+(j\omega\tau)^\alpha)^\beta} - j\left(\frac{\sigma(\omega)}{\varepsilon_0\omega}\right)^N \tag{3-8}$$

3.3. Circuit models for systems with two and more time constants

Very few electrochemical cells can be modeled using a single equivalent circuit element. Instead, EIS models usually consist of a number of elements in a network. Both serial and parallel combinations of elements occur (Figure 3-3). For linear impedance elements in series the equivalent impedance is:

$$Z_{eq} = Z_1 + Z_2 + Z_3 \qquad (3\text{-}9)$$

Total equivalent impedance increases when several resistors are connected in series. When $R_1 = R_2 = R_3$, the equivalent resistance becomes $R_{EQ} = 3R_1$. When several capacitors are connected in series, the total equivalent impedance also increases, but total "equivalent" capacitance C_{EQ} decreases (Eq. 3-10). For example in the case of $C_1 = C_2 = C_3$ the equivalent capacitance becomes $C_{EQ} = C_1/3$. This is a consequence of the inverse relationship between capacitance and impedance:

$$Z_{EQ} = R_1 + R_2 + R_3 = R_{EQ}$$

$$Z_{EQ} = \frac{j}{\omega C_1} + \frac{j}{\omega C_2} + \frac{j}{\omega C_3} = j\frac{C_2C_3 + C_1C_3 + C_1C_2}{\omega C_1C_2C_3} = \frac{j}{\omega C_{EQ}} \qquad (3\text{-}10)$$

For linear impedance elements in parallel the equivalent impedance is:

$$\frac{1}{Z_{EQ}} = \frac{1}{Z_1} + \frac{1}{Z_2} + \frac{1}{Z_3} \qquad (3\text{-}11)$$

Total equivalent impedance decreases when several resistors are connected in series. For example, in the case of $R_1 = R_2 = R_3$ the equivalent resistance becomes $R_{EQ} = R_1/3$. When several capacitors are connected in series, the total impedance also decreases and total equivalent capacitance increases, becoming equal to the sum of all parallel capacitors (Eq. 3-12):

$$Z_{EQ} = \frac{R_1R_2R_3}{R_2R_3 + R_1R_3 + R_1R_2} \qquad Z_{EQ} = \frac{j}{\omega(C_1 + C_2 + C_3)} = \frac{j}{\omega C_{EQ}} \qquad (3\text{-}12)$$

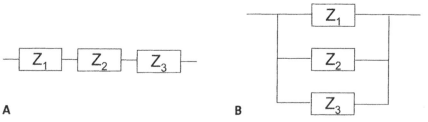

A   **B**

FIGURE 3-3   A. Impedances in series; B. Impedances in parallel

A series combination of two parallel $R\,|\,C$ circuits can be easily analyzed using the principles for parallel and sequential circuit components. The impedance for a single $R_1\,|\,C_1$ circuit becomes:

$$\frac{1}{Z_{EQ1}} = \frac{1}{Z_R} + \frac{1}{Z_C} = \frac{1}{R_1} + j\omega C_1 = \frac{1 + j\omega R_1 C_1}{R_1}$$

$$Z_{EQ1} = \frac{R_1}{1 + j\omega R_1 C_1} = \frac{R_1 - j\omega R_1^2 C_1}{1 + (\omega R_1 C_1)^2} \qquad (3\text{-}13)$$

For a series combination of two elements Z_{EQ1} and Z_{EQ2}, the expression becomes:

$$Z_{EQ} = Z_{EQ1} + Z_{EQ2} = \frac{R_1}{1 + (\omega R_1 C_1)^2} + \frac{R_2}{1 + (\omega R_2 C_2)^2} - j\left[\frac{\omega R_1^2 C_1}{1 + (\omega R_1 C_1)^2} + \frac{\omega R_2^2 C_2}{1 + (\omega R_2 C_2)^2}\right]$$

$$(3\text{-}14)$$

The first two components of the expression correspond to the frequency-dependent real portion of the impedance, and the two later components represent the frequency-dependent imaginary portion of the impedance. The resulting Nyquist plot shown in Figure 3-4 displays two clearly defined time constants: $\tau_1 = R_1 C_1$ and $\tau_2 = R_2 C_2$. "Ohmic resistor" R_{OHM} is also added to the system.

This spectrum can be modeled by any of the equivalent circuits shown in Figure 3-5. Real impedance data can very often be represented by various circuit models, and it is left to the investigator to decide which of these models yields the best description of the process of current traveling between the electrodes through the analyzed media. The models describing the impedance data with two $R\,|\,C$ time constants [1, p. 95–96, 2] (Figure 3-5) are: parallel (Maxwell), hierarchical ladder (Randles), and series (Voigt). The heirarchical ladder network representation of these systems ensures that bulk processes are represented at the highest frequencies, followed by the interfacial impedance arc (representing charge transfer, double layer, diffusion, and adsorption processes) as the frequency decreases. The Voigt circuitry does not impose this limitation; interfacial impedance arc(s) may occur anywhere in a diagram. However, they almost always are featured at the lowest frequencies.

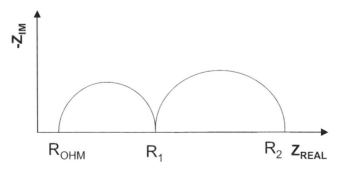

FIGURE 3-4   Two time constants spectrum

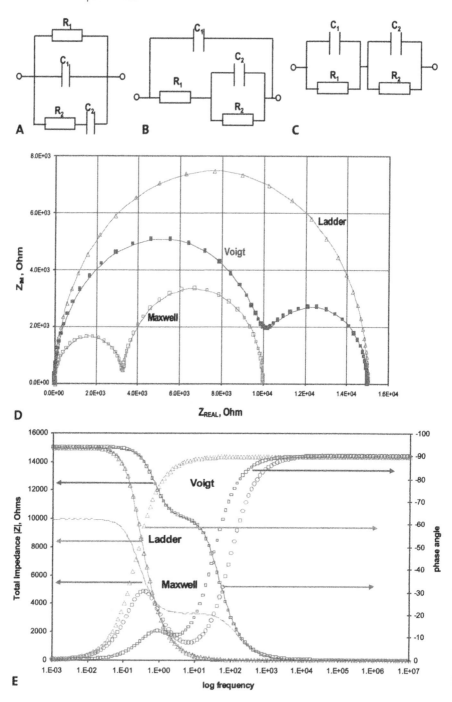

FIGURE 3-5 Equivalent circuit models with two time constants for $R_1 = 5$ kohm, $C_1 = 50\ \mu F$, $R_2 = 10$ kohm, $C_2 = 0.5\ \mu F$. A. Maxwell; B. Ladder; C. Voigt and their representation as D. Nyquist and E. Bode plots

This example shows three types of circuit diagrams for two relaxation time constants with the same R and C values, resulting in a Nyquist plot that usually shows two time constants (in the case of the Ladder model only one time constant was observed due to selected values for R_1, R_2, C_1, C_2 parameters) but with various characteristic fitting parameters. One always has to attempt to select a circuit that represents the geometrical distribution of impedance-related processes that physically occur inside of the system. A criterion of simplicity also has to be applied based on selection of only feasible processes with defined and well-resolved impedance characteristics to model the system. Computer-assisted fitting is necessary to resolve the overlapping arcs in the Z^* and M^* planes, but this method typically is unable to resolve arcs with time constants that are different by less than at least a factor of 100.

References

1. S. Krause, *Impedance methods*, in Encyclopedia of Electrochemistry, A. J. Bard (Ed.), Wiley-VCH, Vol. 3, 2001.

2. E. Barsukov, J. R. MacDonald, *Impedance spectroscopy*, J. Wiley & Sons, Hoboken, New Jersey, 2005.

3. S. Grimnes, O.G. Martinsen, *Bioimpedance and bioelectricity basics*, Academic Press, 2000.

4. C. S. Hsu, F. Mansfeld, *Concerning the conversion of the constant phase element parameter yo into a capacitance*, Corrosion (2001) 57, pp. 747–755.

5. G. J. Brug, A. L. G. van den Eeden, M. Sluyters-Rehbach, J. H. Sluyters, The analysis of electrode impedances complicated by the presence of a constant phase element, J. Electroanal. Chem., 1984, 176, pp. 275–295.

6. B. Hirschorn, M. E. Orazem, B. Tribollet, V. Vivier, I. Frateur, M. Musiani, *Determination of effective capacitance and film thickness from constant-phase-element parameters*, Electrochim. Acta ,2010, 55, 21, pp. 6218–6227.

7. M. E. Orazem, N. Pebere, B. Tribollet, *Enhanced graphical representation of electrochemical impedance data*, J. Electrochem. Soc., 2006, 153, 4, pp. B129–B136.

8. V. Mie-Wen Huang, V. Vivier, M. E. Orazem, N. Pebere, B. Tribollet, *The apparent constant-phase-element behavior of an ideally polarized blocking electrode*, J. Electrochem. Soc., 2007, 154, 2, pp. C81–C88.

9. V. Mie-Wen Huang, V. Vivier, M. E. Orazem, N. Pebere, B. Tribollet, *The global and local impedance response of a blocking disk electrode with local constant-phase-element behavior*, J. Electrochem. Soc., 2007, 154, 2, pp. C89–C98.

10. V. Mie-Wen Huang, V. Vivier, M. E. Orazem, N. Pebere, B. Tribollet, *The apparent constant-phase-element behavior of a disk electrode with faradaic reactions*, J. Electrochem. Soc., 2007, 154, 2, pp. C99–C107.

11. S. –L. Wu, M. E. Orazem, B. Tribollet, V. Vivier, *Impedance of a disc electrode with reactions involving an adsorbed intermediate: local and global analysis*, J. Electrochem. Soc., 2009, 156, 1, pp. C28–C38.

12. B. Hirschorn, M. E. Orazem, B. Tribollet, V. Vivier, I. Frateur, M. Musiani, *Determination of effective capacitance and film thickness from constant-phase-element parameters*, Electrochim. Acta, 2010, 55, pp. 6218–6227.

13. J.–B. Jorcin, M. E. Orazem, N. Pebere, B. Tribollet, *CPE analysis by local electrochemical impedance spectroscopy*, Electrochim. Acta, 2006, 51, pp.1473–1479.

CHAPTER 4

Examples of Simple Equivalent Circuit Models

4.1. Basic R-C circuit

A basic R-C circuit (Figure 4-1A) arrangement does not allow low AC frequency and DC current to pass through, as capacitance C presents to the current a very high impedance $Z = 1/\omega C$ and rejects the current at low frequency where $\omega \to 0$. At high frequency the resistor R limits the passage of the current as $I = V/R$. The time constant for the circuit equals $\tau = RC$. The Nyquist plot of the circuit (Figure 4-1B) is represented by a straight vertical line where the imaginary portion of the impedance goes to infinity at low frequency. The Bode plot at high frequency (Figures 4-1C and 4-1D) reveals the resistive component R and corresponding phase angle $0°$, and at $f < f_c = 1/(2\pi RC)$ the total impedance increases proportionally to the frequency decrease with the phase angle approaching $-90°$.

This circuit is representative of a conductive solution with conductance inversely proportional to the R-parameter. At low frequencies a "blocking" fully capacitive interface emerges where only the double-layer charging represented by the capacitor C is present. Total impedance of the circuit and its real and imaginary components can be expressed as:

$$Z^*(\omega) = R - j\frac{1}{\omega C} \tag{4-1}$$

$$Z_R(\omega) = R \quad Z_{IM}(\omega) = -\frac{1}{\omega C} \tag{4-2}$$

$$|Z_{TOTAL}| = \sqrt{R^2 + \frac{1}{(\omega C)^2}} \quad \phi = \arctan(-\frac{1}{\omega RC}) = \arctan(-\frac{1}{\omega \tau}) \tag{4-3}$$

49

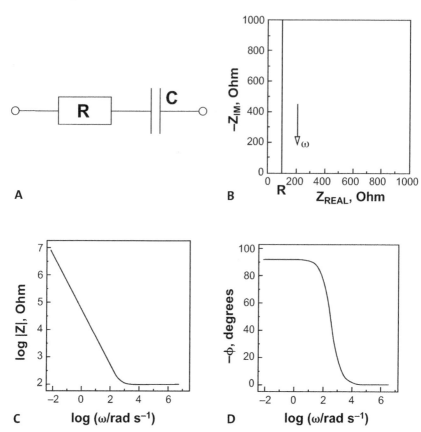

FIGURE 4-1 A. *R-C* circuit diagram and corresponding: B. Nyquist and Bode; C. total impedance; D. phase angle plots

Using Eq. 1-2 for resistance *R* and Eq. 1-3 for capacitance *C*, which are frequency-dependent parameters when analyzed over a broad frequency range, a general impedance equation (Eq. 1-10) can be developed as a function of the frequency-dependent characteristic electrical properties of the analyzed material or system. This expression takes into account the frequency-dependent values of relative permittivity $\varepsilon(\omega)$ and resistivity $\rho(\omega)$ or its inverse, electrical conductivity $\sigma(\omega)$, combined with the geometry factors of the electrode surface area (A) and the sample's thickness between the bounding electrodes (d):

$$Z^*(\omega) = Z_R(\omega) + jZ_{IM}(\omega) = R(\omega) + \frac{j}{\omega C(\omega)} =$$

$$= \rho(\omega)\frac{d}{A} + \frac{jd}{\omega \varepsilon(\omega)\varepsilon_0 A} = \frac{d}{A}\left[\rho(\omega) + \frac{j}{\omega \varepsilon(\omega)\varepsilon_0}\right] = \frac{d}{A}\left[\frac{1}{\sigma(\omega)} + \frac{j}{\omega \varepsilon(\omega)\varepsilon_0}\right] \quad (4\text{-}4)$$

4.2.   Basic R|C circuit

A basic $R \mid C$ circuit (Figure 4-2A) is characterized by a transition from the current flowing entirely through the capacitor at high frequency, where the circuit impedance is $Z = 1/\omega C$ and the phase angle is $-90°$, to a situation at low frequency, where the current flows only through the finite resistor R and the phase angle is $0°$. At high frequency the finite resistor rejects the current, while the capacitive impedance is very low and allows the current to pass through. Alternatively at the low frequencies ($\omega \to 0$) the capacitive impedance $Z = 1/\omega C$ becomes very large and rejects the current flow, which passes entirely through the resistor as $I = V/R$. The time constant for the circuit equals $\tau = RC$, with phase angle transitioning from $-90°$ to $0°$ as frequency decreases, reaching $-45°$ at the critical frequency $f_C = 1/(2\pi RC)$.

The Nyquist plot of the circuit (Figure 4-2B) is represented by a semicircular arc that intersects the x-axis at low frequency, indicating the limiting value of the real impedance $Z_{REAL} = R$. The Bode plot (Figures 4-2C and 4-2D) at low frequency reveals the resistive component R and corresponding phase angle of $0°$, with the phase angle reaching $-90°$ at high frequencies.

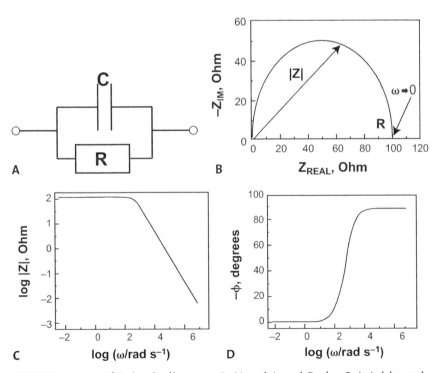

FIGURE 4-2   A. $R|C$ circuit diagram; B. Nyquist and Bode; C. total impedance; D. phase angle plots

The circuit analysis results in the expressions for total, real, and imaginary impedance of the circuit as:

$$Z*(\omega) = \frac{R}{1+(\omega RC)^2} - j\frac{\omega R^2 C}{1+(\omega RC)^2} \tag{4-5}$$

$$Z_R(\omega) = \frac{R}{1+(\omega RC)^2} \qquad Z_{IM}(\omega) = -\frac{\omega R^2 C}{1+(\omega RC)^2} \tag{4-6}$$

$$|Z_{TOTAL}| = \frac{R}{\sqrt{1+(\omega RC)^2}} \qquad \phi = \arctan(-\omega RC) = \arctan(-\omega\tau) \tag{4-7}$$

Interestingly, although R is frequency-independent, Z_{REAL} is frequency-dependent. At very high frequencies $Z_{REAL} \to 0$ and at very low frequencies $Z_{REAL} \to R$. With respect to Z_{IM}, at very high frequencies $Z_{IM} \sim -1/\omega C$ and has some finite value, and at very low frequencies $Z_{IM} \to 0$. These observations in fact represent very important yet simple practical guidance to exercise caution when using commercial equipment software for evaluation of impedance-derived frequency-dependent parameters such as conductivity and permittivity.

The $R|C$ model represents bulk conduction in many nonpolar dielectric materials. The circuit also serves as the initial representation of the processes in the electrochemical double layer where the charge-transfer reaction impedance represented by R_{CT} is coupled with the double-layer capacitance C_{DL}. In the latter case the $R|C$ model allows one to clearly demonstrate how the double-layer charging with the capacitive current occurs at higher frequencies, followed by the resistive charge-transfer process at lower frequencies. The circuit represents a "shift" from conduction through a completely capacitance-dominated charging current at high frequency (as long as $1/\omega C < R$) to a completely resistance-dominated conduction current at low frequency (where $1/\omega C > R$). The corresponding impedance values shown for high frequencies are $Z_{REAL} \to 0$ ohm (where the actual resistance of R rejects any current flow through the resistive component) and for the low frequencies $Z_{IM} \to 0$ ohm (where the actual capacitive impedance of $1/\omega C$ is very high and rejects any current flow through the capacitive circuit branch). A blind analysis of conductivity or permittivity for such a circuit is entirely misleading; if software defines conductivity as $\sim 1/Z_{REAL}$ (instead of $\sim 1/R$), it will show very high real conductivity at high frequency, while in reality the system at high frequency does not conduct current through the high resistive impedance R. The high-frequency current is entirely capacitive, as indicated by phase angle $-90°$. For instance, for a highly resistive system with $R = 1$ Mohm, the imaginary impedance $Z_{IM} \sim 1000$ ohm, relative permittivity ~ 2, and real impedance $Z_{REAL} \sim 0$ ohm will be observed. The resulting "software calculated" conductivity as $1/Z_{REAL}$ at 1MHz will erroneously be very high.

At low frequencies the imaginary impedance $Z_{IM} \to 0$ resulting in very high "estimated" permittivity if $\varepsilon = \dfrac{d}{Z_{IM}\omega\varepsilon_0 A}$ is used instead of a correct $\varepsilon = \dfrac{Cd}{\varepsilon_0 A}$

definition. Such a situation may imply capacitive conduction, while in reality at low frequency the capacitive impedance is very high and rejects the current flow. The instrument shows $Z_{REAL} = 1$Mohm, and $Z_{IM} \rightarrow 0$ (showing very high corresponding permittivity to accommodate for $Z_{IM} \rightarrow 0$). In reality the media permittivity is still ~2 (for insulating oil) or ~80 (in the case of water), and the high capacitance value is caused by the impedance of a nanometer-thick double region and not by the impedance of the whole sample d (Section 5-3). Eq. 1-3 can be applied to determine sample capacitance C using high-frequency impedance parameters. A discussion of appropriate representation and selection of relevant frequency regions and geometry parameters will be offered in Section 6-3. With that in mind, it is imperative to start the analysis with an appropriate impedance model and to select frequency regions appropriate for identification of conductivity, resistivity, and permittivity characteristics of the sample.

4.3.   Randles $R_{SOL} - R_{CT}|C_{DL}$ circuit

A Randles circuit is one of the simplest and most common cell models used for many aqueous, conductive, and ionic solutions [4]. It includes only solution resistance R_{SOL}, a parallel combination of a double-layer capacitor C_{DL},

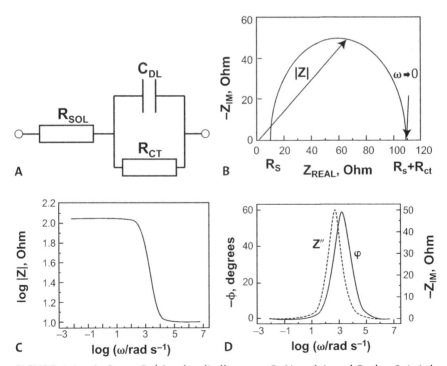

FIGURE 4-3   A. $R_{SOL} - R_{CT}|C_{DL}$ circuit diagram; B. Nyquist and Bode; C. total impedance; D. ZIM and phase angle plots

and a charge transfer or polarization resistance R_{CT} (Figure 4-3A). The Randles circuit is characterized by high-frequency purely resistive solution component R_{SOL} where the phase angle is $0°$ and the current is $I = V/R_{SOL}$. At lower frequencies the current flow is distributed between the double-layer capacitor C_{DL} at medium frequencies with a corresponding increase in the absolute value of the phase angle, and the finite charge-transfer resistor R_{CT} at lower frequencies where the phase angle, approaches $0°$ and the current approaches $I = V/(R_{SOL} + R_{CT})$. Therefore the details of current distribution in the Randles circuit are largely similar to those in the $R|C$ circuit, with the necessary correction for the presence of the additional series resistor R_{SOL} (Section 2-1). The time constant for the circuit equals $\tau = R_{CT}C_{DL}$, with the imaginary impedance reaching maximum absolute value at the critical frequency $f_c = 1/(2\pi R_{CT}C_{DL})$. The simplified Randles cell model is often the starting point for other more complex models, mainly for charge-transfer kinetics analysis in highly conductive solution systems not impeded by migration and diffusion mass-transport effects.

The Nyquist plot for a Randles cell is always a semicircle (Figure 4-3B). The solution resistance R_{SOL} can found by reading the real axis value at the high-frequency intercept where $Z_{REAL} \rightarrow R_{SOL}$ and $Z_{IM} \sim 1/\omega C_{DL} \rightarrow 0$. The real axis value at the low-frequency intercept is $Z_{REAL} \rightarrow R_{SOL} + R_{CT}$, equaling the sum of the charge-transfer resistance and the solution resistance $R_{SOL} + R_{CT}$, where $Z_{IM} \rightarrow 0$. The diameter of the semicircle is therefore equal to the charge-transfer resistance R_{CT}. The Bode impedance magnitude plot shows the solution resistance R_{SOL} at high, and the sum of the solution resistance R_{SOL} and the charge-transfer resistance R_{CT} at low frequencies. The phase angle is resistive $(0°)$ at high frequencies, changes to significant negative values when the impedance becomes partially capacitive at medium frequencies, and becomes again completely resistive at low frequencies.

The circuit analysis results in the expressions for total, real, and imaginary impedance of the circuit as:

$$Z*(\omega) = R_{SOL} + \frac{R_{CT}}{1+(\omega R_{CT}C_{DL})^2} - j\frac{\omega R_{CT}{}^2 C_{DL}}{1+(\omega R_{CT}C_{DL})^2}$$

$$(4-8)$$

$$Z_R(\omega) = R_{SOL} + \frac{R_{CT}}{1+(\omega R_{CT}C_{DL})^2}$$

$$(4-9)$$

$$Z_{IM}(\omega) = -\frac{\omega R_{CT}{}^2 C_{DL}}{1+(\omega R_{CT}C_{DL})^2}$$

$$(4-10)$$

4.4. Debye dielectric relaxation $(R_1 - C_1) | C_2$ circuit

A Debye circuit (Section 1-2) is another example of a common cell model used for many years as a popular representation for the impedance response of many dielectric materials. It includes dielectric material capacitance C_2 in parallel with a series combination of resistor R_1 and capacitor C_1, where $C_1 > C_2$ (Figure 4-4). Typically at high frequency the Debye circuit is characterized by a high-frequency purely capacitive response dominated by C_2, where the phase angle is $-90°$. The finite resistance R_1 at high frequencies prevents current from flowing through the $R_1 - C_1$ branch. As frequency decreases and capacitive impedances $1/\omega C_2$ and $1/\omega C_1$ increase, the magnitude of resistance R_1 becomes less than the difference between $1/\omega C_2$ and $1/\omega C_1$, and the current starts flowing through the $R_1 - C_1$ branch. The phase angle absolute value decreases to reflect the resistive impedance contribution and may approach

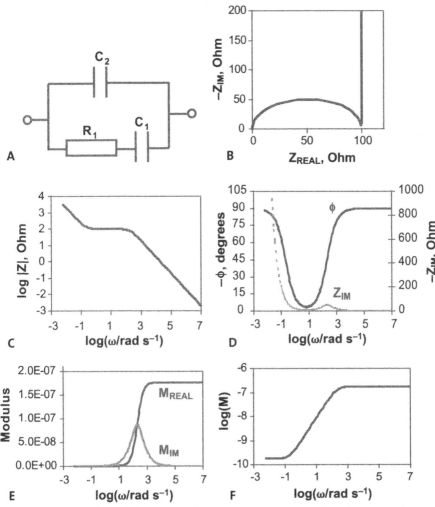

FIGURE 4-4   A. $(R_1\text{-}C_1) \mid C_2$ circuit diagram; B. Nyquist and Bode; C; total impedance; D; Z_{IM} and phase angle; E. modulus plots; F. modulus plots

~ 0°, with the current value approaching $I = V/R_1$. At low frequencies the capacitance C_1 dominates the impedance response of the $R_1 - C_1$ branch, and the phase angle again approaches –90°. The Debye circuit is completely capacitive at high frequencies, largely resistive in medium-frequency ranges, and again becomes completely capacitive (due to the C_1 element) at low frequencies. The time constant for the circuit equals $\tau_{C1} = R_1C_2$, with the imaginary impedance reaching maximum absolute value at the critical frequency $f_{C1} = 1/(2\pi\,R_1C_2)$. This critical frequency can be determined from the peak imaginary impedance and imaginary modulus values on the Bode plot and from the first (downward) intersection of the phase-angle plot with the –45° line.

The Nyquist plot for a Debye circuit is often a severely depressed semicircle (Figure 4-4B), with the resistive response often suppressed by the transitions from high-frequency capacitance to low-frequency capacitance. The resistance R_1 can be found by reading the real axis value at the medium-frequency intercept or from the diameter of the semicircle. The flat medium-frequency segment on the Bode total impedance plot can also be used to determine the resistance R_1 value (Figure 4-4C). However, the modulus notation is more useful for the analysis of the overall circuit structure (Figure 4-4E). The peak in the imaginary modulus corresponds to the critical frequency value. The capacitive values C_2 and C_1 can be easily evaluated from the horizontal segments in the total modulus plot at high and low frequencies, respectively. The circuit analysis results in the expressions for total, real, and imaginary impedance of the circuit as:

$$Z*(\omega) = \frac{\omega^2 R_1 C_1^{\,2}}{(\omega^2 R_1 C_1 C_2)^2 + \omega^2 (C_1 + C_2)^2} - j\,\frac{\omega^3 R_1^{\,2} C_1^2 C_2 + \omega(C_1 + C_2)}{(\omega^2 R_1 C_1 C_2)^2 + \omega^2 (C_1 + C_2)^2} \quad (4\text{-}11)$$

$$Z_R(\omega) = \frac{\omega^2 R_1 C_1^{\,2}}{(\omega^2 R_1 C_1 C_2)^2 + \omega^2 (C_1 + C_2)^2} \quad (4\text{-}12)$$

$$Z_{IM}(\omega) = -\frac{\omega^3 R_1^{\,2} C_1^2 C_2 + \omega(C_1 + C_2)}{(\omega^2 R_1 C_1 C_2)^2 + \omega^2 (C_1 + C_2)^2} \quad (4\text{-}13)$$

At very high frequencies $Z_{REAL} \to 1/\omega^2 R_1 C_2^{\,2} \to 0$ and at very low frequencies $Z_{REAL} \to R_1$. At very high frequencies $Z_{IM} \sim 1/\omega C_2 \to 0$, and at very low frequencies $Z_{IM} \to 1/\omega(C_1 + C_2) \to \infty$.

The Debye circuit has been described in the dielectric literature using the complex permittivity notation. The circuit serves as an expression for a "single dielectric relaxation" system, where transition occurs from high-frequency permittivity ε_∞ (or in this example capacitance C_2) to low-frequency permittivity ε_{LF} (capacitance C_1). As was shown above, in the Debye circuit the response is completely capacitive at high- and low-frequency extremes, making $\varepsilon* = \varepsilon'$ (real permittivity) for both ε_∞ and $\varepsilon_{LF} = \varepsilon_\infty - \Delta\varepsilon$ (Section 1-2).

$$\varepsilon*(\omega) = \varepsilon_\infty' + \frac{\Delta\varepsilon'}{1 + j\omega\tau} \quad (4\text{-}14)$$

It is important to state that the Debye circuit also contains another time constant $\tau_{C2} = R_1 C_1$ representing transition between resistive and capacitive conduction in the $R_1 - C_1$ branch with corresponding critical relaxation frequency f_{C2}. This critical frequency corresponds to the frequency where the loss factor ε'' is at maximum, or to the second (upward) intersection of the phase angle plot with the $-45°$ line.

On the practical level, the Debye transition typically occurs in dielectrics with bulk capacitance C_2 where the second conduction path takes place through abundantly present dipolar species (or particles) with characteristic capacitance C_1 and resistance R_1. These species become polarized in the external electric field at frequencies below the critical relaxation frequency of the circuit $\tau_{C2} = R_1 C_1$. For example, the model is adequate to describe conduction by polarized particles present at high concentrations (dozens of percent) in many highly resistive nonionic organic materials and dispersions, where no interfacial charging takes place even at very low frequencies (Section 12-3). High concentration of the polarized particles allows for the conductive path through the particles $R_1 - C_1$ to become noticeable in the impedance data. However, for the majority of practical colloidal and polymeric systems, the Debye relaxation typically reduces to a simple $R_2 | C_2$ bulk material model, where R_2 is finite resistance to conduction through the bulk solution. It should be remembered that the Debye model is in fact a reduction of a full relaxation circuit $R_2 | C_2 | (R_1 - C_1)$, as there is in principle always a conduction path through a (sometimes) very high but still finite bulk material resistance R_2 allowing for noticeable bulk solution conduction (or "loss") through R_2 at low frequencies. This bulk material conduction through the R_2 element often supersedes the conduction through the $R_1 - C_1$ branch which becomes undetectable in the impedance analysis. At even lower frequencies the effects of the particles' capacitance C_1 is frequently masked by the interfacial processes.

The Maxwell–Wagner dispersion effect due to conductance in parallel with capacitance for two ideal dielectric materials in series $R_1 | C_1 - R_2 | C_2$ can also be represented by Debye dispersion without postulating anything about dipole relaxation in dielectric. In the ideal case of zero conductivity for both dielectrics $(R_1 \rightarrow \infty, R_2 \rightarrow \infty)$, there is no charging of the interfaces from free charge carriers, and the relaxation can be modeled by a single capacitive relaxation-time constant.

CHAPTER 5

Impedance Representation of Bulk Material and Electrode Processes

In Chapters 1 through 3 the concept of impedance representation of a complete electrochemical system composed of electrodes, bulk media, and electrode-solution interface was introduced. In this chapter individual electrochemical segments of analyzed experimental or applied system will be separately discussed. Six typical components of the electrochemical process are considered [1, pp. 45–48]:

- uncompensated impedance in wires, electrodes, and related measurement artifacts
- bulk-media conduction and relaxation phenomena
- electric charging of the electrochemical double layer
- Faradaic electron transfer occurring near the electrode surface
- sorption (adsorption and/or desorption) of species at the electrode surface
- mass transport from the bulk solution to the electrode surface

More details on impedance analysis of practical electrochemical systems composed of these processes and their interrelationships can be found in Chapters 6 and 7. In these chapters an emphasis will be put on developing a complete equivalent circuit representation of an analyzed system and establishing relationships among various components of the circuit.

5.1.   Uncompensated impedance Z_{OHM}

Under most circumstances, the "uncompensated" impedance Z_{OHM} present due to electrodes, connectors, and conducting cables connecting electrochemical analyzers and electrodes is extremely small. The uncompensated impedance is typically represented by ~ 1 ohm range ohmic impedance R_{OHM}. However, all cables develop an inductance in very high (MHz) frequency ranges (Figures 5-1 and 5-2) which becomes particularly noticeable as at these high frequencies the impedance of the analyzed sample becomes comparably small. A more obvious case for cable loading effects is manifested when a multi-electrode probe is constructed or a very long cable (~ several meters) has to be used to connect impedance analyzer and electrodes (Chapter 8).

A general formula for cable's impedance Z_{OHM} in the external field of AC frequency f becomes dependent on the cable's properties—inductance L_{OHM}, capacitance C_{OHM}, and conductance G_{OHM} of insulator material separating the cable's wires. Typically the insulator's conductance is extremely low ($G_{OHM} \rightarrow 0$), and the circuit is represented by a parallel combination of capaci-

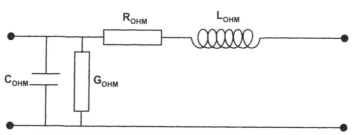

FIGURE 5-1   Equivalent circuit for a cable

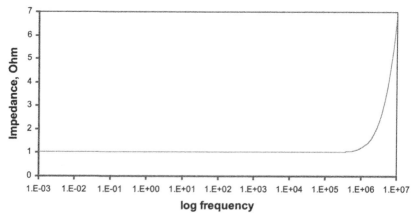

FIGURE 5-2   Representative analysis of cable impedance, for R_{OHM}~1 ohm, C_{OHM}~10^{-10} F and L_{OHM}~10^{-7} H

tive impedance and a series combination of the cable's low electrical resistance R_{OHM} and its inductance as:

$$Z_{OHM} =$$

$$\frac{R_{OHM}(1-\omega^2 C_{OHM}L_{OHM})+\omega^2 R_{OHM}C_{OHM}L_{OHM} + j(\omega L_{OHM}(1-\omega^2 C_{OHM}L_{OHM})-\omega R_{OHM}^2 C_{OHM})}{(1-\omega^2 C_{OHM}L_{OHM})^2 + (\omega R_{OHM}C_{OHM})^2}$$

(5-1)

For a typical cable, the values of $R_{OHM} \sim 1$ ohm, $C_{OHM} \sim 10^{-10}$ F, and $L_{OHM} \sim 10^{-7}$ H can be expected. In practical cables characteristic impedance is determined by the cable geometry (such as length and gauge) and the dielectric properties of the insulating material. At low frequencies the $2\pi f L_{OHM}$ term is insignificant compared to the R_{OHM}, which becomes parallel with the insulator's large low-frequency capacitive impedance. Therefore all current flows through the resistor and $Z_{OHM} = R_{OHM}$. At very high frequencies (over 1 MHz, as one would conclude from Figure 5-2) both frequency terms become large and the total cable impedance becomes:

$$Z_{OHM} \sim \frac{\omega L_{OHM}}{1-\omega^2 C_{OHM}L_{OHM}} \sim \omega L_{OHM}$$

(5-2)

Thus, the inductive effects due to the actual physical impedance loads of the cable' material and of the electrodes are expected at the highest frequencies.

Another source of high-frequency inductive behavior is mutual inductance of cables connecting a potentiostat to a multi-electrode cell. AC currents that flow in cables connecting the electrodes generate magnetic fields surrounding the cables. This time-varying magnetic field induces a voltage in the cables. Because the effect is magnetic, using shielded wires or coax does not eliminate the effect. These effects can be minimized by twisting the wires, and by proper wire placement. Moving the wires will change the observed inductance. The remaining effect is difficult to "calibrate out" unless the positions of the wires are strictly controlled and reproducible. Finally, use of three- and four-electrode probes often results in various instrumentation-related measurement artifacts that may appear as additional relaxations at high, medium and low frequencies. These artifacts will be discussed in more detail in Chapter 8.

5.2. Bulk media impedance—R_{SOL}, R_{BULK}, and C_{BULK}

Impedance analysis is essential in investigation of the motion of ions, polarized particles, and other conducting species in the "bulk" media under an external electric field. As was discussed in Chapter 1, the impedance analysis treats bulk and interfacial processes separately, with separation being effectively achieved on the basis of selective responses of bulk and interfacial processes to sampling AC frequencies. Apart from convection (if present), an electrical current in the bulk liquid and solid materials is transported by migration and charging effects of electrons, ions, conductive particles, and dipoles. The conduction mechanism is different for various investigated systems. For example, bulk-media conduction in polar liquids is typically purely ionic, in many solid conducting polymers it is mixed ionic and electronic, and in composite materials it is often based on percolation and charging of conductive particles dispersed in insulating base media [1, p. 21].

The overall migration current equation is often simplified in terms of its linear dependency on an applied electric field V. Ohm's Law is obeyed as $i = V/R$. If in a bounded sample area with the surface area of the electrodes exposed to the analyzed sample A, and thickness of the sample between the electrodes d, a uniform current is being transported between two electrodes. The "bulk" media resistance is defined as:

$$R_{BULK} = \rho \, \frac{d}{A} \qquad (5\text{-}3)$$

where ρ [ohm m] is resistivity of the analyzed homogeneous material. The material electrical conductivity σ [ohm^{-1} m^{-1}] is an inverse of resistivity and is more commonly used in material science. The standard units for σ are siemens per meter (Sm/m), which is the reciprocal of the ohm, so 1 Sm = 1/ohm. Following that, the concept of bulk-material resistance R_{BULK} can be introduced, which is related to the measured conductivity as:

$$R_{BULK} = \frac{1}{\sigma} \frac{d}{A} = \frac{d}{AF} \, \frac{1}{\sum_i z_i u_i C_i{}^*} \qquad (5\text{-}4)$$

The bulk resistance of a homogeneous material depends on the bulk concentration C_i^* of conducting species, their mobility u_i, charge z_i, sample temperature T, and the electrode geometry of the area A in which current is carried. This equation has been developed for ionic conduction, but it can also be broadly applied to other types of conduction by charged species. For a well-defined ionic system, the total media conductivity σ can be calculated from specific ion conductance σ_i of the ions or other species responsible for the conduction in the sample.

This simplified model of migration-driven ionic conduction in the bulk media is strongly related to a concept of ionic electrophoretic mobility. Under an application of an external electrical field, ions and charged conducting particles start moving and participating in conduction of electrical current. In the first approximation, the finite-sized charged conducting particles can be essentially treated as ions or macro-ions. The resulting current depends on the electrophoretic mobility u_i of the participating ions or conducting particles, which, as will be shown in Section 7-3, is determined by their permittivity, zeta-potential (ζ), and viscosity of the liquid [1, p. 32]. For this simplified analysis, which assumes a linear relationship between the measured current and sample conductivity, the overall bulk conductivity σ is governed by the migration of participating charged species. For a given type of conducting species of molar concentration C_i^*, radius a, charge z_i in solvent of viscosity η, and applied voltage V, their contribution into solution conductivity becomes, according to Stoke's Law:

$$\sigma_i = \frac{z_i^2 e_0 F}{6\pi\eta a} \, C_i^*$$ (5-5)

With a corresponding mobility u_i and velocity v_i of conducting species with sizes much smaller than the Debye length $\lambda_{DEBYE} \sim 10nm$ ($a << \lambda_{DEBYE}$) in an external electric field V:

$$u_i = \frac{v_i}{V} = \frac{\sigma_i}{zFC_i^*} = \frac{ze_0}{6\pi\eta a}$$ (5-6)

For a general case where the particle diameter-to-Debye length ratio is large ($a >> \lambda_{DEBYE}$), the mobility equation becomes:

$$u_i = \frac{\varepsilon\varepsilon_0\zeta}{\eta}$$ (5-7)

Standard chemical handbooks list σ values for specific solutions at certain (typically room) temperatures. The actual conductivity, however, may vary with the electrolyte concentration. At very low concentration conductivity decreases, as there are not enough ions to support the conduction. At very high concentrations the electrostatic interactions between the ions reduce their mobility if the ions are very tightly packed. For electrolytes of low permittivity the electrostatic forces between the particles increase and ionic pairs can be formed. Higher temperatures may lead to formation of a less coherent ionic atmosphere, resulting in higher conductivity.

The presence or absence of supporting electrolyte is responsible for the relative magnitudes of bulk resistance and capacitance. In a bulk solution with free ions there is electroneutrality and the sum of charges is zero as $\sum z_i C_i^* = 0$. This condition does not prevail in space charge regions at the electrode-media boundaries, where migrating charged species are attracted to the electrode interface of the opposite charge, resulting (along with the diffusing species discharging at the electrodes) in the formation of a double layer.

The presented bulk impedance analysis assumes essentially no effects from interfacial electrochemical reactions, electrochemical double-layer charging, or diffusion-driven mass transport in the space-charge regions on the measured electrical parameters such as current and conductivity, and certainly simplifies just about any experimental electrochemical system to an extreme. This simplification is largely valid only at high AC frequencies, where the current must be transported through the bulk media and overcome the bulk impedance. In aqueous systems with excess of "supporting electrolyte" R_{BULK} is typically very small, on the order of several ohms, resulting in small "IR drop" [1, pp. 45–48]. For these systems, the bulk impedance Z_{BULK} is typically expressed in the literature as a purely real impedance component, represented by a solution resistance ($R_{SOL} \sim 1/\sigma$) with associated low ohmic values. However, the solution resistance R_{SOL} often becomes a significant factor in the total impedance of an electrochemical cell, as even a relatively small R_{SOL} complicates the impedance analysis and data representation [2], as was shown in Chapter 2. Any solution-resistance contribution between the electrodes must be considered when an electrochemical cell is modeled. A modern three-electrode potentiostat employs various more or less accurate methods of compensation for the total solution resistance between the working and counter/reference electrodes. However, as will be shown in Chapter 8, in three- and four-electrode electrochemical cells the solution resistance R_{SOL} may still introduce various distortions and instrumentation-related artifacts. The IR drop has to be subtracted from the total impedance by experimental or mathematical means to better quantify the interfacial electrode-related kinetics (Chapter 2).

In conductive aqueous solutions with an excess of supporting electrolyte, R_{SOL} and associated media electrical conductivity σ are independent of the applied potential and sampling AC frequency [1, p. 68]. The supporting electrolyte is not interfering with the electrochemical discharge at the electrodes and is not expected to participate in any charge-transfer reactions under the applied combinations of DC and AC potentials. Therefore, the supporting electrolyte is present at the same constant volumetric concentration during the experiments (as it is not consumed or produced at the electrodes) and is not affected by the concentration-driven diffusion mass transport.

However, in electrochemical impedance analysis one must always consider several factors that may challenge the above premises. First of all, most electrochemical cells do not have uniform current distribution through a definite electrolyte area. The major problem in accurately calculating the solution resistance is concerned with determination of the current flow path and the geometry of the media that carries the current. In a series of recent papers [3, 4, 5, 6, 7], the influence of nonuniform current and potential distributions associated with a macrosized working electrode geometry was explored. These studies have shown that at high frequency noticeable imaginary impedance with an associated CPE element is present due to the radial distribution of the electrochemical potential and resulting localized heterogeneity in the bulk solution impedance Z_{BULK}. However, as a first approximation fitting experimental EIS data into a model allows estimating R_{SOL} values from the intercept of the semicircle with the real axis at high frequencies, where impedance of double-layer capacitance is very small and charge-transfer resistance is effectively shorted out [8].

Secondly, even a conductive aqueous-based solution in principle possesses a capacitive current path in parallel with the purely resistive R_{SOL}. The resulting effects of the imaginary bulk impedance component should often be considered—for instance, in the analysis of tissues, cells, and proteins. The bulk impedance of biological media is often composed of extracellular resistive and intracellular capacitive components, and the analyzed systems can be viewed as aqueous colloids with dipolar particles or micelle-based conduction. Ions of opposite charge surround a charged particle (or dipole) within the Debye length λ_{DEBYE}. When a charge on an ion disappears, it takes about 1μs for the surrounding ions to rearrange themselves, which in itself can be viewed as an example of high-frequency relaxation.

If an aqueous solution is replaced by a tissue or a dielectric medium, a more complex circuit consisting of both resistive and capacitive elements replaces the R_{SOL} resistor. This more complicated circuit is represented by a parallel combination of $R_{BULK} | C_{BULK}$, with impedance response to bulk solution processes dominating the kHz-MHz frequency ranges (Chapter 11). As will also be shown in Chapter 7, in complex multicomponent media several relaxations represented by a combination of several $R_{BULK} | C_{BULK}$ elements may be present. As the first approximation, a single $R_{BULK} | C_{BULK}$ element can be selected to represent the bulk-material relaxation. For the bulk processes in dielectrics the R_{BULK} represents a lossy part of the relaxation mechanism, and C_{BULK} is a dipolar capacitive contribution [1, p. 68].

In highly conductive samples the contribution of C_{BULK} is not observed within a typical impedance measurement range, and the bulk solution resistance R_{BULK} is very low (several ohms) and always presents the path of the least impedance to the current. The capacitive relaxation time constant $\tau = R_{SOL}C_{BULK}$ becomes very small (less than 10^{-7} sec, which is detectable in a frequency range in excess of 10MHz), and the very high-frequency capacitive contribution is still present but is unnoticeable except at very low temperatures [9, p. 14]. Therefore, in conductive aqueous and polar nonaqueous systems the capacitive contribution is typically not considered. C_{BULK} is omitted for supported situations, as experimental measurements rarely extend to high enough frequencies (which are often outside the measurement range of many impedance analyzers) to make C_{BULK} contribution noticeable [9, p. 104], and the bulk solution impedance is represented solely as $Z_{BULK} = R_{SOL}$.

In highly resistive systems the concept of bulk solution impedance composed of resistive R_{BULK} and capacitive C_{BULK} components is introduced. A simple equivalent circuit [10] can be used to represent such a system as a parallel combination of capacitance C and AC conductance $G_{AC} = \sigma/\varepsilon_0$, equivalent to the frequency-dependent $R_{BULK} | C_{BULK}$ coupling derived in Eq. 1-13 for complex permittivity:

$$C_C = C'_C - j\frac{G_{AC}}{\omega} \qquad (5\text{-}8)$$

The above dielectric (or complex capacitance) notation and Debye dispersion (Eq. 1-15) have often been used to describe a single bulk-media dielectric relaxation process in organic and polymeric (lossy) systems where at least two components with resistive and capacitive features exist [9, p. 33]. The permittivity of a lossy dielectric with negligible parallel DC conductance can be expressed on the basis of the Havriliak-Negami model (Eq. 1-16). Equivalent circuits representing a Debye model for lossy dielectric, where $C_{INF} = C_{HF} \sim \varepsilon_\infty$, $C_0 = C_{LF} \sim \varepsilon_{LF}$, $R \sim 1/G$ [1, p. 65, p. 216], are shown in Figure 5-3.

The Debye circuit represents a transition from a conduction mechanism through the first capacitive C_{INF} at high frequency to the second independent higher-value capacitive component C_0 in a series with finite resistor R. At very high frequencies C_{INF}, usually bulk capacitance C_{BULK} of ion-free "matrix" continuous-phase material such as nearly completely insulating oil, dominates the total impedance response. The finite resistor R prevents current from flowing through the secondary resistive-capacitive branch. This branch very often rep-

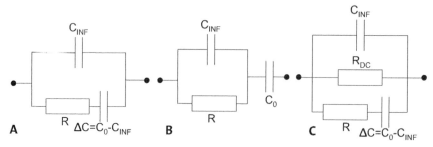

FIGURE 5-3 Parallel (A) and series (B) equivalent circuit versions for a Debye single dispersion in lossy dielectrics. Circuit (C) represents realistic Debye dispersion with lossy relaxation

resents suspended polarized particles, which provide second current path through a series combination $R - \Delta C$. The polarizable particles produce characteristic concentration-dependent capacitance C_0, and at the same time cannot play the role of strong predominant conductor, as in that case the particles' mobility will shorten the circuit through the low-frequency resistance R_{DC} as shown in Figure 5-3C. At very low frequencies the capacitive impedance values become very high and R becomes negligible, and two capacitors C_{INF} and $\Delta C = C_0 - C_{INF}$ are effectively in parallel and are added to each other. Therefore, at both high and low frequencies the circuit is capacitive and can be effectively analyzed in the complex modulus or permittivity plots. The characteristic relaxation frequency f_C appears somewhere between 10MHz and 1 Hz corresponding to $\omega_{C2}\tau = 1$ and the maximum ε'' with $\tau_{C2} = R\Delta C$ (Section 4-4).

Cole and Cole found that frequency dispersion in dielectrics results in an arc in the complex ε plane with its center below the real axis. This dispersion is modeled by a modified equivalent circuit where pure resistor R is replaced by a CPE. The model can be expressed with the following empirical formula derived from the original "ideal" Cole and Cole (Eq. 1-15) for the complex dielectric constant and the CPE parameter Q:

$$\varepsilon^* = \varepsilon_\infty + \frac{\varepsilon_{LF} - \varepsilon_\infty}{1 + (j\omega\tau)^\alpha} \tag{5-9}$$

$$\tau = \left[\frac{C_{LF} - C_{HF}}{Q} \right]^{1/\alpha} \tag{5-10}$$

In realistic highly resistive systems the capacitive response dominates either at the high frequency (kHz-MHz) range only in finite R_{DC} allows current to flow at low frequencies, or in the whole range of the impedance measurements in the case of insulating substrates that do not develop sizeable conductive "loss" even in sub-Hertzian frequencies ($R_{DC} \to \infty$). For instance, the latter cases take place in classical "ideal" dielectric materials, where no ionic migration occurs. This situation occurs in many insulating engineering plastics (such as Teflon®) and transformer oils with Tohm bulk media resistance.

In many other highly resistive materials with associated bulk impedance values above 1 Mohm, the range of frequencies corresponding to the bulk-material impedance can still cover several decades of frequency range—often from MHz to low Hz. When conduction through the bulk is not ionic, there would be no significant accumulation of charges at the interface and essentially no double-layer capacitance develops. The total circuit is represented by C_{BULK}, and the total thickness of the sample d can be used to estimate this capacitive response.

The systems represented by the classical Debye dispersion model at constant temperature exist as multicomponent systems. So-called "electrorheological fluids" (Section 12-3) represent examples of such a system. In many other "real-life" systems the Debye capacitive transition is often not observed due to the presence of measurable ionic conduction ($= 1/R_{DC}$) in parallel with high-frequency C_{INF} bulk capacitance and resulting "lossy relaxation" with time constant $\tau = 1/R_{DC}C_{INF}$. Even in highly resistive "unsupported" systems there is always a possibility for current to flow through the bulk-material resistance, bypassing the capacitive element. In this case, a simple transition occurs from high-frequency capacitive to a resistance-dominated conduction path through the media at low frequencies, with characteristic frequency easily determined at a point where the phase angle becomes equal to $-45°$ (Section 4-2).

In highly resistive solutions, unlike in "supported" electrolytes, electrical conductivity σ and sample resistance show a significant dependence on temperature and applied electric field voltage and frequency. In colloidal systems where conduction is governed by percolation of polarized particles, the application of AC voltage amplitude above a certain threshold may cause an "electrorheological effect" of conductive chain formation, short circuiting the system. Also quite frequently the ionic conduction leads to "interfacial polarization" at the electrodes, producing a low-frequency capacitive response that masks the Debye relaxation process. An application of external electric field in many ionic nonpolar liquids causes the buildup of surface charges near the electrodes without charges crossing the interface due to blocking [11]. The interfacial polarization may result in an additional sizable voltage drop at the interface and diminished effective electric field in the bulk media. Charged particles and ions in the bulk will move more slowly because of a diminished effective electric field and appear to have lower electrophoretic mobility than they actually do, resulting in a measurable increase of R_{BULK} [12]. Polarization of electrodes to large potential values is a problem in low-conductivity media where there are only a few oxidizable or reducible polar species that can eliminate polarization through charge transport on electrodes. Reduction of electrode polarization is usually accomplished by reversing the direction of the electric field every few seconds, which is equivalent to operating in a sufficiently high AC frequency regime.

Commercial conductivity meters rely on operation at often arbitrarily chosen "high" AC frequencies to avoid the effects of interfacial polarization on measured bulk-solution conductivity. The reason behind this selection of operational frequency is based on the assumption that the measurements at frequencies lower than 1 kHz will result in an additional large contribution of interfacial polarization and double layer capacitive effects to the total "measured" impedance, producing values that have little in common with pure bulk-solution properties. Another experimental alternative for eliminating the effects of interfacial polarization is a four-electrode setup with two independent reference electrodes. The employment of a four-electrode probe often results in significant improvements in the accuracy of conductivity measurements for highly conductive samples. However, as will be demonstrated in Chapter 8, this option is very difficult to implement appropriately for even moderately higher bulk-impedance media, and sometimes even for highly conductive fluids, as many experimental artifacts may influence the measured conductivity value [11].

Traditional dielectric analysis of highly resistive materials, such as ceramics, plastics, polymers, and colloids, is frequently more concerned with purely bulk-material effects and often relies on non-electrochemical analysis methods, such as modulation of the sample temperature (Chapter 1). In addition, the temperature effects on bulk-solution conductivity values must be considered in any experimental setup. Most referenced solution conductivities are measured and reported at room temperature, and for a standard laboratory aqueous-solution analysis this information is appropriate to consider as the first approximation. However, the bulk-solution conductivity σ of aqueous media increases at ~2% per degree °C as the result of decrease in viscosity of water with increase in the temperature [1, p. 17].

For many nonaqueous systems temperature modulation is one of the most effective methods of perturbation-based analysis. Conductivity-temperature dependence in various systems with ionic conductivity typical of activated mechanisms can usually be described by the Arrhenius equation, derived from the Nernst-Einstein and Fick equations describing DC conductivity based on ion hopping through a structure [13]:

$$\sigma = \frac{A_R}{T} e^{-\frac{E_A}{K_B T}} \qquad (5\text{-}11)$$

where A_R = constant proportional to the number of carrier ions, K_B = Boltzman constant, and E_A = activation energy. A modified expression for solution conductivity-temperature dependence for polymers with ionic conductivity typical of activated mechanisms can be described by the Vogel-Tammann-Fulcher (VTF) equation:

$$\sigma = \frac{A_R}{\sqrt{T}} e^{-\frac{E_A}{K_B(T-T_0)}} \qquad (5\text{-}12)$$

where T_0 = equilibrium glass transition temperature below which no ionic conduction is detectable [14]. A more complex generalized Vogel equation was derived [15] based on ion hopping between the defects and accounting for both temperature and pressure dependence of conductivity:

$$\sigma = \frac{A_R}{T(1-\delta)} e^{-\frac{E_A T_0^{1.5}}{(T-T_0)^{1.5}(1-\delta)}} \qquad (5\text{-}13)$$

Where $(1-\delta)$ = fractional volume change of the material due to the pressure effects.

For a majority of liquid and solid experimental systems with ionic conduction, an increase in temperature leads to an increase in media conductivity σ according to Eqs. 5-11 through 5-13. A small decrease in the bulk-media capacitance and an increase in conductivity with temperature may be observed in dipolar or particle-based conduction systems such as organic colloids and liquid polymers. These effects are due to the redistribution of species' electrical properties with temperature and an increase in the number of conducting and a decrease in the number of capacitive (or dipolar) species [1, p. 17]. In the majority of solid polymers and plastics the temperature increase leads to an increase in both capacitance and conductivity of the material.

The bulk-solution resistance R_{BULK} was also found to be proportional to a square root of viscosity-density product (Eq. 5-14). This empirical relationship is being used in rheological analysis and development of surface acoustic-wave sensors for viscosity measurements for systems where an assumption of constant density holds [16]:

$$R_{BULK} \sim \sqrt{\rho_D \eta} \qquad (5\text{-}14)$$

5.3. Electrochemical double layer capacitance C_{DL}

An electrical double layer exists at the interface between an electrode and its surrounding electrolyte. This double layer is formed as ions and other charged species from the solution "stick on" the electrode surface. Charges in the electrode are separated from the charges of these ions resulting in a potential drop across the separation distance, establishing a double layer capacitor C_{DL}. This charge separation is very small, on the order of several nanometers. As was discussed in Section 3-2, the measured double-layer parameter is typically not an ideal capacitor, as some heterogeneity should be expected due to porosity of solid electrodes, electrode material, atomic-scale inhomogeneity, applied voltage, and current distribution. The impedance diagram often reveals a depressed semicircular feature, and an ideal double-layer capacitor C_{DL} is replaced by a CPE_{DL} representation in parallel with a charge-transfer resistance R_{CT}. The double-layer capacitance charging typically occurs in a Hertzian frequency range (~10–1000 Hz in conductive aqueous solutions, and at ~ 1Hz in resistive colloids) at room temperatures.

The actual value of the double-layer capacitance depends on many variables including electrode type, electrochemical potential, oxide layers, electrode surface heterogeneity, impurity adsorption, media type, temperature, etc. [1, pp. 45–48]. Capacitance of the double layer also largely depends on the intermolecular structure of the analyzed media, such as the dielectric constant (or high-frequency permittivity), concentration and types of conducting species, electron-pair donicity, dipole moment, molecular size, and shape of solvent molecules. Systematic correlation with dielectric constant is lacking and complex, due to ionic interactions in the solution. In ionic aqueous solutions with supporting electrolyte ("supported system") the values of C_{DL} ~10–60 µF/cm^2 are typically experimentally observed for thin double layers and solution permittivity $\varepsilon \sim 80$. The double-layer capacitance values for nonpolar dielectrics ("unsupported systems") are usually ~1 $\mu F/cm^2$, or ~10–100 less that those for water [17, p. 170]. In organic and polymeric systems many differences compared to polar liquids are apparent, such as smaller values of dielectric constants ($\varepsilon \sim 2 - 5$), low values of potential gradient $\Delta\phi/\Delta x$ that may extend further away from the electrode across the whole sample thickness, resulting in thicker double layers [18], lower ionic concentration C^*, smaller electrosorption valency, weak interaction between the adsorbate and electrode, and often impeded charge transfer [19, 20, 21].

The double-layer capacitance is composed of several contributions. In a geometrical sense the double layer in "supported" systems is represented by the compact "Helmholtz" or "Stern" layer. The electrostatically attracted solvated species reside in the "outer Helmholtz plane" (OHP), and specifically adsorbed species reside closer to the electrode in the "inner Helmholtz plane" (IHP). The double-layer structure is completed by a "diffuse" layer, composed of electrostatically attracted species at some distance from the electrode surface. The full thickness of the double layer can be defined as the external boundary of the diffuse layer separating it from the bulk solution, where the measured potential becomes equal to that of the bulk solution and no local potential gradient driven by the difference between the electrode potential ϕ_{EL} and the solution potential ϕ_{SOL} can be determined (Figure 5-4).

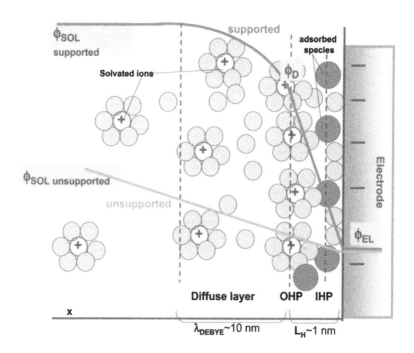

FIGURE 5-4 The electrode-sample interface

The double-layer thickness influencing the C_{DL} is different than the "diffusion layer" ($\sim 10^{-2}$–10^{-4} cm thick), which plays a role in the diffusion-limited mass transport and will be discussed in Section 5-6. The principles behind the two parameters are also very different—the diffusion layer is driven by a gradient of concentration that extends far into solution as the species are being consumed in the electrochemical reaction. The double-layer thickness results from an electrochemical potential difference between the media and the electrode, which decays over a much smaller distance from the electrode surface.

Total double-layer capacitance is composed of a series combination of the compact Helmholtz layer and the diffuse-layer capacitances as:

$$\frac{1}{C_{DL}} = \frac{1}{C_{HELMHOLTZ}} + \frac{1}{C_{DIFFUSE}} \qquad (5\text{-}15)$$

The Helmholtz capacitor $C_{HELMHOLTZ}$ (F/cm^2) is potential-independent and is expressed by a capacitor formula:

$$C_{HELMHOLTZ} = \frac{\varepsilon\varepsilon_0}{L_H} \qquad (5\text{-}16)$$

The thickness of the compact Helmholtz layer ($L_H \sim 1$–2 nm) is approximately equal to the length of closest proximity where the discharging ions can approach the interface. For aqueous media with relative permittivity $\varepsilon = 80$ and a ~ 2 nm thick Helmholtz layer, $C_{HELMHOLTZ} \sim 35\ \mu F/cm^2$. While the Helmholtz layer capacitance is generally considered to be a constant parameter that is equal to ~ 1–$60\ \mu F/cm^2$ for different media, the diffuse-layer capacitance is dependent on electrochemical potential and concentration of electrolyte. The diffuse-layer thickness depends on the ionic concentration in the sample, with samples with a high presence of supporting electrolyte ($\sim 0.1M$) having thin diffuse layers of ~ 10 nm. Unsupported samples with low ionic presence ($10^{-7}M$) can have thicker diffuse layers up to $\sim 1\mu m$. The thickness of diffuse layer can be determined from the formula for Debye length λ_{DEBYE}, which is the measure of the distance in the electrolyte over which a small potential perturbation decays by $1/e$—effectively a region of space charge with an excess of charges where electroneutrality no longer holds, derived from the Poisson-Boltzman equation:

$$\lambda_{DEBYE} = \sqrt{\frac{\varepsilon\varepsilon_0 K_B T}{8\pi e_0^2 \sum z_i^2 C_i{}^*}} \qquad (5\text{-}17)$$

$C_{DIFFUSE}$ (F/cm^2) increases for higher applied voltages and higher concentrations of charged species C^*. For higher electrolyte concentrations (>0.1 M) $C_{DIFFUSE}$ becomes constant, and total double layer capacitance C_{DL} is dominated by $C_{HELMHOLTZ}$. The correction for the value of electrochemical potential ϕ_D at the point of separation between the Helmholtz and the diffuse layers is introduced as the second multiplier in Eq. 5-18 as:

$$C_{DIFFUSE} = \frac{\varepsilon\varepsilon_0}{\lambda_{DEBYE}}\cosh\frac{z_i e_0 \phi_D}{K_B T} = \sqrt{\frac{8\pi\varepsilon\varepsilon_0 e_0^2 \sum z_i^2 C_i{}^*}{K_B T}}\cosh\frac{z_i e_0 \phi_D}{K_B T} \qquad (5\text{-}18)$$

In diluted solutions the double-layer capacitance C_{DL} increases rapidly when external DC potential V is applied [22]. As presented in Eq. 5-18, only the diffuse-layer capacitor $C_{DIFFUSE}$ (or "differential capacitance") depends on the electrode potential ϕ_D which is controlled by the externally applied DC potential V. This equation, a derivation from the Guoy-Chapman theory based on Poisson-Boltzmann equation [23] for the double layer differential capacitance, was also presented for a two-electrode experimental arrangement as [24]:

$$C_{DIFFUSE} = \frac{\varepsilon\varepsilon_0}{4\pi\lambda_{DEBYE}} \frac{1}{\tanh\dfrac{d}{2\lambda_{DEBYE}}} \cosh\frac{ze_0 V}{4K_B T} \qquad (5\text{-}19)$$

with d- electrode gap. For most cases $d/\lambda_{DEBYE} >3$ (that is, the separation gap between the electrodes is more than three times larger than the thickness of the diffusion layer), and $[\tanh(d/\lambda_{DEBYE})]^{-1} \to 1$, while $\cosh(ze_0 V/K_B T)$, and $C_{DIFFUSE}$ linearly increase with the applied DC potential V. The previous equation reduces to:

$$C_{DIFFUSE} = \frac{\varepsilon\varepsilon_0}{4\pi\lambda_{DEBYE}} \cosh\frac{ze_0 V}{4K_B T} \qquad (5\text{-}20)$$

The double-layer capacitance can be also represented by a kinetic notation [1, p. 115], where C_{DL} is treated as a combination of $C_{DIFFUSE}$, $C_{HELMHOLTZ}$ (which is composed of the capacitances of the outer and inner Helmholtz layers C_{OHP} and C_{IHP}), reaction capacitance C_R, and the adsorption pseudocapacitance C_{ADS} (if it is present). C_R accounts for the capacitive contribution of ions located in both IHP and OHP that are participating in charge-transfer reactions. When adsorption is present, the adsorption-capacitance element C_{ADS} can be added in parallel with C_{IHP} (or with both C_{OHP} and C_{IHP}), assuming that both C_{OHP} and C_{IHP} originate from supporting electrolyte species. Typically it is assumed that these contributions are minor, as migrating ions cannot fully penetrate the compact portion of the double layer. The total representation for the double-layer capacitance becomes $C_{DIFFUSE} - \{C_R \mid [C_{OHP} - C_{IHP} \mid C_{ADS}]\}$ or $C_{DIFFUSE} - \{C_R \mid C_{ADS} \mid [C_{OHP} - C_{IHP}]\}$.

The famous Mott-Schottky relationship [25, 26] in Eq. 5-21 represents a different potential-dependent surface capacitive case. This relationship was derived to express the electronic properties of passive capacitive films of constant thickness formed on metals. The methods based on the Mott-Schottky equation have been widely used as a valid tool to determine semiconductive character and dopant density of the surface films in the semiconductor industry and in corrosion studies. The change of the space-charge layer capacitance C_{SC} of the passive film (or space charge distribution) depends on the difference between the applied DC potential V and flat band potential V_{FB} characteristic of the surface film, where N_D = concentration of donors (or acceptors) or "doping density" ($\sim 10^{12} - 10^{22}$ cm^{-3}), and $e_0 = 1.6 \, 10^{-19}$ C electron charge:

$$\frac{1}{C_{SC}^2} = \frac{2}{\varepsilon\varepsilon_0 e_0 N_D}\left(V - V_{FB} - \frac{K_B T}{e_0}\right) \qquad (5\text{-}21)$$

The donor density N_D can be calculated from the slope of the $1/C^2$ vs. V curve, and the flat band potential V_{FB} can be determined by extrapolation to $C = 0$. For a p-type semiconductor response $1/C^2$ vs. V should be linear with a negative slope that is inversely proportional to the acceptor concentration N_D. On the other hand, for an n-type semiconductor the slope is positive and is inversely proportional to the concentration of donors in the film.

This calculation model is based on an assumption that total capacitance of the system C estimated from the inverse relationship to the Z_{IM} is composed of two capacitances—that of the space-charge region C_{SC} and that of the double layer C_{DL}. Since capacitances are in series, the total measured capacitance is the sum of their reciprocals. As the space-charge capacitance C_{CD} is typically ~ 100 times smaller than the double-layer capacitance C_{DL}, the double-layer capacitance contribution to the total capacitance is negligible. Therefore, the capacitance value calculated from this model is assumed to be the value of the space-charge capacitance, C_{SC}, which decreases with the applied potential. This model is adequate in particular if the sampling frequency is high enough (on the order of kHz). There is an alternative expression relating capacitance of a passive film to an inverse overpotential as C_{SC}~$1/V$, which covers cases where film thickness changes occur.

There is another capacitive element in the electrochemical charge-transfer process, which is called pseudocapacitance $C_{PSEUDOCAP}$ [17]. Pseudocapacitance is Faradaic in nature and therefore is different from C_{DL}. Pseudocapacitance usually originates from electrosorption (specific adsorption) processes and related partial electron transfer or surface charging (Section 5-5). On metal oxide "supercapacitor" types of electrodes (such as iridium and ruthenium oxides) possessing several oxidation states, pseudocapacitance originates from potential dependence of multiple oxidation/reduction couples that are active on the surfaces of the materials. Another common source of pseudocapacitance is charging-discharging of redox-active polymers such as polyaniline and polypyrrole.

Pseudocapacitance can be 5 to 30 times larger than C_{DL}, sometimes reaching ~ 700–800 $\mu F/cm^2$. When present, pseudocapacitance should be placed in parallel with C_{DL}, resulting in an equivalent circuit $R_{SOL} - C_{DL} | \{R_{CT} - (C_{PSEUDOCAP} | R_{CT2})\}$, where R_{CT2} as "leakage resistance of pseudocapacitor" may not be present. Such a system in principle illustrates pseudocapacitance originating from specific adsorption, similar to what was discussed above for "kinetic" representation of C_{DL} composition with the adsorption C_{ADS} and the double-layer C_{DL} capacitors in parallel. In multicomponent systems several combinations of $C_{ADS} | R_{ADS}$ elements may be present. For such complex cases it is fundamentally impossible to accurately distinguish between C_{DL} and parallel Faradaic pseudocapacitance $C_{PSEUDOCAP}$ [17], apart from cases where separate capacitive elements can be identified in the impedance diagrams.

5.4. Electrochemical charge-transfer resistance R_{CT}

When an interface is perturbed from its equilibrium by means of an external energy source, a permanent flow of charge and matter takes place. This is due to:

1. the existence of electrochemical reactions allowing the electric charge transfer between the electronic conductor (metal or semiconductor)

electrode and the conductor of ionic or other charges (liquid or solid electrolyte)

2. the gradients of electric and chemical potentials that charge the interfaces and allow the transport of the reacting species between the bulk of the electrolyte and the interfacial reaction zone

Electrode reaction is usually composed of charge transfer, adsorption/desorption, and mass transport parts that are present at the low Hz-μHz range of the impedance spectrum. Electrode "polarization" occurs whenever the potential of an electrode V is forced away from its equilibrium value at an open circuit V_{EQ}. When an electrode is polarized, it can cause current to flow via electrochemical reactions that occur at the electrode surface at characteristic redox potential V_0 or charge the interface with overabundant species that cannot discharge due to kinetic restrictions, such as sluggish electrode reaction, adsorption, or diffusion limitations [9, p. 100].

The charge-transfer resistance R_{CT} is related to the activation energy and the distance between the species and the electrode so the charge transfer may occur. The charge-transfer reaction is controlled by the so called "Faradaic" response, which is the rate of electron transfer to the electroactive species in the media occurring near the electrode surface in the double layer. The charge transfer may occur by tunneling over 1–2 nm distances separating the electroactive species from the interface (Figure 5-4). This charge-transfer process has a certain speed reflected in Faradaic current which is dependent on concentration of discharging (electroactive) species, type of reaction, applied electrochemical AC and DC field, temperature, pressure, surface area of the electrodes, and convection (by the rotation speed of the working electrode, bubbling of nitrogen, or external magnetic field). R_{CT} parameter dominates the total impedance response at low Hz and mHz regions and at DC.

The amount of current is controlled by the kinetics of the electrochemical reactions and the diffusion of reactants both toward and away from the electrode. The general relation between the potential and the current is [23, p.99]:

$$i = i_0 \left[\frac{C_{OX}}{C_{OX}{}^*} e^{\frac{\alpha z F(V-V_{EQ})}{R_G T}} - \frac{C_{RED}}{C_{RED}{}^*} e^{\frac{-(1-\alpha)zF(V-V_{EQ})}{R_G T}} \right] \tag{5-22}$$

Where:

i_0 = exchange current density
C_{OX} = concentration of oxidant at the electrode surface
$C_{OX}{}^*$ = concentration of oxidant in the bulk
C_{RED} = concentration of reductant at the electrode surface
$C_{RED}{}^*$ = concentration of reductant in the bulk
F = 96500 C/mol (Faraday constant)
T = absolute temperature in K
R_G = 8.31 J/(K mol) = gas constant
α = reaction order ("transfer coefficient")
z = number of electrons involved
V = applied potential
V_{EQ} = equilibrium electrode potential

When the concentration in the bulk is the same as at the electrode surface, $C_{OX} = C_{OX}{}^*$ and $C_{RED} = C_{RED}{}^*$, yielding the Butler-Volmer equation (Eq. 5-23):

$$i = i_0 \left[e^{\frac{\alpha z F(V - V_{EQ})}{R_G T}} - e^{-\frac{(1-\alpha)zF(V-V_{EQ})}{R_G T}} \right] \tag{5-23}$$

Charge-transfer resistance R_{CT}, represented in Eq. 5-24, is a non-linear element controlled by the Butler-Volmer relationship [1, pp. 45–48], with strong dependence on the electrochemical potential. External convection will enhance mass transport and minimize diffusion effects, keeping the assumptions of $C_{OX} = C_{OX}{}^*$ and $C_{RED} = C_{RED}{}^*$ valid:

$$R_{CT} = \frac{R_G T}{z F i_0} \tag{5-24}$$

The kinetics rate constant k_0 can be expressed as a function of charge-transfer resistance at the electrode with surface area A [23, p. 292]:

$$k_0 = \frac{R_G T}{z^2 F^2 R_{CT} A} \frac{1}{\sqrt{\frac{D_{OX}^\alpha}{D_{RED}^\alpha} C_{OX}^\alpha C_{RED}^{1-\alpha}}} \frac{1+e^\alpha}{e^{\alpha(1-\alpha)}} \tag{5-25}$$

$$k_0 \sim \frac{R_G T}{z^2 F^2 R_{CT} A} \frac{1}{C_{OX}^\alpha C_{RED}^{1-\alpha}} \tag{5-26}$$

Where: exchange current [9, p. 69] is:

$$i_0 = zFAk_0 C^*{}_{OX} \, e^{-\frac{\alpha n F}{R_G T}(V - V_{EQ})} = zFAk_0 C^*{}_{OX}^{1-\alpha} C^*{}_{RED}^\alpha \tag{5-27}$$

When the "overpotential" $V - V_{EQ}$ is very small and the electrochemical system is at equilibrium, the expression for the charge-transfer resistance of species i present in bulk solution concentration $C_i{}^*$ discharging with a heterogeneous kinetic rate constant k_{heter}, is expressed as [23, p. 381]:

$$R_{CT} = \frac{R_G T}{F^2 A k_{heter} \sum z_i^2 C_i{}^*} \tag{5-28}$$

The charge-transfer resistance becomes well defined only when the polarization depends on the electrochemical charge-transfer kinetics and current (Eq. 5-27). If there is no electron transfer, charge-transfer resistance R_{CT} becomes very large, and the electrode is polarizable with poorly defined potential. If there is an electron-transfer reaction, R_{CT} becomes smaller and connects in parallel with the double-layer capacitance C_{DL} [1, pp 45–48]. It is a common principle to add charge-transfer resistance (R_{CT}) or ionic conductance ($\sigma = 1/R_{DC}$) in parallel with double-layer capacitance to describe a complete electrode interface [1, p. 67]. Semicircles corresponding to charge-transfer processes are often depressed [9, p. 121], mostly due to current distribution and electrode inhomogeneity, roughness, and porosity. The resulting impedance Z_{ARC} will become a function combining a CPE (Section3-2) representing the double layer capacitance and ideal DC charge-transfer resistance in parallel [9, p. 121]:

$$Z_{ARC} = \frac{R_{CT}}{1 + (j\omega\tau)^{\alpha}} \qquad (5\text{-}29)$$

The charge transfer impedance shows strong dependence on the difference between the applied electrochemical DC potential V and the "standard redox reaction potential" V_0. When $V = V_0$ the electrochemical charge transfer between the electrode and media species in the double layer occurs. General charge-transfer impedance can be expressed as an expansion of Eq. 5-28 [27, p. 16]:

$$R_{CT} = \frac{R_G T}{z^2 F^2 A k_{heter}} \left(\frac{1}{\alpha C_{RED} e^{\frac{\alpha z F(V - V_0)}{R_G T}} + (1 - \alpha) C_{OX} e^{\frac{(\alpha - 1) z F(V - V_0)}{R_G T}}} \right) \qquad (5\text{-}30)$$

For 0 DC current [27, p. 20] the equation simplifies as (compare with Eq. 5-28):

$$R_{CT} = \frac{R_G T}{z^2 F^2 A k_{heter}} \frac{1}{C_{OX}^{*\alpha} C_{RED}^{*1-\alpha}} \qquad (5\text{-}31)$$

For applied DC potential V for reversible reaction, the expression includes provisions for diffusion limitations through diffusion coefficients D_{OX} and D_{RED} for both oxidized and reduced species. The expression becomes [27, p. 20]:

$$R_{CT} = \frac{R_G T}{z^2 F^2 A k_{heter}} \frac{e^{\frac{(1 - \alpha) z F(V - V_0)}{R_G T}} + \sqrt{\frac{D_{RED}}{D_{OX}}} e^{\frac{-\alpha z F(V - V_0)}{R_G T}}}{\sqrt{\frac{D_{RED}}{D_{OX}}} C_{RED}^* + C_{OX}^*} \qquad (5\text{-}32)$$

The separate equations for charge-transfer and diffusion-system impedances $Z = R_{CT} + Z_{DIFF}$ response to the sum of DC and AC potential perturbations $V = V_0 - vt - V_A \sin \omega t$ were derived as [28-31]:

$$R_{CT} = \frac{R_G T}{z^2 F^2 A k_{heter} C_{OX}^*} \left(\frac{D_{RED}}{D_{OX}} \right)^{\alpha/2} \frac{1 + e^{\frac{-z F(V - V_0)}{R_G T}}}{e^{\frac{-\alpha z F(V - V_0)}{R_G T}}} \qquad (5\text{-}33)$$

For applied DC potential V for irreversible reaction [27, p. 23] where $C_{RED} = 0$ further simplifications (Eq. 5-32) take place:

$$R_{CT} = \frac{R_G T}{z^2 F^2 A k_{heter} C_{OX}^*} \left[e^{\frac{(1 - \alpha) z F(V - V_0)}{R_G T}} + \sqrt{\frac{D_{RED}}{D_{OX}}} e^{\frac{-\alpha z F(V - V_0)}{R_G T}} \right] \qquad (5\text{-}34)$$

And for metal-metal ion electrodes where $C_{RED} = 1$ [27, p. 32]:

$$R_{CT} = \frac{R_G T}{z^2 F^2 A k_{heter}} \frac{1}{\alpha e^{\frac{\alpha z F(V - V_0)}{R_G T}} + (1 - \alpha) C_{OX} e^{\frac{(\alpha - 1) z F(V - V_0)}{R_G T}}} \qquad (5\text{-}35)$$

And for a reversible process (where $C_{OX} = exp(zF(V - V_{EQ})/R_GT)$):

$$R_{CT} = \frac{R_GT}{z^2F^2 Ak_{heter}} e^{\frac{-\alpha zF(V-V_0)}{R_GT}} \tag{5-36}$$

Eqs. 5-30 through 5-36 indicate an increase in the charge-transfer and diffusional components of the Faradaic impedance when applied potentials V is lower than standard electrochemical reaction potential V_0. With the applied potential increase, and $V \rightarrow V_0$, and R_{CT} reaches its minimum and stabilizes at a relatively constant low value, which depends on exact values of α parameter, k_{heter}, diffusion coefficients, and the concentrations. It is apparent from Eq. 5-36 that at very high overpotentials ($V > V_0$) the charge-transfer resistance R_{CT} becomes independent of external potential V and the electrode process will not be limited by the electrochemical kinetics. Kinetic-rate constants governing electron transfer depend exponentially on the applied potential [22, 32], with kinetics parameters determined by a ratio of kinetic-rate constants for forward and backward reactions $(k_f + k_b)^2 / k_f k_b$, independent of standard heterogeneous constant k_{heter}, having a minimum at $V = V_0$, where $k_{heter} = k_f = k_B$. In realistic experimental situations the diffusion impedance starts playing a major role [27, p. 26], and the total measured current becomes mass-transport (diffusion)-limited.

5.5. Electrochemical sorption impedance Z_{SORP}

The sorption impedance Z_{SORP} represents the impedance to current resulting from kinetics of specifically adsorbed or desorbed species in the electrochemical double layer (Figure 5-4). The adsorbed species typically do not exchange electrons directly with the electrode and do not produce "pure" Faradaic current (otherwise such adsorption processes can be treated like most charge transfer reactions), but they change the surface charge density and capacity and cause a pure AC current path, resulting in interfacial charging [1, p. 46]. Sorption processes are usually facile, and other processes (such mass and charge transport) typically determine the interfacial kinetics [9, p. 70].

The sorption impedance depends on charge associated with specific adsorption and/or desorption of charged species at the interface, primarily in the compact portion of the double layer. Hence the adsorption process results in an additional "pseudocapacitive effect" that can be broadly presented as $C_{ADS}^{-1} = C_{ADS OHP}^{-1} + C_{ADS IHP}^{-1}$ with $C_{ADS IHP}$ and $C_{ADS OHP}$ being adsorption capacitances of inner and outer Helmholtz layers. Modeled as a series combination of current-limiting resistor R_{ADS} with a CPE_{ADS} or an inductor L_{ADS}, Z_{SORP} may dominate the circuit in high Hz and low kHz regions.

Pseudocapacitance due to formation of the adsorption layer depends on electrochemical potential, media temperature, and electrode condition (surface area, roughness, electrode material, etc.). Z_{SORP} may depend on externally applied DC voltage V in nonlinear fashion as adsorption/desorption may occur abruptly at some sharply defined potentials. However, it is frequently constant with applied potential and temperature within a certain range if no change in the status of adsorption/desorption coverage occurs or if the charged specifically adsorbed species do not exchange charges with the electrode. If it is found experimentally that in certain range the low-frequency impedance is nearly independent of applied DC voltage and/or temperature, it is likely that it is dominated by the adsorption capacitance C_{ADS} rather than a charge-transfer process R_{CT} or diffuse capacitance $C_{DIFFUSE}$ [4, p. 120]. Specific adsorption at potentials opposite to purely electrostatic attraction and resulting potential profile may lead to current decreasing with voltage and appearance of "negative capacitance" loops. CPE_{ADS} may become negative for particular kinetic conditions (Section 7-4) and in purely electrical terms may be represented by an "ideal" inductor L_{ADS} [9, p. 75]. Also noncharged specifically adsorbed components may reduce available electrode area, reducing both measured Faradaic current and double-layer capacitance.

A derivation for electron-transfer kinetics in electroactive monolayers with an amount of adsorbed species Γ (mol/cm^2) resulted in expressions for the adsorption capacitance (C_{ADS}) and adsorption resistance (R_{ADS}) as:

$$C_{ADS} = \frac{F^2 A \Gamma}{4 R_G T} \tag{5-37}$$

$$R_{ADS} = \frac{2 R_G T}{F^2 A \Gamma k_f} \tag{5-38}$$

where R_G = the gas constant, F = Faraday's constant, A = the electrode surface area, and k_f = the rate constant for the adsorption-driven kinetics. With temperature T increase, the adsorption capacitance often increases and resistance decreases due to the acceleration of the adsorption kinetics rate constant k_f [21, 22] and an increase in the amount (Γ) of the adsorbed species. These effects exceed the direct decrease in C_{ADS} and increase in R_{ADS} with temperature implied in Eqs. 5-37 and 5-38 as a consequence of enhancement of "disorganizing" Brownian-motion effects at higher temperatures.

5.6. Mass transport impedance

The mass transport impedance reflects the amount of species transported to the reaction site from the bulk of the electrolyte and the amount of reactants transported from the electrode to the bulk solution. The general mass transport occurs by migration and convection in the bulk solution, and by migration, convection, and diffusion in the diffusion layer (Section 1-3). Making the variation of the mass transport impedance dependent on controlled diffusion process, which is driven by a concentration gradient developed as a result of consumption of electroactive species in the process of charge-transfer reactions, is always the preferred method in electrochemistry to manage the complexity of the experimental impedance dependence on the mass transport. The diffusion process in fact occurs as a result of the charge transfer and/or adsorption—two processes that can create a gradient between the bulk and interfacial concentrations of electroactive (discharging) species. The relevant diffusion impedance is typically seen at the lowest frequency range of the impedance spectrum.

The mass transport impedance concept is closely connected with adsorption, charge transfer, and double-layer formation processes. A simplified "geometrical" approach assumes that all potential difference occurs across the compact double layer where the charge transfer and adsorption reactions take place (Figure 5-4). In principle one has to realize that the mass transport of species in the double layer occurs by migration, diffusion, and homogeneous reactions (in the absence of convection effects). Migration carries only species of the charges opposite to the charge on the electrode, and is not responsible for the delivery of neutral species and species of the same charge as the electrode. However, all these types of species can carry charge or adsorb, and therefore influence the double layer capacitance. In situations where the media has an excess of supporting electrolyte, a facile migration transport is predominantly responsible for electrostatic charging of the double layer, with corresponding often negligible migration impedance. In that case the diffusion effect on the double layer charging may be disregarded. On the other hand, the controllable supply of electroactive species participating in the charge-transfer and adsorption reactions occurs almost exclusively from the concentration gradient -driven diffusion from the bulk solution. The migration-driven transport of the electroactive species reacting at the electrodes is minor (or, which is even more important, constant and independent of their consumption at the electrode), as the supporting electrolyte ions constitute the overwhelming majority of migration-transported species. In a typical system charge-transfer process with measurable R_{CT} and/or R_{ADS} consumes the discharging species at the interface, and a gradient between their interfacial and the bulk solution concentrations is developed. The resulting concentration gradient-driven diffusion continues to supply the discharging and adsorbing species to the reaction zone at the interface. In that regard the diffusion impedance has to be overcome to provide an adequate supply of species for these two processes. The "full" Randles circuit representation can be applied to describe this system, with the double-layer capacitance C_{DL} being placed in parallel with the Faradaic process represented by a series of charge-transfer and diffusion impedances $R_{CT} - Z_{DIFF}$ (Figure 1-6).

In the case of several competing charge-transfer and adsorption reactions resulting in concentration gradients between the interface and the bulk solution, several parallel diffusion impedances may manifest. A simplified approach would treat the situation as being heavily dominated by a single type of electroactive species that diffuses to the electrode and discharges. Indeed, under most experimental conditions only one diffusion-related impedance feature is observed. The discharging species should also influence the double-layer capacitance, but for steady-state conditions they develop an equilibrium concentration in the double layer, which therefore remains constant with respect to the discharging species, while supporting electrolyte migration is largely responsible for the actual double-layer capacitance value. An excess of supporting electrolyte essentially eliminates migrational contribution to the "purely diffusive" mass transfer of electroactive species, decreases migration-driven bulk solution resistance, and compresses the double layer. That situation again results in a conventional Randles structure for a supported situation with $Z_{DIFF} - R_{CT}$ for electroactive species supplied by diffusion and C_{DL} developed as a result of the (primarily) migrating supporting ions in series with R_{SOL} (Figure 1-6).

For the concept of diffusion it is important to identify Nernst diffusion layer thickness or diffusion zone. The diffusion zone with a dimension L_D is the length of solution away from the electrode at which the concentration wave is reduced to $1/e$ of its value at the electrode. A frequency-dependent diffusion length L_D depends on the ratio of the sampling frequency f_D and diffusion coefficient D, as $L_D = \sqrt{D/2\pi f_D} = \sqrt{D/\omega_D}$. The thickness of the diffusion layer is concentration-independent as long as the diffusion coefficient D is constant. For uniformly stirred solution the diffusion layer becomes thicker at longer experimental times (at lower frequencies f) as the concentration wave spreads out further into the bulk media as the applied frequency decreases. The diffusion layer thickness L_D can reach from 0.1 mm at 100Hz to ~ 1 mm at 0.01 Hz, and for thin samples may extend all the way to the counter electrode and become comparable with the overall sample thickness d. A relatively thick Nernst diffusion layer can be developed in particularly in the absence of significant external convection, at high overpotentials, low concentrations of analyte and negligible migration. The diffusion process occurs over a much larger portion of a sample adjacent to the electrodes than the double layer thickness (or $L_D \gg L_H + \lambda_{DEBYE}$). In [33] the diffusion zone was in fact separated into two spatial impedance segments. The first segment represents diffusion inside of a few nanometers thick diffuse "Debye" layer (λ_{DEBYE}), where the diffusion component Z_{DIFF1} is placed in parallel with the double layer capacitance C_{DL}. The second segment contains term Z_{DIFF2} representing the diffusion impedance of diffusion waves propagating ahead of the compact electrode region inside the bulk media. Z_{DIFF2} is placed in parallel with the diffusion "second layer" capacitance $C_{DD} = \varepsilon\varepsilon_0 A(L_D - \lambda_{DEBYE})^{-1}$. The effect of this second segment on the overall impedance response becomes especially prominent at higher concentrations of supporting electrolyte.

Diffusion resistance Z_{DIFF} to current flow carried by electroactive species can create impedance, frequently known as the Warburg element [23, p. 376]. If the diffusion layer L_D is assumed to have an unlimited thickness within the experimental AC frequency range, than a "semi-infinite" diffusion may become the rate-determining step in the Faradaic kinetic process. In the "semi-infinite" diffusion model the diffusion layer thickness L_D is assumed to be always much smaller than the total thickness of the sample d ($L_D \gg d$. The equation for the "semi-infinite" Warburg impedance $Z_W(\omega)$ is a function of concentration-driven potential gradient dV/dC^*. The "semi-infinite" diffusion limitation is modeled by characteristic resistance R_W and a Warburg infinite diffusion component Z_W that can be derived [8] as:

$$Z_W(\omega) = \frac{dV}{dC^*}\frac{(\omega^{-1/2} - j\omega^{-1/2})}{zFA\sqrt{2D}} = \frac{dV}{dC^*}\frac{1-j}{zFA\sqrt{2D\omega}} = \frac{R_G T(1-j)}{z^2 F^2 AC^*\sqrt{2\omega D}} = \frac{\sigma_D(1-j)}{\sqrt{\omega}} = \frac{R_W}{\sqrt{j\omega}} \quad (5\text{-}39)$$

Where σ_D = the Warburg coefficient, which can be determined from a slope of a plot of Z vs. $\omega^{-1/2}$ for a frequency range where the diffusion process occurs. The Warburg coefficient is defined as [23, p. 381]:

$$\sigma_D = \frac{R_G T}{z^2 F^2 A \sqrt{2}} \left(\frac{1}{C^*_{OX} \sqrt{D_{OX}}} + \frac{1}{C^*_{RED} \sqrt{D_{RED}}} \right) \text{ or [32]:}$$

$$\sigma_D = \frac{R_G T}{z^2 F^2 A C^* \sqrt{2D}} \frac{(k_f + k_b)^2}{k_f k_b} \quad (5\text{-}40)$$

If the diffusion coefficients D are known, a plot of Z vs. $\omega^{-1/2}$ can be developed to determine unknown bulk concentration C^* and the differential of potential per concentration for ideal and nonideal solutions [9, p. 59] is:

$$\frac{dV}{dC^*} = \frac{R_G T}{zFC^*} \quad (5\text{-}41)$$

Where:

D_{OX} and D_{RED} = diffusion coefficients of the oxidant and the reductant

A = surface area of the electrode

z = number of transferred electron charges per diffusing species

F = 96500 C/mol = Faradays constant

T = absolute temperature in K

R_G = 8.31 J/(K mol) = gas constant

C^*, C^*_{OX}, C^*_{RED} = bulk concentration of the diffusing species ("overall," reduced, oxidized) (moles/cm³)

k_f, k_b = forward and backward rate constants for heterogeneous charge transfer

The semi-infinite total Warburg impedance is inversely proportional to the square root of frequency [9, p. 58]. The resulting CPE (Section 3-2) impedance can be represented as two separate elements—a "diffusional capacitance" and "diffusional resistance," each of them depending on a square root of frequency [17, p. 530]. For a case with a series solution resistance R_{SOL} = 20 ohm as the only other cell impedance component [23, pp. 384–386], on a Nyquist plot the infinite Warburg impedance appears as a line with a slope of 0.5 (Figure 5-5A). On a Bode plot the Warburg impedance exhibits a constant phase shift of −45° at low frequencies (Figure 5-5B). The Warburg coefficient can be calculated to be about 120 ohm sec$^{-1/2}$ at room temperature for a two electron transfer (z = 2), diffusion of a single species with a bulk concentration of C^* = 100 µM and a typical aqueous solution diffusion coefficient $D = 1.6*10^{-5}$ cm²/sec.

Diffusion resistance R_W can be generally expressed as a function of a slope of the coulometric titration curve dV/dx of the host material with molar share x of diffusing ions [34]:

$$R_W = \frac{L_D}{zFADC^* e^{\frac{zF(V-V_0)}{R_G T}}} = \frac{1}{zFA} \frac{L_D}{D} \frac{dV}{dx} \quad (5\text{-}42)$$

which for the Nernstian reversible case takes a simple form:

$$R_W = \frac{R_G T L_D}{z^2 F^2 ADC^*} \quad (5\text{-}43)$$

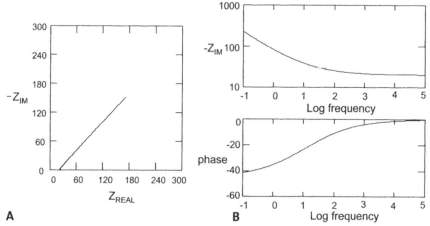

FIGURE 5-5   Warburg Impedance representation as A. Nyquist plot; B. Bode plot

This impedance depends on the DC potential V and AC frequency of the potential perturbation. The equation for a diffusion-limited system impedance Z_W response to the sum of DC and AC potential perturbations $V = V_0 - vt - V_A \sin \omega t$ was derived for oxidized species as [28-31]:

$$Z_W = \frac{4R_G T}{z^2 F^2 A C^* \sqrt{2D}} \cosh^2 \left[\frac{zF(V - V_0)}{2R_G T} \right] \tag{5-44}$$

Analogous to the case of charge-transfer resistance, the mass-transport parameter $Z_W = R_W / \sqrt{j\omega}$ also displays a minimum at the reaction standard potential V_0 where $V - V_0 = 0$.

Due to the assumption of semi-infinite diffusion made by the Warburg impedance for the derivation of the diffusion impedance, it predicts that the impedance diverges from the real axis at low frequencies. The DC impedance of the diffusion-limited electrochemical cell would be infinitely large. The Warburg impedance can be represented by a semi-infinite transmission line (TLM) composed of capacitors and resistors (Figure 5-6) [1, p. 59].

However, in many practical cases a "finite" diffusion-layer thickness has to be taken into consideration. Spatially restricted situations often emerge for "thin" samples, where at low frequencies a diffusion zone L_D thickness ap-

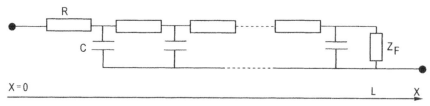

FIGURE 5-6   Transmission line representation of infinite diffusion layer Warburg Impedance terminated in a load Z_F

proaches that of the sample d. The diffusion takes place in a layer of thickness $L_D \sim d$, driven by a diffusion overvoltage at $X = 0$, and the observed diffusion behavior deviates from the pure Warburg impedance. For many problems thin samples dictate the use of a finite-length boundary with "finite" thickness of diffusion layer L_D restricted by the sample thickness d [1, p. 59]. The finite-layer diffusion-impedance model is more realistic than the "semi-infinite" model. This situation is relevant for thin layers of liquid and solid electrolytes, and many biological and separating membranes.

For such situations a characteristic diffusion frequency $\omega_D = D / L_D^2$ separates two regimes [34], with pure Warburg response occurring at higher frequencies, while at low frequencies the impedance response depends on whether the diffusing species are reflected or extracted at the end of the diffusion region ($X = L_D$) (at the electrode surface). For $\omega \gg \omega_D$ the species will not sense the boundary at $L = L_D$, and the system will be effectively semi-infinite [35, 36]. For $\omega \leq \omega_D$ two boundary conditions describe mass transport at the end of the diffusion zone ($X = L_D$). The first condition is a reflective or totally blocking boundary where the boundary cannot be penetrated by electroactive species, with a resulting "blocking effect" or "reflecting boundary," such as encountered with a thin film of conducting polymer (also called a "finite-space boundary") $\left(dC^* / dX \right)_{X=L_D} = 0$. The other boundary condition is an "absorbing" or "transmitting" boundary where the excess concentration of diffused species is instantly drained out $C^*(X = L_D, t) = 0$ [36]. These two conditions arise from elementary considerations about a random walker, who is either reflected or absorbed upon encountering a wall at $X = L_D$.

A finite-length diffusion impedance of charged particles is represented by Z_O parameter. The resulting "finite length" diffusion-impedance response does not have the –45° line, instead displaying a depressed semicircle or a vertical –90° line. The circuit is represented by a parallel combination of a CPE and an ideal resistor R_D, which also strongly depends on the electrochemical potential. The universal expression for finite diffusion impedance $Z_O(j\omega)$ was proposed in [37]:

$$Z_O(\omega) = R_D \frac{1 + \dfrac{Z_F}{R_D} \sqrt{\dfrac{j\omega}{\omega_D}} \coth \sqrt{\dfrac{j\omega}{\omega_D}}}{\dfrac{Z_F}{R_D} \dfrac{j\omega}{\omega_D} + \sqrt{\dfrac{j\omega}{\omega_D}} \coth \sqrt{\dfrac{j\omega}{\omega_D}}} \tag{5-45}$$

where Z_F is essentially a boundary-transfer function with dimensions of impedance that can under mixed mass transport–charge transfer control be analogous to Faradaic impedance.

For a totally blocking boundary condition $\left(dC^* / dX \right)_{X=L_D} = 0$ the last expression can be reduced to a situation where the Faradaic process does not occur or $Z_F(j\omega) \rightarrow \infty$, and the equation becomes [36]:

$$Z_{O\_REFLECTING}(\omega) = R_D \frac{\coth \sqrt{\dfrac{j\omega}{\omega_D}}}{\sqrt{\dfrac{j\omega}{\omega_D}}} \tag{5-46}$$

or [8, 9, p. 59]:

$$Z_{O\_REFLECTIVE} = R_D \frac{\coth\sqrt{\dfrac{j\omega L_D^2}{D}}}{\sqrt{\dfrac{j\omega L_D^2}{D}}} = \frac{R_C T}{z^2 F^2 DAC*} \frac{\coth\sqrt{\dfrac{j\omega L_D^2}{D}}}{\sqrt{\dfrac{j\omega}{D}}} = \frac{L_D^2}{DC_D} \frac{\coth\sqrt{\dfrac{j\omega L_D^2}{D}}}{\sqrt{\dfrac{j\omega L_D^2}{D}}}$$

(5-47)

with C_D = limiting differential capacitance at $\omega \to 0$:

$$C_D = \frac{z^2 F^2 AC* L_D}{R_C T}$$

(5-48)

The diffusion with reflecting boundary produces a transmission line terminated by an open circuit (as $Z_F = \infty$), revealing a straight $-90°$ capacitive line at low frequencies following the high frequency $-45°$ Warburg line (Figure 5-7A). At low frequencies expansion of Eq. 5-47 results in $Z_O = \dfrac{R_D}{3} + \dfrac{1}{\omega C_D}$ and the circuit effectively becomes a serial combination of a resistance $R_D/3 = \dfrac{L_D}{3zAD}(\dfrac{dV}{dC})_0 \sim \dfrac{L_D R_C T}{3z^2 F^2 ADC*}$ and limiting diffusion capacitance $C_D = \dfrac{L_D^2}{R_D D}$ [37].

The impedance for an absorbing or transmitting boundary where concentration of species outside of the diffusion layer at distance X from the electrode surface is constant as $C(X > L_D, t) = C*$ and the species are instantly consumed at the electrode surface as $C(X = L_D, t) = 0$ [1, p. 59, 3]. The situation can be represented by negligible charge-transfer impedance where $Z_F(j\omega) = 0$, and Eq. 5-45 becomes [36]:

$$Z_{O\_TRANSMITTING}(\omega) = R_D \frac{\tanh\sqrt{\dfrac{j\omega}{\omega_D}}}{\sqrt{\dfrac{j\omega}{\omega_D}}}$$

(5-49)

or [8, 9, p. 59]:

$$Z_{O\_TRANSMITTING} = \frac{R_C T L_D}{z^2 F^2 ADC*} \frac{\tanh\sqrt{\dfrac{j\omega L_D^2}{D}}}{\sqrt{\dfrac{j\omega L_D^2}{D}}} = R_D \frac{\tanh\sqrt{\dfrac{j\omega L_D^2}{D}}}{\sqrt{\dfrac{j\omega L_D^2}{D}}} =$$

$$= \frac{\sigma_D(1-j)}{\sqrt{\omega}}\tanh(L_D\sqrt{\dfrac{j\omega}{D}}) = \frac{\dfrac{2\sigma_D}{\sqrt{\omega}}\tanh\sqrt{\dfrac{\omega L_D^2}{D}}\left(1 - j\left(\tanh\sqrt{\dfrac{\omega L_D^2}{D}}\right)^2\right)}{1 + \left(\tanh\sqrt{\dfrac{\omega L_D^2}{D}}\right)^4}$$

(5-50)

where R_D = limiting diffusion resistance for $\omega \to 0$ $R_D = \dfrac{R_G T L_D}{z^2 F^2 D A C^*} \sim \dfrac{1}{C_D} \dfrac{L_D^2}{D}$ [3].

For high frequencies the *tanh* term $\to 1$ and the expression becomes an equivalent of a forward-only semi-infinite diffusion (Eq. 5-39):

$$Z_W = \frac{R_G T}{z^2 F^2 A C^* \sqrt{j\omega D}} \qquad (5\text{-}51)$$

The transmitting or absorbing boundary diffusion impedance will lead to a high-frequency $-45°$ Warburg line [9, p. 60] followed by a "teardrop" line bending over to the Z_{REAL} axis at low frequencies, displaying a distorted semi-circle that indicates finite DC resistance R_D (Figure 5-7B). At low frequencies, where the values of parameter $L_D^2 \omega / D \ll 3$ are effectively at $\omega \to 0$, the diffusion impedance becomes a parallel combination of a resistance $R_W = R_D$ and a limiting differential capacitance $C_D / 3$. When this data is plotted in a complex plane, the initial $-45°$ line reaches its peak when $Z_W^{''} \sim 0.417 R_D$ at $L_D^2 \omega / D = 2.53$ and then decreases toward the Z_{REAL} axis, finally approaching it vertically.

Thus, the diffusion impedance expressions depend on the electrode separations d at low frequencies. One way to identify the finite Warburg impedance is to use measurements at various values of the electrodes' separation d. When $L_D^2 \omega / D \gg 3$ (at $\omega \to \infty$), the *tanh* term approaches unity, the diffusion length is negligible compared to the whole region available for diffusion d, and Z_{DIFF} approaches infinite length Warburg Z_W:

$$Z \to Z_W = \frac{R_W}{\sqrt{\dfrac{j\omega L_D^2}{D}}} = \frac{R_W(1-j)}{L_D\sqrt{\dfrac{2\omega}{D}}} \qquad (5\text{-}52)$$

In a practical situation it is also possible for two types of diffusion processes for two different diffusion species displaced in frequency to be pres-

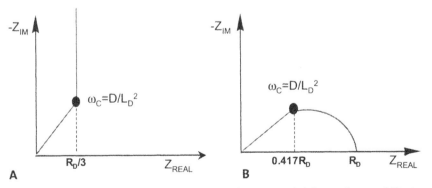

FIGURE 5-7 Complex plots of the impedance model for ordinary diffusion in a layer of thickness L: A. reflecting boundary condition $dC/dX = 0$ at $X = L$; B. absorbing boundary condition $C = 0$ at $X = L$

ent—an infinite diffusion at higher frequencies, followed by a finite diffusion at lower frequencies. In unsupported systems there is a limited supply of conducting species, and migration to control the bulk-solution process is insufficient. Additional types of diffusion impedance may be present, primarily in low frequency regions where the more mobile and abundant charged species have time to rearrange positions and to screen the less mobile and less abundant charges from the electric field [9, p. 115]. This diffusional process occurs in a bulk solution overlapping with a thick diffusion layer and can be placed in series with R_{BULK} [9, p. 115]. In addition, other diffusion limitations may occur, such as diffusion for uncharged adsorbing species or discharging oxygen that may hinder discharge of other electroactive species. Several diffusion impedance-related features can be observed if more than one type of species is discharged or if there is significant product diffusion back into the electrolyte. Distinguishing among various often parallel diffusion mechanisms can be determined by changing electrochemical potential, the electrode separation d (forcing finite boundary diffusion), the electrode surface area A, and the species bulk solution concentration C^* to determine which species are responsible for diffusion [9, p. 116].

The Warburg impedance description of diffusion has its limitations [35]. The Warburg impedance expression cannot be evaluated for consistency using the Kramers-Kronig transformations (Section 8-5). The mechanism assumes exclusively one-dimensional diffusion normal to the electrode surface; the analytical expression fails at low and high frequency extremes. At low frequencies the Warburg semi-infinite expression for impedance approaches infinity and assumes that no current is flowing through an electrode, therefore it is often unable to explain the DC behavior of an electrochemical system.

The semi-infinite and finite diffusion impedance $Z_{DIFF} \sim \omega^{-1/2}$ often gives rise to anomalous power laws that deviate from the 0.5 exponent as a result of secondary reactions or trapping ("subdiffusion") in fractal, amorphous, or porous media [37]. For both absorbing and reflecting cases of finite diffusion in porous media the initial rising portion of the impedance response may not necessarily reach $-45°$ [38]. This situation leads to heterogeneity, where the system can be considered as a two-phase and not a single-phase region. Sometimes a phase angle close to $-22.5°$ is observed for diffusion in pores and coatings, where, according to the de Levie model, the impedance varies as $Z^{1/2}$, with Z being a corresponding impedance of plane electrode (Section 7-5). In polymeric electrolytes, where diffusion of conducting particles (such as Li^+ in polymeric membranes) is complicated by a predominant adsorption, a nearly $0°$ line parallel to the Z_{REAL} axis can be observed.

External convection, for example with rotating disc working electrode, also plays a role in the expression for the diffusional impedance. Solution convection produces a mass-transport boundary layer of finite thickness next to the electrode. At high frequencies ($\omega \to \infty$) or for an infinite thickness of the diffusion layer, the diffusion is still represented by a semi-infinite Warburg impedance model. The absence of significant convection gives rise to a relatively thick Nernst diffusion layer, in particular at high overpotentials and low concentrations of analyte [39]. However, the infinite Warburg impedance measured in the absence of external convection may change to the finite diffusion with transmitting boundary at rotating disk electrode with an increase in the rotation rate ω_{ROT}. Convection influences Warburg mass transport phenomenon, by causing a $-45°$ line to bend over to the Z_{REAL} axis, just like in the case of the finite transmitting boundary diffusion in thin samples where $d{\sim}L_D$ [27, p. 50]. With the increase in the rotation rate a continuous decrease in corresponding diameter of diffusion impedance arc is observed with the corresponding current estimated from the Levich equation:

$$I_{lim} = 0.62FD^{2/3}\eta^{-1/6}C^* \sqrt{\omega_{ROT}}$$ (5-53)

The diffusion layer thickness L_D in a fluid of viscosity η will become effectively dependent on the rotation rate:

$$L_D = \frac{1.6D^{1/3}\eta^{1/6}}{\sqrt{\omega_{ROT}}}$$ (5-54)

5.7. Mixed charge-transfer, homogeneous, and diffusion-controlled kinetics

Diffusion and charge-transfer kinetics are usually coupled. A typical electrochemical (or "Faradaic") reaction is composed of both mass-transport processes of charged species to the electrode surface and their redox discharge at the interface. The Faradaic impedance can be represented by a series combination of Warburg diffusion impedance and charge-transfer resistance:

$$Z_F(j\omega) = R_{CT} + Z_{DIFF} = \frac{R_G T}{zFi_0} + \frac{2R_G T}{z^2F^2} \frac{1-j}{C^*A\sqrt{2D\omega}}$$ (5-55)

The equation points out that as frequency increases ($\omega \to \infty$), the Faradaic impedance approaches R_{CT}. At low frequencies ($\omega \to 0$Hz) the Faradaic impedance can be viewed as two resistances connected in series—one related to electron-transfer kinetics, the other to mass transport toward the electrode.

For reversible charge-transfer kinetics studied on microelectrodes with radius r, the diffusion-layer thickness L_D, depending on D and ω ($\omega_D = D/L_D^2$), is modified as $L_D \sim r$, with the critical frequency becoming $\omega_D \sim 2.53D/r^2$ [35]. When $r^2\omega/D \to 0$ (and $\omega \to 0$Hz), the Warburg impedance Z_w approaches the value of the diffusive transport limited resistance R_D and:

$$Z_F(\omega \to 0) = R_{CT} + R_D = \frac{R_G T}{zFi_0} + \frac{R_G Tr}{z^2F^2DAC^*}$$ (5-56)

The value of R_D is often several orders of magnitude higher than R_{CT}. For bulk-solution concentration $C^* \sim 10^{-6}$ mol/cm^3 and diffusion coefficient $D \sim 10^{-5}$ cm/sec at 298K, the $R_D \sim 8$Kohm ($r = 1$cm) or $R_D \sim 8$Mohm ($r = 10\mu$m) [35].

A typical electrochemical cell where polarization is due to a combination of kinetic and diffusion processes is modeled by a "full" Randles equivalent circuit composed of a parallel addition of a double-layer capacitance C_{DL} and Faradaic impedance [23, p. 385] (Figure 5-8). This circuit also includes solution resistance element R_{SOL}. An example of the Nyquist and Bode plots for this circuit

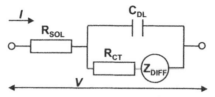

FIGURE 5-8 Equivalent circuit with mixed kinetic and charge-transfer control

is shown in Figure 5-9. As in the example for a pure diffusion process presented in Figure 5-5, the Warburg coefficient is assumed to be $\sigma_D = 150$ ohm sec$^{-1/2}$. Other parameters of the circuit are $R_{SOL} = 20\ \Omega$, $R_{CT} = 250\ \Omega$, and $C_{DL} = 40\ \mu F$.

The impedance of the Randles circuit is derived in Eq. 5-57. In a more general case the expression for CPE replaces C_{DL}. The Faradaic impedance can be expressed as a series combination of the Warburg impedance (Eq. 5-39) and charge-transfer resistance (Eq. 5-24). Solution resistance R_{SOL} completes the circuit:

$$Z(\omega) = R_{SOL} + \frac{R_{CT} + Z_W}{1 + j\omega C_{DL}(R_{CT} + Z_W)} = R_{SOL} + \frac{\dfrac{R_G T}{zFi_0} + \sigma_D \omega^{-1/2}(1-j)}{1 + j\omega C_{DL}(\dfrac{R_G T}{zFi_0} + \sigma_D \omega^{-1/2}(1-j))} =$$

$$= R_{SOL} + \frac{R_{CT} + \sigma_D \omega^{-1/2} - j\left[\omega C_{DL}(R_{CT} + \sigma_D \omega^{-1/2})^2 + \sigma_D \omega^{-1/2}(\sigma_D \omega^{1/2} C_{DL} + 1)\right]}{(\sigma_D \omega^{1/2} C_{DL} + 1)^2 + \omega^2 C_{DL}^2 (R_{CT} + \sigma_D \omega^{-1/2})^2} \tag{5-57}$$

A derivation [40] for semi-infinite diffusion resulted in another closed-form expression for the Faradaic impedance:

$$Z_F(\omega) = R_{CT} + R_{CT}\left(\frac{\dfrac{k_f}{\sqrt{D_{OX}}} + \dfrac{k_b}{\sqrt{D_{RED}}}}{\sqrt{j\omega}}\right) =$$

$$R_{CT}\left(1 + \frac{\dfrac{k_f}{\sqrt{D_{OX}}} + \dfrac{k_b}{\sqrt{D_{RED}}}}{\sqrt{2\omega}}\right) - R_{CT}^2 C_{DL}\left(\frac{k_f}{\sqrt{D_{OX}}} + \frac{k_b}{\sqrt{D_{RED}}}\right)^2 - jR_{CT}\left(\frac{\dfrac{k_f}{\sqrt{D_{OX}}} + \dfrac{k_b}{\sqrt{D_{RED}}}}{\sqrt{2\omega}}\right) \tag{5-58}$$

An extrapolation of the $-45°$ straight line to high frequencies ($\omega \to \infty$) representing the Warburg impedance in the complex plane intersects the real axis at the value R_{WO}, allowing us to calculate diffusion coefficients D_{OX} and D_{RED} from known rate constants k_f and k_b (Figure 5-10A):

$$R_{WO} = R_{SOL} + R_{CT} - R_{CT}^2 C_{DL}\left(\frac{k_f}{\sqrt{D_{OX}}} + \frac{k_b}{\sqrt{D_{RED}}}\right)^2 \tag{5-59}$$

However, depending upon the relative values of the charge-transfer and the diffusion parameters, various shapes can be obtained for the impedance diagram. Hence, obtaining electrochemical quantities by simple extrapolation of the $-45°$ straight line may become difficult. Analogous Faradaic impedance

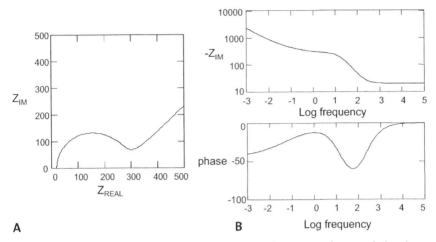

FIGURE 5-9 A. Nyquist diagram; B. Bode plot for a mixed control circuit

derivation for a case of finite diffusion with absorbing boundary [40] resulted in Eq. 5-60, graphically represented in Figure 5-10B:

$$Z_F(\omega) = R_{CT} + R_{CT}\left(\frac{k_f \tanh L_D\sqrt{\dfrac{j\omega}{D_{OX}}}}{\sqrt{j\omega D_{OX}}} + \frac{k_b \tanh L_D\sqrt{\dfrac{j\omega}{D_{RED}}}}{\sqrt{j\omega D_{RED}}}\right) \qquad (5\text{-}60)$$

For conditions of infinite diffusion-layer thickness $L_D \to \infty$, as well as for high frequencies $\omega \to \infty$, Eq. 5-60 transforms into Eq. 5-58 for semi-infinite diffusion. At low frequencies $\omega \to 0$ and only one reaction (for instance, oxidation) taking place $\tanh L_D\sqrt{\dfrac{j\omega}{D_{OX}}} \to L_D\sqrt{\dfrac{j\omega}{D_{OX}}}$ and the impedance approaches the value of $Z(\omega \to 0) = R_{SOL} + R_{CT} + R_{CT}\dfrac{k_f L_D}{D_{OX}}$.

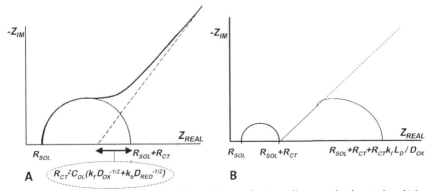

FIGURE 5-10 Electrochemical impedance for Randles equivalent circuit in the complex plane for A. diffusion layer of infinite thickness; B. diffusion layer of finite thickness

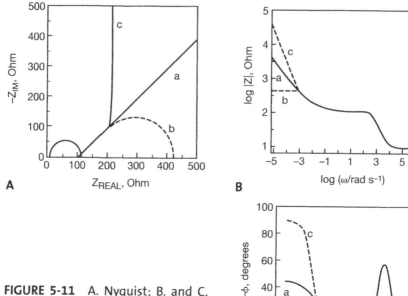

FIGURE 5-11 A. Nyquist; B. and C. Bodeplots for: a) linear semi-infinite, b) finite transmissive, and c) finite reflective boundaries, with $R_{SOL} = 10$ ohm, $R_{CT} = 100$ ohm

Previous discussions and the data in Figure 5-10 can be further generalized for the spatially restricted finite-diffusion situations that were discussed in Section 5-6. The semi-infinite diffusion Warburg equation is replaced by appropriate finite-diffusion expressions, such as Eqs. 5-46 and 5-49 [1, p. 76]. The resulting Nyquist and Bode plots are presented in Figure 5-11 [28].

In general, three characteristic frequency regions can be identified [40] when an electrochemical system represented by a Randles circuit is characterized by EIS (Figure 5-11A). The low-frequency region is identified for characterization of diffusion-limited mass transport through its characteristic frequency:

$$\omega_D = D / L_D^2 \tag{5-61}$$

The second (typically higher) frequency is determined by the ratio of the Faradaic (mainly charge transfer) and capacitive double layer charging currents:

$$\omega_{RC} = 1 / (R_{CT} C_{DL}) \tag{5-62}$$

Above this critical frequency a high-frequency impedance region exists ($\omega > \omega_{RC}$) where the impedance is determined by the double-layer capacitance as well as the bulk-solution impedance, resistance of the wires, film thickness of bulk polymer, etc. In the second, medium-frequency ($\omega_{RC} > \omega > \omega_D$) region, the impedance characteristic depends on Faradaic charge transfer and infinite diffusion impedances. The nonstationary diffusion layers are generated near the interface, while the concentration perturbations are zero outside these layers (at $\omega > \omega_D$). For $\omega > \omega_D$ the species will not sense the boundary at $x = L_D$, and the system will be characterized by semi-infinite diffusion, with Warburg $-45°$ response. At the low frequencies ($\omega_D > \omega$) the concentration distributions are quasistationary, and the diffusion impedance may become finite. Z_{IM} decreases to the real axis at the onset of finite diffusion for transmitting or absorbing boundaries or rises at $-90°$ as a purely capacitive response for blocking or reflecting boundaries [40].

The problem of combining various types of diffusion processes, including diffusion with finite boundaries and homogeneous reactions were addressed earlier [34, 36]. One of the studied cases was finite reflecting boundary diffusion, which also shows a capacitive dispersion [37]. A total expression for diffusion impedance was derived as a rearrangement of Eq. 5-45 for a case where the capacitive dispersion can be described by a constant phase element $Z_F(j\omega) = \dfrac{1}{Q(j\omega)^\alpha}$. Substituting this expression into Eq. 5-45 yields:

$$Z_O(j\omega) = R_W \frac{R_W Q(j\omega)^\alpha + \sqrt{\dfrac{j\omega}{\omega_D}}\coth\sqrt{\dfrac{j\omega}{\omega_D}}}{\dfrac{j\omega}{\omega_D} + R_W Q(j\omega)^\alpha \sqrt{\dfrac{j\omega}{\omega_D}}\coth\sqrt{\dfrac{j\omega}{\omega_D}}} \qquad (5\text{-}63)$$

In this expression the parameter $R_W Q(j\omega)^\alpha$ essentially represents a ratio of diffusion and capacitive dispersion contributions to the overall impedance process. This type of process can be represented by a parallel combination of a CPE and diffusion resistance R_W. At high frequencies a familiar $-45°$ semi-infinite-diffusion Warburg impedance line is observed as a function of $\omega^{-1/2}$. At low frequencies the shape of the impedance plot depends on the value of the parameter $R_W Q(j\omega)^\alpha$. When $R_W Q(j\omega)^\alpha \ll 1$ and $Z_O(j\omega) \sim 1/(Q\omega^\alpha) \to \infty$ a $-90°$ totally blocking capacitive line typical of finite reflection boundary diffusion is observed. When $R_W Q(j\omega)^\alpha \sim 1$ a deviation from the $-90°$ vertical line is observed. If $R_W Q(j\omega)^\alpha \gg 1$ than $Z_O(j\omega) \to R_W$ and the impedance plot reveals a distorted semicircle typical of finite transmission boundary diffusion with a finite R_W value at Z_{REAL} axis following the initial $-45°$ Warburg line.

Other more complicated forms of impedance-frequency dependencies can be revealed when a combination of diffusion process and homogeneous bulk solution reaction or recombination of electroactive species is considered [34]. In the case of homogeneous bulk solution and electrochemical reaction, the corresponding charge-transfer impedance Z_R can be simplified by a kinetic resistance R_K. Another parameter, ω_K—critical frequency of reaction or recombination—can also be introduced. The resulting impedance of homogeneous reaction or recombination R_K coupled with semi-infinite or finite diffusion R_W can be expressed as:

$$R_K = \frac{\omega_D}{\omega_K} R_W \qquad (5\text{-}64)$$

If R_K is infinitely large with reflecting boundary diffusion ($R_W / R_K \ll 1$), a simple diffusion model takes place (Figure 5-12A):

$$Z = R_W \sqrt{\frac{\omega_D}{j\omega}} \coth \sqrt{\frac{j\omega}{\omega_D}} \tag{5-65}$$

For a case with reflecting boundary diffusion and finite reaction resistance R_K, two cases may occur (Figure 5-12B), depending on a value of R_W / R_K:

$$Z = \sqrt{\frac{R_K R_W}{1 + j\omega / \omega_K}} \coth \left[\sqrt{\frac{\omega_K}{\omega_D}(1 + \frac{j\omega}{\omega_K})} \right] \tag{5-66}$$

If $R_K \gg R_W$ ($R_W / R_K \ll 1$) then at high frequencies ($\omega \gg \omega_D$) the species will not sense the boundary at $X = L_D$, and the impedance becomes completely dominated by semi-infinite diffusion:

$$Z = R_W \sqrt{\frac{\omega_D}{j\omega}} \tag{5-67}$$

At low frequencies ($\omega \ll \omega_D$) the impedance becomes:

$$Z = \frac{1}{3} R_W + \frac{R_K}{1 + j\omega / \omega_K} \quad \text{and} \quad R_{DC} = \frac{1}{3} R_W + R_K \tag{5-68}$$

The resulting impedance has a small Warburg part at high frequencies and a large, slow reaction-related arc at low frequencies (Figure 5-13A). At medium frequencies ($\omega \leq \omega_D$) a deviation from -90^0 vertical capacitive line will be first noted (Figure 5-13B). If $R_W \gg R_K$ ($R_W / R_K \gg 1$), the impedance becomes:

$$Z = \sqrt{\frac{R_K R_W}{1 + j\omega / \omega_K}} \quad \text{and} \quad R_{DC} = \sqrt{R_W R_K} \tag{5-69}$$

In that case the reaction time is shorter than the diffusion time ($\omega_K \gg \omega_D$); the concentration profile decays before the species can reach the boundary at $X = L_D$; and reaction and diffusion occur in semi-infinite space. (The same behavior can occur in very thin layers with rapidly reacting diffusing particles.) This type of phenomenon is called Gerisher impedance (Figure 5-13A). If R_K is infinitely large with absorbing boundary diffusion, a simple diffusion model takes place (Figure 5-12C):

$$Z = R_W \sqrt{\frac{\omega_D}{j\omega}} \tanh \sqrt{\frac{j\omega}{\omega_D}} \tag{5-70}$$

For a case with absorbing boundary diffusion and finite reaction resistance R_K, two cases occur (Figure 5-12D), depending on the ratio of R_W / R_K:

$$Z = \sqrt{\frac{R_K R_W}{1 + j\omega / \omega_K}} \tanh \left[\sqrt{\frac{\omega_K}{\omega_D}(1 + \frac{j\omega}{\omega_K})} \right] \tag{5-71}$$

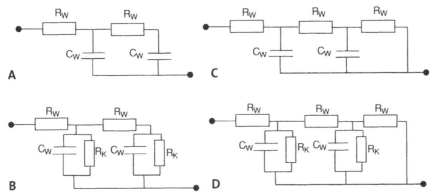

FIGURE 5-12 Transmission-line representation of diffusion impedance: A. simple diffusion with reflecting boundary; B. reflecting boundary diffusion coupled with homogeneous reaction; C. simple diffusion with absorbing boundary; D. absorbing boundary diffusion coupled with homogeneous reaction

If $R_K \gg R_W$ ($R_W / R_K \ll 1$), the effect of the reaction impedance is negligible, and the total impedance becomes completely dominated by semi-infinite diffusion (Figure 5-13C):

$$Z = R_W \sqrt{\frac{\omega_D}{j\omega}} \tanh \sqrt{\frac{j\omega}{\omega_D}} \text{ and } R_{DC} = R_W \qquad (5\text{-}72)$$

If $R_W \gg R_K$ ($R_W / R_K \gg 1$), the Gerischer impedance with boundary conditions becomes irrelevant and impedance becomes:

$$Z = \sqrt{\frac{R_K R_W}{1 + j\omega / \omega_K}} \text{ and } R_{DC} = \sqrt{R_W R_K} \qquad (5\text{-}73)$$

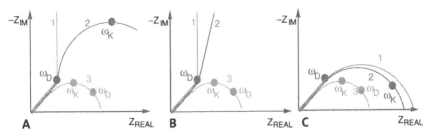

FIGURE 5-13 Impedance model for diffusion coupled with homogeneous reaction:

A. reflecting boundary (1. no reaction, 2. $R_K \gg R_W$, 3. $R_W \gg R_K$);
B. reflecting boundary (enlargement of A);
C. absorbing boundary (1. no reaction, 2. $R_K \gg R_W$, 3. $R_W \gg R_K$)

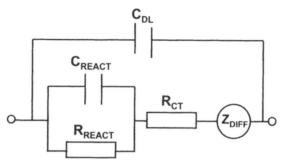

FIGURE 5-14   Circuit diagram for mixed kinetics with preceding reaction

At very low frequencies the Gerischer impedance reaches a constant real value inversely proportional to a value of kinetic rate constant k.

Both Warburg impedance and R_{CT} decrease rapidly and at a similar rate with an increase in potential [41], with their values becoming the lowest at $V = V_0$. With further increase in polarization, both Z_{DIFF} and R_{CT} stabilize, and the diffusion impedance starts increasing due to mass-transport limitation. This limitation is particularly evident in linear-scan voltammetry studies, where a peak at $V = V_0$ for "macro-electrodes" and a plateau in the case of "microelectrodes" are typically observed. Preceding and following homogeneous and heterogeneous reactions introduce a slight correction in the values of $V_0 = V_{MIN}$ for diffusion and charge-transfer processes, with $V_{MIN\,CT}$ becoming unequal to $V_{MIN\,DIFF}$ [30, 31]. As was shown above, preceding and following homogeneous (Gerischer), heterogeneous, and adsorption reactions may also introduce additional impedance semicircular features before the mass-transport Warburg impedance [27, p. 71–72, 28, 29, 30, 41]. The resulting equivalent circuit diagram is displayed in Figure 5-14.

References

1.   S. Grimnes, O.G. Martinsen, *Bioimpedance and bioelectricity basics*, Academic Press, 2000.

2.   M. E. Orazem, N. Pebere, B. Tribollet, *Enhanced graphical representation of electrochemical impedance data*, J. Electrochem. Soc., 2006, 153, 4, pp. B129–B136.

3.   V. Mie-Wen Huang, V. Vivier, M. E. Orazem, N. Pebere, B. Tribollet, *The apparent constant-phase-element behavior of an ideally polarized blocking electrode*, J. Electrochem. Soc., 2007, 154, 2, pp. C81–C88.

4.   V. Mie-Wen Huang, V. Vivier, M. E. Orazem, N. Pebere, B. Tribollet, *The global and local impedance response of a blocking disk electrode with local constant-phase-element behavior*, J. Electrochem. Soc., 2007, 154, 2, pp. C89–C98.

5.   V. Mie-Wen Huang, V. Vivier, M. E. Orazem, N. Pebere, B. Tribollet, *The apparent constant-phase-element behavior of a disk electrode with faradaic reactions*, J. Electrochem. Soc., 2007, 154, 2, pp. C99–C107.

6.   S.-L. Wu, M. E. Orazem, B. Tribollet, V. Vivier, *Impedance of a disc electrode with reactions involving an adsorbed intermediate: local and global analysis*, J. Electrochem. Soc., 2009, 156, 1, pp. C28–C38.

7. J.-B. Jorcin, M. E. Orazem, N. Pebere, B. Tribollet, *CPE analysis by local electrochemical impedance spectroscopy*, Electrochim. Acta, 2006, 51, pp. 1473–1479.

8. S. Krause, *Impedance methods*, Encyclopedia of electrochemistry, A. J. Bard (Ed.), Wiley-VCH, Vol. 3, 2001.

9. E. Barsukov, J. R. MacDonald, *Impedance spectroscopy*, John Wiley & Sons, Hoboken, New Jersey, 2005.

10. K. Asami, *Evaluation of colloids by dielectric spectroscopy*, HP Application Note 380-3, 1995, pp. 1–20.

11. V. F. Lvovich, M. F. Smiechowski, *AC impedance investigation of conductivity of industrial lubricants using two- and four-electrode electrochemical cells*, J. Appl. Electrochem., 2009, 39, 12, pp. 2439–2452.

12. R. E. Kornbrekke, I. D. Morrison, T. Oja, *Electrophoretic mobility measurements in low conductivity media*, Langmuir, 1992, 8, pp. 1211–1217.

13. S. Deabate, F. Henn, S. Devautour, J. C. Giuntini, *Conductivity and dielectric relaxation in various $Ni(OH)_2$ samples*, J. Electrochem. Soc., 2003, 150, 6 pp. J23–J31.

14. H. Nakagawa, S. Izuchi, K. Kuwana, T. Nikuda, Y. Aihara, *Liquid and polymer gel electrolytes for lithium batteries composed of room-temperature molten salt doped by lithium salt*, J. Electrochem. Soc., 2003, 150, 6, pp. A695–A700.

15. R. Larsson, O. Andersson, *Properties of electrolytes under pressure: PPG400 and PPG4000 complexed with $LiCF_3SO_3$*, Electrochim. Acta, 2003, 48, pp. 3481–3489.

16. B. Jacoby, A. Ecker, M. Vellekoop, *Monitoring macro- and microemulsions using physical chemosensors*, Sens. And Act. A, Physical, 2004, 115, 2–3, pp. 209–214.

17. B. E. Conway, *Electrochemical supercapacitors*, Kluwer Academic, New York, 1999.

18. P. Georen, J. Adebahr, P. Jacobsson, G. Lindberg, *Concentration polarization of a polymer electrolyte*, J. Electrochem. Soc., 2002, 149, 8, pp. A1015–A1019.

19. J. W. Schultze, D. Rolle, *The electrosorption valency and charge distribution in the double layer. The influence of surface structure on the adsorption of aromatic molecules*, J. Electroanal. Chem., 2003, 552, pp. 163–169.

20. V. F. Lvovich, M. F. Smiechowski, *Electrochemical impedance spectroscopy analysis of industrial lubricants*, Electrochim. Acta, 2006, 51, pp. 1487–1496.

21. V. F. Lvovich, M. F. Smiechowski, *Non-linear impedance analysis of industrial lubricants*, Electrochim. Acta, 2008, 53, pp. 7375–7385.

22. K. Darowicki, P. Slepski, *Dynamic electrochemical impedance spectroscopy of the first order electrode reaction*, J. Electroanal. Chem., 2003, 547, pp. 1-8.

23. A. J. Bard, L. R. Faulkner, *Electrochemical methods, fundamentals and applications*, J. Wiley & Sons, New York, 2001.

24. J. R. MacDonald, *Static space-charge effects in the diffuse double layer*, J. Chem. Phys., 1954, 22, 8 pp. 1317–1322.

25. W. S. Li, S. Q. Cai, J. L. Luo, *Chronopotentiometric responses and capacitance behaviors of passive films formed on iron in borate buffer solution*, J. Electrochem. Soc., 2004, 151, 4, pp. B220–B226.

26. W. S. Li, S. Q. Cai, J. L. Luo, *Pitting initiation and propagation of hypoeutectoid iron-based alloy with inclusions of martensite in chloride-containing nitrite solutions*, Electrochim. Acta, 2004, 49, pp. 1663–1672.

27. M. Sluyters-Rehbach, J. H. Sluyters, *Sine wave methods in the study of electrode processes*, Electroanalytical chemistry, Vol. 4, A. J. Bard (Ed.), Marcel Dekker, New York, 1970, pp. 1–127.

28. A. Lasia, *Electrochemical impedance spectroscopy and its applications, in modern aspects of electrochemistry*, B.E. Conway, J. Bockris, R. White (Eds.), vol. 32, Kluwer Academic/ Plenum Publishers, New York, 1999, pp. 143–248.

29. A. Lasia, *Applications of the electrochemical Impedance spectroscopy to hydrogen adsorption, evolution and absorption into metals*, B.E. Conway, R. White (Eds.), Modern Aspects of Electrochemistry, Vol. 35, Kluwer Academic/ Plenum Publishers, New York, 2002, pp. 1–49.

30. G. A. Ragoisha, A. S. Bondarenko, *Potentiodynamic electrochemical impedance spectroscopy*, Electrochim. Acta, 2005, 50, pp. 1553–1563.

31. F. Prieto, I. Navarro, M. Rueda, *Impedance study of thallous ion movement through gramicidin-dioleoylphosphatidylcholine self-assembled monolayers supported on mercury electrodes: the C-(C)-CE mechanism*, J. Electroanal. Chem., 2003, 550-551, pp. 253–265.

32. T. Komura, G. Y. Niu, T. Yamahuchi, M. Asano, *Redox and ionic-binding switched fluorescence of phenosafranine and thionine included in Nafion films*, Electrochim. Acta, 2003, 48, pp. 631–639.

33. R. Sandenberg, J. M. A. Figueiredo, *Relationship between spatial and spectral properties of ionic solutions: the distributed impedance of an electrolytic cells*, Electrochim. Acta, 2010, 55, pp. 4722–4727.

34. J. Bisquert, *Theory of the impedance of electron diffusion and recombination in a thin layer*, J. Chem. Phys. B., 2002, 106, pp. 325–333.

35. Navarro-Laboulais, J. J. Garcia-Jareno, F. Vicente, *Kramers-Kronig transformation, dc behaviour and steady state response of the Warburg impedance for a disk electrode inlaid in an insulating surface*, J. Electroanal. Chem., 2002, 536, pp. 11–18.

36. J. Bisquert, A. Compte, *Theory of the electrochemical impedance of anomalous diffusion*, J. Electroanal. Chem., 2001, 499, pp. 112–120.

37. J. Bisquert, G. Garcia-Belmonte, F. Fabregat-Santiago, P. R. Bueno, *Theoretical model of ac impedance of finite diffusion layers exhibiting low frequency dispersion*, J. Electroanal. Chem., 1999, 475, pp. 152–163.

38. A. A. O.Magalhaes, B. Tribollet, O. R. Mattos, I. C. P. Margarit, O. E. Barcia, *Chromate conversion coatings formation on zinc studies by electrochemical and electrohydrodynamic impedances*, J. Electrochem. Soc., 2003, 150, 1, pp. B16–B25.

39. M. V. ten Kortenaar, C. Tessont, Z. I. Kolar, H. van der Weijde, *Anodic oxidation of formaldehyde on gold studied by electrochemical impedance spectroscopy: an equivalent circuit approach*, J. Electrochem. Soc., 1999, 146, 6, pp. 2146–2155.

40. C. Gabrielli, *Identification of electrochemical processes by frequency response analysis*, Solartron Analytical Technical Report 004/83,1998, pp. 1–119.

41. E. H. Yu, K. Scott, R. W. Reeve, L. Yang, R. G. Allen, *Characterization of platinized ti mesh electrodes using electrochemical methods: methanol oxidation in sodium hydroxide solutions*, Electrochim. Acta, 2004, 49, pp. 2443–2452.

CHAPTER 6

Distributed Impedance Models

6.1. Distributed $R_{BULK}|C_{BULK} - R_{INT}|CPE_{DL}$ circuit model

As the first approximation, a circuit combining both bulk media and interfacial processes can be represented by a sequence of two parallel resistance-capacitance loads (Figure 6-1). This simplified equivalent circuit serves as a good initial model for many realistic experimental systems [2, 3], taking into account both high-frequency processes represented by a parallel $R_{BULK}|C_{BULK}$ combination of bulk-media resistance and capacitance and low-frequency processes at the electrochemical interface. The interfacial kinetics can be simplified as a parallel combination of Faradaic interfacial resistance R_{INT}, representing combined effects of charge transfer, adsorption, and diffusion-related resistances, and nonideal double-layer capacitive charging effects modeled by a CPE_{DL} element (Section 3-2). In most cases it is assumed that the impedances of contacts and cables to the measuring cell Z_{OHM} are negligible and/or already subtracted [1, p. 16].

The Nyquist plot reveals two semicircles (Figure 6-2). At the highest frequencies the circuit is characterized by a purely capacitive response, reflecting the effect of the capacitor C_{BULK}, with corresponding –90° phase angle. At medium frequencies there is a region separating the bulk media and the interfacial impedance segments. There the current flow becomes dominated by the resistor R_{BULK}, the phase angle changes to low absolute values approaching 0°, and the current value approaches $I = V/R_{BULK}$. The high frequency "bulk" media impedance analysis in the complex Nyquist plane (Figure 6-2A) reveals an ideal semicircle, where the resistance $R_{BULK} = 10^6$ ohm can be found by reading the real axis value at medium frequency at the boundary between the high-frequency bulk and low-frequency interfacial impedance regions.

The Bode plot (Figure 6-2B) also provides an easy estimate for this parameter from a horizontal section of the total impedance plot at medium frequencies. The time constant for the bulk solution equals $\tau_{HF} = R_{BULK}C_{BULK}$, with the

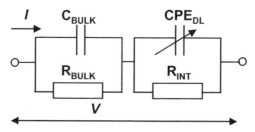

FIGURE 6-1   An equivalent circuit impedance model for a realistic experimental system

characteristic frequency $f_{C\,HF} = 1/(2\pi\,R_{BULK}C_{BULK})$, corresponding to maximum value of imaginary impedance and a phase angle of $-45°$.

At lower frequencies the interfacial impedance contributes to the measured impedance. The diameter of the low-frequency semicircle equals the interfacial resistance R_{INT}; the semicircle is not ideal due to the CPE_{DL} dispersion effects. At medium to low frequencies the double-layer contribution through CPE_{DL} is the primary contributor to the impedance, with a corresponding increase in the absolute value of the phase angle (Figure 6-2C). At the lowest frequencies the finite charge-transfer resistor R_{INT} becomes pronounced where

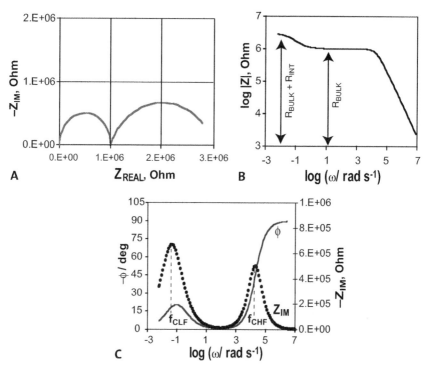

FIGURE 6-2   A. Nyquist and B., C. Bode impedance plots for the $R_{BULK}|C_{BULK}$ – $R_{INT}|CPE_{DL}$ circuit, where $R_{BULK} = 10^6$ ohm, $C_{BULK} = 50$ pF, $R_{INT} = 2 \cdot 10^6$ ohm, $C_{DL} = 5\mu F$, $\alpha = 0.75$.

the phase angle again approaches $0°$ and the current approaches $I = V/(R_{BULK}+R_{INT})$. The Bode plot shows the sum of the bulk resistance and the interfacial resistance $R_{BULK} + R_{INT}$ from the magnitude plot at low frequencies. The time constant for the interfacial impedance region equals $\tau_{LF} = R_{INT}C_{EFF\,DL}$ with $C_{EFF\,DL}$ calculated from experimental CPE_{DL} values using Eqs. 3-3 through 3-6. As was noted previously, the phase-angle maximum does not accurately indicate the critical relaxation frequency value $f_{C\,LF}$ for the lower-frequency interfacial processes. The $f_{C\,LF}$ parameter can usually be determined from the maximum value of the imaginary impedance (Figure 6-2 C).

The expression for the high-frequency ("bulk media") impedance response represented by the parallel combination of R_{BULK} and C_{BULK} becomes:

$$Z_{HF} = \frac{R_{BULK}}{1+\omega^2 C_{BULK}^2 R_{BULK}^2} - j\left(\frac{\omega R_{BULK}^2 C_{BULK}}{1+\omega^2 C_{BULK}^2 R_{BULK}^2}\right) \qquad (6\text{-}1)$$

A similar equation can be solved for the low-frequency "interfacial region" impedance, taking into consideration the substitution of a constant phase element for a capacitor as $Z_{CPE} = \dfrac{1}{(j\omega)^\alpha Q_{DL}}$ placed here in parallel with the resistor R_{INT}. Typical α values are $0 < \alpha < 1$ (for the example in Figure 6-2 $\alpha = 0.75$), resulting in a "depressed semicircle" on the complex impedance Nyquist diagram. Solving for the low-frequency impedance response yields:

$$Z_{LF} = \frac{R_{INT} + R_{INT}^2 Q_{DL}\omega^\alpha \cos\left(\dfrac{\pi\alpha}{2}\right) - j\left(R_{INT}^2 Q_{DL}\omega^\alpha \sin\left(\dfrac{\pi\alpha}{2}\right)\right)}{1+\omega^\alpha Q_{DL}R_{INT}\left(2\cos\left(\dfrac{\pi\alpha}{2}\right)\right)+\omega^{2\alpha}Q_{DL}^2 R_{INT}^2} \qquad (6\text{-}2)$$

The total impedance of the model will be the sum of Eqs. 6-1 and 6-2. Using these equations, expressions for critical relaxation frequencies for the high- and low-frequency impedance features can be developed:

$$f_{CHF} = \frac{1}{2\pi R_{BULK}C_{BULK}} \qquad (6\text{-}3)$$

$$f_{CLF} = \frac{1}{2\pi\left(R_{INT}Q_{DL}\right)^{1/\alpha}} \qquad (6\text{-}4)$$

Let's define first the actual geometry of the sample corresponding to the "bulk media" phenomenon. The bulk media resistance is determined largely by migration (and convection if present) processes, as defined by Eq. 5-4. The diffusion effects are present in the thick diffusion layer and overlap with the migration zone. In the presence of supporting electrolyte the diffusion contribution to the bulk solution resistance is negligible. Migration and convection processes take place in the bulk media outside of the compact Helmholtz part of the double layer with a thickness $L_H \sim 1$–2 nm, where migrating ions cannot fully penetrate. The migrating charged species can penetrate, however, into the diffuse portion of the double layer. The full conduction via migration occurs in a region of length that equals the difference between the geometrical thickness of the sample d, and $2L_H$ [1, p. 110], as we would assume that the Helmholtz layer exists on both working and counter electrodes. Nevertheless, the thickness of the Helmholtz layer is significantly smaller than the geometrical thickness of the sample ($L_H << d$), as it can be safely assumed that the bulk-solution resistance R_{BULK} is determined by the full "geometrical" thickness of the analyzed sample d.

The same logic applies to the geometrical definitions used for estimating the bulk media capacitance as $C_{BULK} = \dfrac{\varepsilon \varepsilon_0 A}{d}$, which serves as an indicator of the dielectric properties of the media. However, total measured sample capacitance is composed of the bulk media C_{BULK} and the double layer C_{DL} capacitances, and their spatial distribution is different from that for the bulk resistance R_{BULK} and interfacial resistance R_{INT}. The double-layer capacitance is composed of the compact Helmholtz layer and the diffuse-layer capacitances (Eq. 5-15). Therefore, the actual "bulk capacitance" effective distance is reduced by both thicknesses of the compact layer L_H ~1 nm and by the diffuse-layer thickness λ_{DEBYE}~10 nm. However, as the condition for the sample thickness distribution $d >> (2L_H + 2\lambda_{DEBYE})$ practically always holds, the bulk capacitance can be estimated as an inverse function of the total sample thickness d.

The bulk material capacitance C_{BULK} can be placed in parallel only with the bulk material resistance R_{BULK} (as shown in Figure 6-1) or with the whole circuit (such as that shown, for example, in Figure 6-3). In theory the choice can be made on the basis of geometrical representation of charging and ionic migration processes occurring in the bulk material and consideration of the geometrical boundaries separating the "bulk" and the "interfacial" segments in the sample. As was discussed above, the effective thickness of the bulk region used to estimate $R_{BULK} | C_{BULK}$ components should be reduced by the thickness of the Helmholtz compact portion L_H for the bulk resistance and by the thickness of both compact and diffuse layers $L_H + \lambda_{DEBYE}$ for the bulk capacitance. However, these segments are yet again very small compared to the typical thickness of the sample, and the total thickness of the sample d can be used to estimate the bulk impedance responses. In practice there is no real difference between the two position choices for C_{BULK}. As the first approximation, we may consider that the "bulk" region ends at the Helmholtz outer layer [1, p. 100], resulting in the representation of $R_{BULK} | C_{BULK}$ in series with $C_{DL} | R_{INT}$ (Figure 6-1). Inside the double layer (the Helmholtz layer and to a lesser extent the diffuse layer) the charging of the interfaces occurs, resulting in the appearance of large double-layer capacitance $C_{DL} >> C_{BULK}$.

In moderately resistive samples with ionic conduction migration process controls the bulk-media conduction and the bulk resistance exceed the bulk capacitive impedance only at the highest frequencies. In practically all "real life" systems there is at least some "lossy path" through a bulk-media resistance R_{BULK} in parallel with the C_{BULK}, resulting in predominant conduction through a resistive element at the frequencies where $1/\omega C_{BULK} > R_{BULK}$. The bulk migration process results in movement of charges through the bulk and their accumulation at the interface and formation of the double-layer capacitance C_{DL}. If there is no discharge ($R_{INT} \to \infty$), than C_{DL} can be placed in series with R_{BULK} alone (emphasizing that R_{BULK} represents ionic mobility conduction leading to accumulation of charges at the interface and formation of the double layer), resulting in $C_{BULK} \mid (R_{BULK} - C_{DL})$ representation. Alternatively C_{DL} can be positioned in series with the combined $R_{BULK} \mid C_{BULK}$ component, resulting in a ($R_{BULK} \mid C_{BULK}) - C_{DL}$ combination. Both alternatives result in a "double layer" version of the Debye circuit (Figure 5-3).

If C_{BULK} is placed in parallel only with R_{BULK}, then for an ideal "nonlossy" dielectric with $R_{BULK} \to \infty$ C_{DL} will be in series with C_{BULK}. The total impedance of the circuit is represented by $1 / j\omega(C_{DL} + C_{BULK})$, and the total measured impedance become dependent only on the relative magnitudes of C_{DL} and C_{BULK}. Typically $C_{DL} \gg C_{BULK}$ and the total impedance become dependent primarily on the bulk-solution capacitance contribution C_{BULK}. If there is no accumulation of charges at the interface, the sample impedance becomes inversely dependent exclusively on C_{BULK}.

If the bulk-solution capacitance is placed in parallel with the rest of the circuit and determined by the total thickness of the sample d, the combination can be reduced to the Debye circuit as $C_{BULK} \mid (R_{BULK} - C_{DL})$. For this circuit, an infinitely large R_{BULK} will result in rejecting current flow through the resistive portion of the circuit, and the total impedance of the circuit will again become dependent on the bulk-media capacitance C_{BULK}. For a realistic situation with a finite value of R_{BULK} component, the circuit impedance becomes dominated by the C_{DL} double-layer capacitance at the lowest frequencies and by C_{BULK} at high frequencies, with R_{BULK} dominating the response in the medium-frequency range. At low frequencies C_{BULK} and C_{DL} are effectively in parallel, and the measured capacitance is proportional to the sum of the two. With $C_{DL} \gg C_{BULK}$ the total measured capacitance at low frequency becomes equal to that of the double layer. Therefore, C_{DL} can be safely placed either in series with R_{BULK} only or with $R_{BULK} \mid C_{BULK}$ together.

The interfacial impedance described by a parallel combination of interfacial resistance R_{INT} and the double-layer capacitance C_{DL} dominates the circuit at low frequencies. The nature of the response depends on several factors, such as electrochemical potential, media temperature, and types of electrodes. The impedance interfacial response can be studied in both reversibly and irreversibly polarizable electrodes. A completely irreversible (or ideally polarizable) electrode is an electrode where no charge transfer due to the electrode's internal reactions occurs across the interface and interfacial resistance to such processes is infinitely large ($R_{INT} \to \infty$). Upon application of external voltage or passing externally driven current, the electrochemical potential of such an electrode changes very substantially and depends entirely on the media species capable of reducing or oxidizing at the electrode's surface. Since the charge due to the internal reactions cannot cross the interface, the ideally polarizable electrode behaves as an ideal capacitor, represented by C_{DL}. In the absence of charge transfer reactions due to the media redox species and controlled by the externally applied potential, the impedance response of the electrode becomes largely capacitive. Ideally polarizable electrodes represent an extreme case that does not happen in real life. Real electrodes with some degree of reversibility allow steady-state Faradaic current to pass (for instance due to electron exchange between the electrode material and its oxides at the surface). However, "inert metal" electrode materials such as platinum and gold are nearly ideally polarizable. These electrodes have very low exchange current and show a very large change in potential per application of even a small external current, which allows them to be used as non-interfering "working" electrodes in studies of electro-chemical potential-dependent kinetics exclusively for species contained in analyzed sample.

Reversible (or ideally nonpolarizable) electrodes typically possess very high exchange current maintained by a redox couple with highly reversible and facile electron exchange that is a part of the electrode chemistry itself. This current significantly exceeds a current due to the media species electrochemical processes and reactions that proceed as a result of an external electric field application. For the ideally nonpolarizable electrode the interfacial resistance is infinitely small ($R_{INT} \rightarrow 0$), and the impedance response of the electrode is governed by this very small interfacial resistance and therefore is also very small. When an external electric field is applied to the reversible electrode, the current due to the internal electrode charge exchange always exceeds the current due to the media species reactions; and equilibrium is constantly maintained in the vicinity of the interface. No essential modification in the electrode potential takes place, and the potential drop across the interface remains independent of the current across the interface. Ideally nonpolarizable electrodes, such as Ag/AgCl, serve as "reference" electrodes maintaining constant potential in electrochemical experiments. Such electrodes are useful as external voltage reference for redox kinetics studies on polarizable working electrode as a function of externally applied voltage.

Real-life electrodes are situated between the two extreme cases. Applied external potential results in generation of current response, with some "charging" current passing in the electrochemical double layer and some "Faradaic" current being generated in the case of charge-transfer processes. Strictly speaking, there is a close coupling between the Faradaic current I_F and the charging current I_C of the double layer. Full calculation of the impedance with no separation of the electrode impedance into Faradaic and double-layer components has been attempted, but the analytical expression of the impedance is particularly complicated. Hence, in most models it is assumed that the two current components can be separated [4].

6.2. General impedance models for distributed electrode processes

A full cell with two identical electrodes of surface area A can always be represented by an equivalent circuit, as shown in Figure 6-3 [1]. The electrodes are separated by a uniform bulk material of thickness d with a geometric capacitance $C_{BULK} = \dfrac{\varepsilon \varepsilon_0 A}{d}$. The bulk-material resistance R_{BULK} is determined by ionic mobility (u_i) of the bulk-solution species i with valence number z_i and bulk concentration C^* as $R_{BULK} = \dfrac{d}{AF} \dfrac{1}{\sum\limits_i z_i u_i C^*_i}$. Both electrodes have the electro-

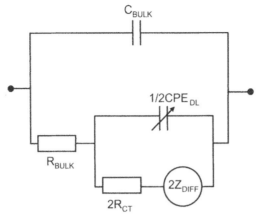

C_{BULK}

$1/2CPE_{DL}$

R_{BULK}

$2Z_{DIFF}$

$2R_{CT}$

FIGURE 6-3   Equivalent circuit for a full cell with two identical electrodes

chemical double layer represented by identical CPE_{DL} parameter, the identical interfacial process represented by a combination of identical charge-transfer resistances R_{CT}, and identical mass-transport impedances Z_{DIFF}. After these separate contributions are added together, general impedance for the full cell circuit with two identical electrodes can be written as a series combination of common bulk-media impedance with two interfacial impedances. The combined interfacial impedance resulting from the addition of interfacial impedances for each of the electrodes results in "doubled" interfacial parameters, accounting for two separate identical double layers on both electrodes, or a combined double layer with a total CPE of $Q_{DL}/2$, charge transfer resistance $2R_{CT}$ and diffusion impedance $2Z_{DIFF}$. The resulting expression for the total impedance of the full cell with two identical electrodes becomes:

$$Z = \cfrac{1}{j\omega C_{BULK} + \cfrac{1}{R_{BULK} + \left[\frac{1}{2}(j\omega)^\alpha Q_{DL} + \cfrac{1}{2R_{CT} + 2Z_{DIFF}}\right]^{-1}}} \qquad (6\text{-}5)$$

The above electrode system is frequently used as a "parallel plate" arrangement in many experimental and application-specific impedance probes, both in highly conductive and highly resistive media. "Parallel plate" is a simple and convenient term that can be broadly applied to sets of evenly spaced impedance electrodes of various geometries where a surface area of primary "working" electrode is close (not necessarily exactly equal) to that of a secondary "counter" electrode. For instance, many probes of that type are constructed as a popular mechanically robust "concentric cylinders" arrangement. The impedance analysis, as was shown above, is simple and straightforward. Typically a high-frequency relaxation due to a parallel combination of $R_{BULK} | C_{BULK}$ is present in the high-frequency range, and at the low-frequency range the interfacial impedance due to the double-layer capacitance in parallel with the

Faradaic process can be observed. Individual double-layer and Faradaic impedance parameters for the working electrode only can be assessed by taking into consideration a factor of two difference between the measured values of $CPE_{DL}/2$, $2R_{CT}$, and $2Z_{DIFF}$ and those of Faradaic impedance components for the working electrode. This analysis can be further simplified in many conductive solution systems, where it is possible to make the counter electrode essentially kinetically reversible with a surface area significantly larger that that of the working electrode. These modifications result in very low relative impedance contributions for the counter electrode into the total measured impedance response for the circuit. All applied voltage falls across the working-electrode interface, and counter-electrode impedance becomes negligible. The experimentally determined Faradaic impedance parameters are largely indicative of those for the working electrode only. In highly resistive systems counter-electrode reversibility may be difficult to achieve.

A complete equivalent circuit for the electrode process [2, p. 45] is presented in Figure 6-4 and Eq. 6-6. An expanded representation for low-frequency impedance includes, in addition to charge-transfer resistance R_{CT} and diffusion impedance Z_{DIFF}, possible (ad)sorption impedance, represented by a serial combination of sorption resistance R_{SORP} and inductive (or capacitive) element L_{SORP}:

$$Z = Z_{OHM} + \left(\frac{R_{BULK}}{1 + j\omega R_{BULK}C_{BULK}}\right) + \cfrac{1}{\left(\cfrac{1}{R_{SORP} + j\omega L_{SORP}}\right) + Q_{DL}(j\omega)^\alpha + \left(\cfrac{1}{R_{CT} + Z_{DIFF}}\right)}$$

$$(6\text{-}6)$$

An alternative representation [1, p. 110] can be considered where the circuit is developed from the "reaction kinetics" standpoint (Figures 6-5 and 6-6),

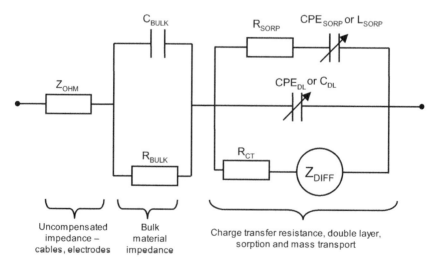

FIGURE 6-4 Equivalent circuit for full electrode process

which results in a different system of representation and a different distribution of reacting species and processes than those illustrated in Figures 6-1 and 6-4. Both representations contain an uncompensated impedance element Z_{OHM}. However, the capacitance of the bulk solution (which is the capacitance due to dielectric properties of supporting and electroactive species that occur outside of the Helmholtz and diffuse layers) is replaced by a better-defined "geometrical capacitance" C_{GEOM} that bridges the whole sample and includes the portions of Helmholtz and diffuse-layer capacitances due to the presence of supporting electrolyte but not electroactive species discharged at the electrode. The bulk-resistance parameter due to migration remains the same in both representations. In the kinetic representation the double-layer capacitance (that is, the capacitance between the electrode and both supporting electrolyte and electroactive species in diffuse and Helmholtz layers) and the charge-transfer resistance (due to electroactive species in the compact Helmholtz layer) are replaced by the reaction capacitance C_{REACT} in parallel with the reaction resistance R_{REACT}. These parameters represent capacitance and resistance to charge transfer for only the electroactive discharging species in the Helmholtz compact layer. The only mechanism of transporting electroactive reactants to the electrode is by diffusion of discharging species represented by impedance Z_{DIFF}, which is placed in series with R_{BULK}. The reaction impedance $C_{REACT} \mid R_{REACT}$ should therefore be placed in series with the diffusion-impedance element representing mass-transport limitations of supplying discharging agents to the reaction zone. The reaction may also be limited by partial adsorption of nonreacting species at the interface, which reduces the surface area available for the reaction in the inner Helmholtz layer. There is a parallel diffusion path for the adsorbing species, represented in most cases by a separate finite Warburg impedance $Z_{DIFF\ ADS}$ [1, p. 115]. $Z_{DIFF\ ADS}$ hinders the supply of adsorbed species; therefore it usually is placed in the adsorption-related part of the circuit (Figure 6-5).

Both Figures 6-4 and 6-6 represent kinetic processes in the bulk and at the interface for various practical applications. Let's consider a system composed of electroactive species of Type A, supporting electrolyte Type B, and adsorbing species Type C. The presence of all types of species can vary. The time-dependent response of the system to an external AC and DC voltage input can be visualized as changing from short input times at high AC frequencies to very long input times at low AC frequencies.

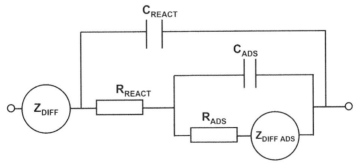

FIGURE 6-5   A kinetic representation of the electrode-sample interface

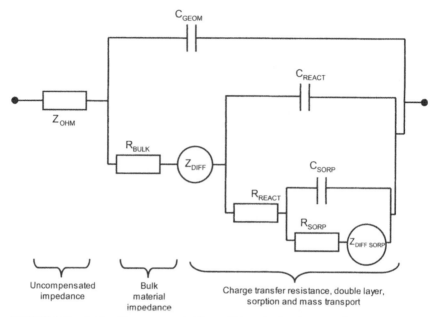

FIGURE 6-6   A kinetic representation of the electrode-sample system

Type A and Type B are both present in "supported" conductive systems. In that case the bulk-media capacitance showing charging of the sample bulk will be visible only in the upper MHz range. At lower frequencies the conduction mechanism is driven by ionic migration of A and B species, resulting in completely resistive bulk impedance, represented by R_{BULK}. If supporting electrolyte is in large excess, than the bulk resistance becomes small, and the corresponding phase angle is $0°$. If the supporting electrolyte concentration is low, then a larger bulk resistance appears. In that case an investigator should also pay attention to changes in the phase angle in the MHz-kHz region. Phase angle may be significantly different from $0°$, as with some aqueous colloidal suspensions, and bulk-media dispersion effects may occur (Chapter 11). As the frequency decreases, the migrating species A and B of the charge opposite to that of an electrode approach the interface. A difference in the accumulated charges between the species A and B in the diffuse and Helmholtz layers and the electrode surface results in the establishment of the electrochemical double layer with capacitance C_{DL}. This effect of double-layer charging becomes noticeable in the low kHz range. The capacitive charging at lower frequencies can result in continuous accumulation of charged species A and B at the double layer until the physical limit of saturation is reached, with the imaginary impedance increasing to infinity ("blocking interface"), as it is inversely proportional to the decreasing frequency. However, as the applied DC potential V approaches standard potential V_0 for species A to discharge, the double-layer capacitance C_{DL} becomes shortened through this Faradaic discharge process occurring in the Helmholtz layer. This process is limited by resistance of the charge-transfer kinetics R_{CT} and by the diffusion process (Z_{DIFF}) supplying species A from the bulk solution to the reaction zone to replace the species consumed in the discharge process. Alternatively, Figure 6-6 represents the process as the reaction capacitance C_{REACT} and resistance R_{REACT}, both representing species A in the Helmholtz layer supplied through a diffusion process Z_{DIFF}. C_{REACT} does not represent the C_{DL} completely, as it ignores the contribution from species B in the diffuse layer. This remaining capacitive segment of the double layer due to the presence of migrating species B is represented by a (very small) portion of the geometrical capacitance C_{GEOM}. Lastly, the effect of the adsorbing species C is represented differently in both models. The sorption impedance Z_{SORP} is typically modeled as a series or a parallel combination of current-limiting resistor R_{SORP} with a C_{SORP} (or CPE_{SORP}) or an inductor L_{SORP} (Figures 6-4 through 6-6) associated with the kinetics of the adsorption process. Diffusion impedance due to mass-transport delivery limitations for the adsorbing species from the bulk solution to the interface can also be present in series with the adsorption impedance. In principle the notation in Figure 6-6 is somewhat less ambiguous, as the classical sorption (representing species C adsorbing at the interface) takes place in the inner Helmholtz layer, has a diffusion component, and should be placed in series or in parallel with the reaction process for species A. In Figure 6-4 the sorption impedance is placed in parallel with the double layer and charge transfer/diffusion process. This representation can be used as long as there is an understanding that the sorption impedance represents essentially an electrochemical process for species C that is similar to the charge transfer/diffusion process for species A.

For highly resistive solutions where no discharge or adsorption processes take place $R_{CT} \rightarrow \infty$. (It is assumed that only species B in small concentrations are present; these species are dipolar or ionic in nature and can respond to the AC field but cannot discharge at the interface at applied DC potentials.) The attention switches to capacitive effects in the bulk media. At high frequencies the response is entirely capacitive, and C_{BULK} (or C_{GEOM}) represents the charging of the sample. In case of particle-based conductance a parallel path can be present, resulting in the phase angle absolute value decreasing at lower frequencies. With further decrease in frequencies the phase angle approaches $-90°$ levels again with a corresponding slightly higher bulk capacitance. At even low frequencies the bulk capacitance-associated impedance becomes higher than the bulk resistance of the sample, and the phase angle transitions to $0°$. Some ionic conduction and resulting migration of species A and B may then occur, and the circuit becomes dominated by the bulk resistance R_{BULK}. This process will eventually (at very low frequencies) result in double-layer formation with charging capacitance C_{DL} (often represented by CPE_{DL}). Typically there are no discharge reactions and hence no diffusion with the circuit represented by $R_{BULK} | C_{BULK} - C_{DL}$, with Figure 6-4 being a better model. The difference with Figure 6-6, however, is minor, as the diffusion element can be considered to be $Z_{DIFF} = 0$, and $C_R = C_{DL}$, with the understanding that C_{DL} represents the potential drop across the whole double-layer region (both diffuse and compact). That typically occurs in unsupported systems ($B \rightarrow 0$) where the double layer is created primarily by the diffusing and migrating discharging species A, and the double-layer capacitance becomes identical with the reaction capacitance. As double-layer capacitance is larger than the bulk (or geometrical) capacitance, placing the two capacitors in parallel (Figure 6-6) or in series (Figure 6-4) resulting in identical impedance responses at a low frequency depending mostly on C_{DL}. For higher applied DC potentials the discharge may take place at very low frequencies, and the double-layer capacitance is shortened by the Faradaic process, as was shown above for supported conductive systems.

6.3.   Identification of frequency ranges for conductivity and permittivity measurements

An important task of practical impedance measurements is to identify the frequency ranges for correct evaluation of characteristic parameters of an analyzed sample, such as bulk-media resistance R_{BULK}, capacitance C_{BULK}, and interfacial impedance. These parameters can be respectively evaluated by measuring the current inside the cell of known geometry, especially in the presence of uniform electric field distributions. For instance, many practical applications often report "conductivity" of materials (σ), the parameter inversely proportional to the bulk-material resistivity ρ and resistance R_{BULK} (or R_{SOL}). A permittivity parameter ε, determined from capacitance measurements and Eq. 1-3, is another important property of analyzed material.

For accurate measurements of media conductivity, it is necessary to realize that the measured resistance value of a sample at an arbitrarily chosen AC sampling frequency may not be a correct representation of the media bulk resistance. The measured total resistance may contain contributions from electrode polarization, Faradaic impedances, lead cables, and other artifacts. To make accurate measurements of the bulk resistive properties of a material, it is necessary to know the measurement frequency range where both capacitive interference from the double layer (and other electrode interfacial impedance effects such as adsorption/desorption) and the bulk capacitance are absent [5]. A sampling frequency has to be chosen that is within the frequency region where the impedance spectrum is dominated by the bulk-material resistance. This task essentially involves the development of a concept of spatially distributed impedance.

The following analysis will be shown for a realistic system equivalent circuit model (Figure 6-1), further simplified by replacing CPE_{DL} with C_{DL} as shown in Figure 2-6A. The resistance of the material dominates the lower cutoff frequency f_{LO}. At lower frequencies the double-layer capacitance and other interfacial processes will cause the impedance to decrease with increasing frequency. This will continue until the impedance from the double-layer capacitor C_{DL} becomes lower than the impedance representing the bulk-material resistance R_{BULK}, which occurs at the frequency:

$$f_{LO} > \frac{1}{2\pi R_{BULK} C_{DL}}$$

(6-7)

This assessment was originally made in [6] and was further expanded in [7] for systems with interfacial resistance R_{INT} lower than the bulk-solution resistance R_{BULK}, where Eq. 6-7 has to be combined with Eq. 6-8:

$$f_{LO} > \frac{1}{2\pi R_{INT} C_{DL}} \tag{6-8}$$

The high-frequency limit f_{HI} is defined as the frequency where the impedance of the cell becomes lower than R_{BULK}, indicating a largely capacitive current; it can be determined with the following equation, a derivation of Eq. 6-3:

$$f_{HI} \approx \frac{1}{2\pi R_{BULK} C_{BULK}} = \frac{\sigma}{2\pi \varepsilon \varepsilon_0} \tag{6-9}$$

At frequencies above f_{LO} and below f_{HI} resistive impedance dominates the circuit, and material conductivity σ can be calculated for the known sample geometry—thickness d and surface area contacting the electrodes A—as:

$$Z_{f_{LO}<f<f_{HI}} \sim R_{BULK} = \frac{1}{\sigma} \frac{d}{A} \tag{6-10}$$

At frequencies above f_{HI} capacitive impedance dominates the circuit, and material permittivity ε can be determined from:

$$Z_{f>f_{HI}} \sim \frac{1}{2\pi f C_{BULK}} = \frac{1}{2\pi f \varepsilon \varepsilon_0} \frac{d}{A} \tag{6-11}$$

For higher-resistance dielectric materials, the suitable frequency range used to determine R_{BULK} (or conductivity σ) shifts to lower frequencies. At high frequencies the dielectric-material resistance rejects the current flow, and the current becomes almost entirely capacitive, with the high-frequency impedance Z related to the material permittivity (Eq. 6-11). At frequencies $f_{LO} < f < f_{HI}$ conductivity can be correctly evaluated as $1/R_{BULK}$ (Eq. 6-10), measured before the onset of the interfacial polarization effects. For conductive ionic materials, this conductivity region shifts towards higher frequencies, where the impedance response is represented solely by small $R_{BULK} = R_{SOL}$, with high-frequency capacitive impedance being relatively high and rejecting the current flow. For a 1 cm² working electrode and 0.1 cm separation between the working and the counter electrodes and measured $|Z| = 1$ ohm, the conductivity from Eq. 6-10 becomes $\sigma \sim \dfrac{d}{|Z|*A} = \dfrac{0.1cm}{1Ohm * 1cm^2} \sim 0.1 Sm/cm$. If a solution conductivity is determined from a single frequency measurement (as is typically done with conductivity application cells), then frequencies around 10 kHz are suitable for materials with conductivity of ~10 mS/m, while 100 Hz to 1 kHz are suitable for materials with ~0.5 mS/m conductivity levels.

At lower frequencies ($f < f_{LO}$) the double-layer capacitance and interfacial impedance dominate, and estimated conductivity decreases dramatically as total impedance increases. Similar to the high-frequency impedance used in

Eq. 6-11, in conductive aqueous solutions the double-layer capacitance $C_{DL} \sim 10\text{–}60 \ \mu F/cm^2$ directly determined from the impedance measurements can also be utilized to estimate the media permittivity ε. For this permittivity evaluation it is important, however, to use the correct value for the double-layer region thickness $L_{DL} \sim 5\text{–}10$ nm (Figure 5-4). The correct permittivity for an aqueous solution with $L_{DL} \sim 7$ nm thick and $C_{DL} \sim 10 \ \mu F/cm^2$ is:

$$\varepsilon = \frac{C_{DL} L_{DL}}{\varepsilon_0} = \frac{10 * 10^{-6} F / cm^2 * 7 * 10^{-7} cm}{8.85 * 10^{-14} F / cm} \sim 79 \ . \quad \text{The low-frequency software-}$$

calculated permittivity is estimated from the total sample thickness d, which is often several millimeters, resulting in extremely high (in millions) erroneous ε values. The incorrect software-estimated low-frequency permittivity originates from a sample geometrical factor, but it stems from the fact that the analyzed portion of the circuit changed from $R_{BULK} | C_{BULK}$ at high frequency to $R_{INT} | C_{DL}$ at low frequency, with a corresponding switch to a different characteristic thickness parameter L_{DL}.

As is shown in these simple examples, appropriate data representation and model selection are essential for correct estimations of geometry-dependent parameters. Permittivity and conductivity cannot be measured at some arbitrarily preset frequency without previous knowledge of the system and correct selection of its impedance model. Figure 6-7 shows an example of frequency distribution for the measurement ranges for a system represented in Figure 2-6 with values $R_{BULK} = 10^7$ ohm, $C_{BULK} = 10^{-10}$ F, $R_{INT} = 10^6$ ohm, and $C_{DL} = 10^{-6}$ F.

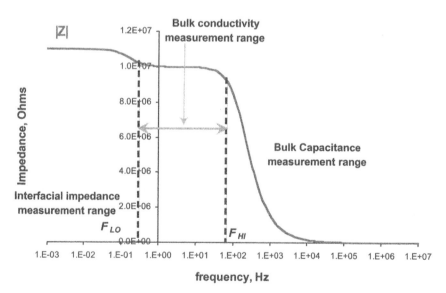

FIGURE 6-7 Frequency ranges for bulk capacitance, resistance, and interfacial measurements in highly resistive material

References

1. E. Barsukov, J. R. MacDonald, *Impedance spectroscopy*, J. Wiley & Sons, Hoboken, New Jersey, 2005.

2. S. Grimnes, O.G. Martinsen, *Bioimpedance and Bioelectricity basics*, Academic Press, 2000.

3. V. F. Lvovich, C. C. Liu, M. F. Smiechowski, *Optimization and fabrication of planar inter-digitated impedance sensors for highly resistive non-aqueous industrial fluids*, Sensors and Actuators, 2007, 119, 2, pp. 490–496.

4. M. E. Orazem, B. Tribollet, *Electrochemical impedance spectroscopy*, J. Wiley and Sons, Hoboken, New Jersey, 2008.

5. R. Sandenberg, J. M. A. Figueiredo, *Relationship between spatial and spectral properties of ionic solutions: the distributed impedance of electrolytic cells*, Electrochim. Acta, 2010, 55, pp. 4722–4727.

6. W. Olthuis, W. Steekstra, P. Bergveld, *Theoretical and experimental determination of cell constants of planar-interdigitated electrolyte conductivity sensors*, Sens. Act. B: Chem., 1995, 24–25, pp. 252–256.

7. V. F. Lvovich, C. C. Liu, M. F. Smiechowski, *Optimization and fabrication of planar inter-digitated impedance sensors for highly resistive non-aqueous industrial fluids*, Sensors and Actuators, 2007, 119, 2, pp. 490–496.

CHAPTER 7

Impedance Analysis of Complex Systems

7.1. Dielectric analysis of highly resistive composite materials with particles conduction

The investigation of mechanisms of deformation, disturbance, and restoration of composite highly resistive liquid and solid materials is a major task of rheology. In a classical composite system conductive particles are dispersed in a continuous-phase "base" material, most commonly an insulating fluid or solid matrix. Examples of composite media include organic nonpolar colloid dispersions, gels, liquid polymers, and many polycrystalline solid-state materials. Composite materials are widely used in industrial applications such as resistors, sensors, and transducers. Despite the significance of rheological measurements, they alone are insufficient for a comprehensive structural analysis of solid composites and liquid dispersions. Many peculiarities of composite materials can be revealed by combining rheology and dielectric studies [1, p. 123].

Electrochemical behavior of conductor-insulator composite materials is a result of several factors, such as the presence, character, and concentration of conductive particles and the composition and dispersive nature of the continuous insulating phase. It is well known that the conductivity of electrical materials is strongly affected by their atomistic composition (charge carrier mobility and concentration, hopping rate, etc.) and their microstructure (grain size distribution, grain morphology, porosity, etc.). The conducting particles often have different shapes and sizes, which can be represented by a probability density function. In the medium both electrochemically active and inactive conducting particles surrounded by a continuous insulating phase are present. There is a potential distribution on the particles' surface, as the ohmic drop of each conducting particle could not be the same.

In composite materials the percolation process of conductive particles typically governs the bulk-material electrical properties [2]. In addition to the percolation process, where the charged particles are essentially viewed as migrating "macro ions" with associated surface diffusion-related relaxation processes, the particles may agglomerate and develop a resistive or capacitive "conductive path." In a composite medium these phenomena very often result in a complicated and nonlinear pattern of dependency of bulk-material conduction on experimental parameters such as concentration of conductive particles, temperature, age of sample, applied electrochemical potential, and AC frequency. For example, in composite polymers an enhancement in electrical conductivity is often achieved by the addition of a nonconductive filler additive, leading to the formation of an amorphous highly conductive grain shell on the polymer-filler interface (composite grain). For low concentrations of the filler, little changes in conductivity are observed. For higher filler concentrations a large increase in conductivity (sometimes by several orders of magnitude) may occur, which is characteristic of amorphous highly conductive grain boundaries. Afterwards, with the increase in the filler concentration, conductivity starts to decrease due to the dilution effect, followed by a large drop in conductivity at the point when the polymer matrix loses its continuity and conductivity of the system approaches that of the nonconductive filler [3].

Composite materials can often be viewed as being composed of "grains" and "grain boundaries." Impedance analysis usually reveals at least two bulk-solution relaxation processes—one related to "grains" and another to "grain boundaries." The ability to express the impedance data in different notations, such as a complex impedance Z^* and complex modulus M^*, is of great value in being able to distinguish among various bulk solution-related processes in a composite medium. For continuous phase of low conductivity into which highly conductive particles are suspended, either the impedance or the modulus spectra can resolve the microstructural components by showing at least two arcs [4, p. 200]. In both modulus and impedance planes the low-frequency arc corresponds to relaxations in the low-conductivity continuous phase and the high-frequency arc to relaxations in the the conductive suspended phase. Both arcs are often nonideal due to various shapes and sizes of the particles, with their centers being located below the Z_{REAL} axis [4, p. 204].

The two phases have representative time constants $\tau_1 = R_1 C_1$ and $\tau_2 = R_2 C_2$; if these time constants differ as a result of differences in capacitance, then the arcs will be well resolved in the impedance spectrum. If two time constants are significantly different as a result of resistance differences, the modulus spectrum will resolve the arcs [2]. However, in practice good resolution is difficult to obtain. For example, when high-conductivity suspended matter reaches a certain volume fraction Φ, the impedance spectrum is often unable to resolve the time constants, and the complex modulus spectrum is preferred in order to see two arcs [4, p. 200].

The Maxwell-Wagner model describes dispersions and composites as conducting spheres suspended in a continuous insulating medium [4, pp. 192–198]. Several different models for a two-phase microstructure are possible:

1. "Series layer model" represents two phases stacked in layers parallel to the electrodes, represented by a series equivalent circuit model

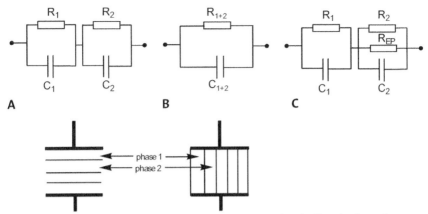

FIGURE 7-1 Equivalent circuit models representing bulk solution of a two-phase microstructure A. series layer; B. parallel layer; C. "easy path"

$R_1 | C_1 - R_2 | C_2$ (Figure 7-1A). As was discussed above, this situation should result in the appearance of two arcs.

2. "Parallel layer model" has two phases stacked across the electrodes, represented by a parallel equivalent circuit model $(R_1 | C_1) | (R_2 | C_2) = R_{1+2} | C_{1+2}$. This circuit shows only one relaxation (Figure 7-1B).

3. Realistic "brick layer model" represents system as "grains" (discontinuous phasing) separated by "grain boundaries" (continuous phasing). For a system where R_1, C_1 = resistance and capacitance of grains and R_2, C_2 – resistance and capacitance of grain boundaries:

   a. If conduction along the grain boundaries is negligible and grains themselves are highly conductive, the series equivalent circuit (Figure 7-1A) represents the system. In that case the current must travel across the insulating grain boundaries, but the current path may be somewhat "shortened" by the conductive grains.

   b. If current is conducted by the grain boundaries, the parallel-layer model (Figure 7-1B) represents the system.

   c. A possible modification of the conductive grains–nonconductive grain boundaries model is an ""easy" or "conductive path" model. Sometimes an "easy path" with resistance $R_{EP} < R_2$ can be found for the current traveling from one grain to the next across a "thin portion" of the boundary. For example, in colloidal systems this situation occurs when a significant agglomeration of conducting particles takes place. The presence of the "easy path" modifies the series equivalent circuit (Figure 7-1C).

The electrical properties of a composite material composed of insulating media and conductive particles can always be modeled as a function of concentration or fraction of the conducting particles Φ. It was demonstrated in a modeling study of composite polycrystalline media composed of two types of closed packed hard spheres [2], where the impedance structure of the system can be represented by either parallel or series circuits with characteristic R and C values. For example, this type of system can be represented by an insulating continuous medium with characteristic resistance $R_1 = 10^9$ ohm, a variable concentration Φ of conductive particles with characteristic resistance $R_2 = 10^3$ ohm, and characteristic capacitances of the continuous medium and the particles equal at $C_1 = C_2 = 1*10^{-10}$ F. As follows from the preceding discussion, the two relaxations for this "layer model" structure (Figure 7-1A) can be better resolved in the modulus plane rather than in the impedance plot. The ratio of diameters of two semicircles is equal to the ratio of two volume fractions as $R_2/R_1 = \Phi / (1 - \Phi)$.

The data in Figure 7-2A (the author's calculations, unpublished data) demonstrate the effect of the conductive particles fraction Φ on the impedance

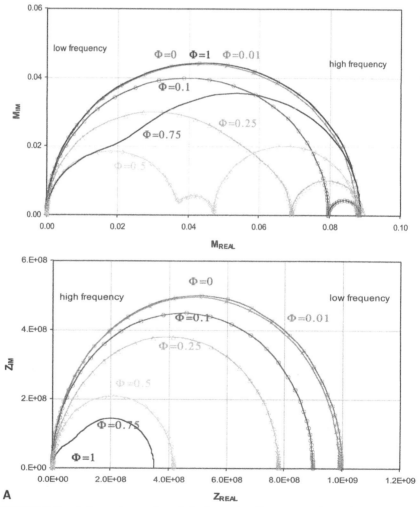

A

FIGURE 7-2 A. Impedance (modulus) spectra for composite polycrystalline media with random distribution of conductive spheres (Φ is a fraction of conductive medium) *(continued on next page)*

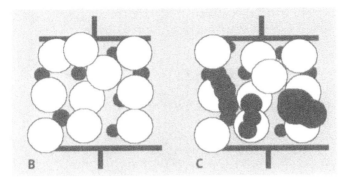

FIGURE 7-2 *(continued)* B. representation of dispersed and C. agglomerated particles

and modulus diagrams. When a very small (~1%) amount of conductive particles is added to an insulating medium, the current path maintained through the conductive particles is negligible and the measured resistance of the sample is heavily dominated by the conduction through the insulating medium with a single low-frequency relaxation and high value of the measured impedance. For a random distribution pattern where no agglomeration of conductive particles into organized orientated clusters occurs (Figure 7-2B) and the low value of the fraction of conductive particles $\Phi < 0.25$, the impedance spectrum can be clearly analyzed as a two-component $R_1 | C_1 - R_2 | C_2$ system. For high values of $\Phi > 0.5$, the system slowly develops into a single time-constant parallel model as electrodes become effectively shortened by continuous clusters of conductive particles. For intermediate values of conductive medium fraction $0.25 < \Phi < 0.5$ and random distribution of conductive spheres (which also includes agglomeration of some of these spheres into clusters, as in Figure 7-2C), an additional time constant emerges in medium frequencies. This relaxation corresponds to agglomeration of conductive particles and formation of clusters. The Nyquist complex impedance plot reveals only a single relaxation for all studied situations, unlike the complex modulus plot, which resolves all major relaxations within the system (Figure 7-2A). However, plotting the impedance data as a modified Bode plot of $\log Z_{REAL}$ vs. $\log (f)$ and $\log Z_{IM}$ vs. $\log (f)$ sometimes allows to resolve most of the relaxations.

Figure 7-3 (author's unpublished experimental data) demonstrates the results of the impedance analysis for a medium composed of insulating-base silicon oil with characteristic resistance $\sim 10^{12}$ ohm and capacitance $\sim 5*10^{-11}$ F with added 25% urea-coated barium titanyl oxalate conductive particles with characteristic resistance $\sim 10^{6}$ ohm and capacitance $\sim 10^{-7}$ F at 25 °C. The pure base oil shows a single relaxation, and particle-containing samples show relaxations at high frequency due to the conductive particles and at low frequency due to the base fluid in the complex modulus plot.

With an increase in the sample temperature to 80°C, a third relaxation at medium frequency emerges (shown by the arrows on Figure 7-3A), corresponding to the conductive particles' agglomeration. In many systems containing conductive particles accelerated agglomeration takes place at elevated temperatures [5], resulting in the appearance of a medium-frequency

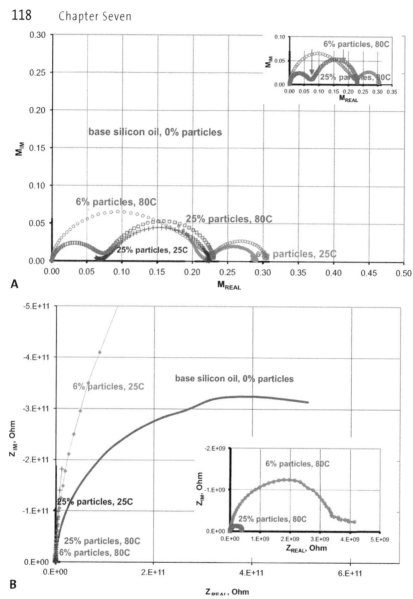

FIGURE 7-3 Complex modulus (A) and impedance (B) plots for experimental data for silicon (5cSt) insulating oil with added BaTiOxalate urea-coated conductive particles with random distribution of conductive spheres at 25 °C

agglomerate-induced relaxation above certain sample temperatures. The appearance of the additional time constant $\tau_3 = R_3 C_3$ and the medium-frequency dispersion is a result of the formation of elongated agglomerates of conductive particles, capable of capacitive dispersion and alignment along the preferred current flow lines perpendicular to the electrodes. These elongated agglomerates typically have higher characteristic capacitance C_3 and higher resistance R_3 than capacitance C_2 and resistance R_2 of the individual conducting particles responsible for the cluster formation. The clusters composed of the conducting particles have larger size, lower electrophoretic mobility, and higher characteristic resistance than those of the individual conducting particles. Elongated clusters usually take longer to align with an external AC field, while they possess higher overall charge-storage capacity, reflected in higher characteristic capacitance and permittivity. The external electric field effectively causes an arrangement of the elongated agglomerates in parallel with both individual (non-agglomerated) conductive particles and with the layer of insulating continuous medium. The current lines "bend" towards the agglomerate, resulting in formation of a new system component with a medium-frequency time constant $\tau_3 = R_3 C_3 > \tau_2$ and corresponding relaxation frequency $f_3 < f_2$ (Figure 7-4) [2]. The complex impedance plots (Figure 7-3B) revealed only one poorly resolved relaxation for all investigated samples, including two samples analyzed at elevated temperatures (shown on a separate inset in Figure 7-3B).

It is important to realize, however, that the additional medium frequency relaxation is caused only by formation of clusters that are elongated and aligned along the current lines. Figure 7-5 demonstrates an effect of shape and orientation of clusters of agglomerated particles on the complex modulus diagram. Agglomeration of particles in hexagonal, cubic, or spherical structures (Figure 7-5, A1 and B1) does not produce the medium frequency relaxation and reveals only two well-resolved time constants (Figure 7-5, A2 and B2). Development of vertically elongated clusters (Figure 7-5, C1 and D1) oriented along the electric field/current flow lines leads to the appearance of the third semicircle in the medium-frequency range (Figure 7-5, C2 and D2), while their orientation perpendicular to the field (Figure 7-5, E1) yet again does not reveal the medium-frequency semicircle (Figure 7-5, E2).

The formation of agglomerates leads to "shortening" of the system only after a certain concentration threshold is reached, after which the conductive

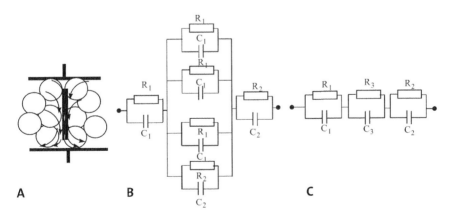

FIGURE 7-4 Distribution of current inside elongated clusters arranged parallel to an electric field (A) and simulated equivalent circuit diagram for the medium (B), resulting in appearance of a third time constant (C)

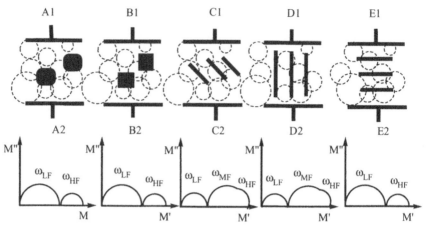

FIGURE 7-5 Distribution of clusters and corresponding Impedance spectra

particles agglomeration results in formation of sufficient numbers of uninterrupted chains of elongated agglomerates connecting the electrodes, presenting the current with sufficient "easy path of least resistance." That condition again requires the initial presence of large amounts of conductive particles in the system, allowing them to become polarized and form the aggregates. The intermediate situation, described in Figure 7-5D, is related to partial formation of large elongated clusters capable of aligning themselves with the current lines but due to their large sizes requiring longer relaxation times τ_3 than individual conductive particles.

Changes in the concentration of dispersed conducting particles also result in apparent changes in measured total high-frequency permittivity ε_{HF} and low-frequency conductivity σ_{LF} of composite media. The Maxwell-Wagner equation for dilute monodispersed emulsion, where ε_M and ε_p = real permittivities ("dielectric constants") of the dispersion "insulating" media and the disperse conducting particles phase, σ_M and σ_p = conductivities of the dispersion media and the disperse phase, respectively, and Φ = volume fraction of the disperse phase, yields total measured system permittivity as [1, p. 91]:

$$\varepsilon_{HF} = \varepsilon_M + 3\varepsilon_M \frac{\varepsilon_p - \varepsilon_M}{2\varepsilon_M + \varepsilon_p}\, \Phi = \varepsilon_M(1 + 3[\beta_R]\Phi) \qquad (7\text{-}1)$$

where the "real" Clausius-Mossotti factor $[\beta_R] = \dfrac{\varepsilon_p - \varepsilon_M}{\varepsilon_p + 2\varepsilon_M}$ becomes $[\beta_R] \to 1$ for conductive particles in insulating fluids where $\varepsilon_p \gg \varepsilon_M$. For $\Phi < 0.1$ the Bruggeman equations yield [1, p. 100]:

$$\frac{\varepsilon_{HF} - \varepsilon_p}{\varepsilon_M - \varepsilon_p}\left(\frac{\varepsilon_M}{\varepsilon_{HF}}\right)^{1/3} = 1 - \Phi \qquad (7\text{-}2)$$

$$\frac{\sigma_{LF} - \sigma_p}{\sigma_M - \sigma_p}\left(\frac{\sigma_M}{\sigma_{LF}}\right)^{1/3} = 1 - \Phi \qquad (7\text{-}3)$$

For spherical particles with high dielectric constant (for example, water with $\varepsilon_p = 80$) suspended in insulating oil ($\varepsilon_M \sim 2$) where $\Phi < 0.2$ and $[\beta_R] \to 1$, Eq. 7-1 high-frequency sample permittivity ε_{HF} becomes:

$$\varepsilon_{HF} = \varepsilon_M(1 + 3[\beta_R]\Phi) \sim \varepsilon_M(1 + 3\Phi) \tag{7-4}$$

This approximation shows that for small Φ, absence of agglomeration, and $\varepsilon_M \ll \varepsilon_p$, the absolute permittivity of the colloidal medium ε is independent of ε_p. For these conditions every 1% increase in concentration of conducting particles Φ increases the total system high-frequency permittivity by 3%. A similar derivation by Schwarz [6] based on the Maxwell-Wagner model [7] resulted in an alternative expression:

$$\varepsilon_{HF} = \varepsilon_M \frac{1 + 2[\beta_R]\Phi}{1 - [\beta_R]\Phi} \sim \varepsilon_M \frac{1 + 2\Phi}{1 - \Phi} \quad \text{and} \quad \sigma_{LF} \sim \sigma_M \frac{1 + 2\Phi}{1 - \Phi} \tag{7-5}$$

This expression is generally better supported by experimental data for $0.1 < \Phi < 0.2$ values than Eq. 7-4, while for $0 < \Phi < 0.1$ both Eqs. 7-4 and 7-5 converge. For nonspherical particles a form factor coefficient α_A was introduced [8]:

$$\varepsilon_{HF} = \varepsilon_M(1 + 3\alpha_A\Phi) \tag{7-6}$$

Factor α_A is unity when particles are spherical and $\alpha_A > 1$ for nonspherical dispersions producing the elongated agglomerated clusters (Figure 7-5D). Other extreme cases, such as platelike or thread-like elongated structures perpendicular to the electric field (Figure 7-5E), may result in values of $\alpha_A < 1$ (Figure 7-5E). When particle agglomeration occurs in the quiescent (nonshear) state, the additional agglomeration factor δ_A is introduced [8]:

$$\varepsilon_{HF} = \varepsilon_M(1 + 3\alpha_A\delta_A\Phi) \tag{7-7}$$

Factor δ_A depends on shear, particles-to-media interactions, particles concentration, and the type of dispersing media. This factor changes from $\delta_A = 1$ for conditions of no agglomeration and/or high values of shear stress to high values (~ 20 for some examples) when agglomeration occurs over time in a quiet solution. Agglomerates of suspended particles can be considered an extreme case of nonspherical particles in a randomly oriented state, leading to higher measured permittivity values according to Eq. 7-7. In a quiet solution increase in conductive particles concentration above a certain threshold will result in formation of continuous particles chains, and the segment will be short-circuited [9, p. 83].

In water-in-oil colloidal suspension with increased shear stress, agglomerated colloidal particles may be broken into separate particles, leading to a decrease in permittivity ε_{HF} [1, p. 116]. It was noted that high-frequency total measured high frequency permittivity ε_{HF} decreased significantly when shear was applied to a system containing higher loading of conductive particles (5%–7%) due to the breaking up of their agglomerates [10]. For oil-in-water emulsions, which represent the case of insulating particles dispersed in conducting media, no interfacial medium-frequency polarization is observed, and permittivity is not affected by shearing or flow.

For water-in-oil emulsions where $\sigma_M \ll \sigma_p$ and $\varepsilon_M \ll \varepsilon_p$ for high concentration of conductive particles ($\Phi > 0.2$), Eqs. 7-2 and 7-3 for low-frequency conductivity σ_{LF} and high-frequency permittivity ε_{HF} take more general forms [1]:

$$\sigma_{LF} = \frac{\sigma_M}{(1-\Phi)^3} \text{ and } \varepsilon_{HF} = \frac{\varepsilon_M}{(1-\Phi)^3} \tag{7-8}$$

For example, for the system presented in Figure 7-3 for 25% particles dispersed in silicon oil with permittivity $\varepsilon_M = 2.2$ and conductivity $\sigma_M = 10^{-14}$ Sm/cm, the measured 1MHz permittivity (above the particles critical relaxation frequency) of the system becomes $\varepsilon_{HF} \sim \dfrac{\varepsilon_M}{(1-\Phi)^3} = \dfrac{2.2}{(1-0.25)^3} = 5.21$ and conductivity at low frequency $\sigma_{LF} \sim \dfrac{\sigma_M}{(1-\Phi)^3} = \dfrac{10^{-14}}{(1-0.25)^3} = 3 \cdot 10^{-14} Sm/cm$, both in complete agreement with the experimental data.

For water-in-oil emulsions where low-frequency permittivity (ε_{LF}) is measured at the frequency below the particles' relaxation frequency f_c, all particles become polarized and fully oriented in the external AC field. With every revolution of the AC field all conductive particles have sufficient time to rotate and become fully aligned with the field. The capacitors of the particles will increase the capacitance of the system, which will be ultimately determined by the permittivity of the particles with respect to the medium. The double layer on each particle adds capacitance to the whole system, and the measured voltage decreases with frequency and lags the current [9, p. 82]. The measured low frequency permittivity of the system ε_{LF} increases, approaching a weighted average of dielectric constants of suspending and suspended phases ε_M and ε_p as:

$$\varepsilon_{LF} \sim \Phi\varepsilon_p + (1-\Phi)\varepsilon_M \to \Phi\varepsilon_p \tag{7-9}$$

Combined permittivity vs. frequency plot for a colloidal suspension with agglomerates is shown in Figure 7-6A. The system can be modeled as an insulating medium with bulk resistance $R_{BULK} = R_1 = R_M = 10^9$ ohm and capacitance $C_{BULK} = C_1 = C_M = 80$ pF, suspended particles with resistance $R_2 = R_p = 10^5$ ohm and capacitance $C_2 = C_p = 200$ pF, and particles' agglomerates with resistance $R_3 = R_{AGG} = 5*10^7$ ohm and capacitance $C_3 = C_{AGG} = 600$ pF (Figure 7-6B). In the external electric field the elongated agglomerates can be arranged in parallel with both individual (nonagglomerated) conductive particles and with the insulating continuous medium, increasing permittivity at both high and low frequencies (Figure 7-6A).

For oil-in-water emulsions where the dispersed oil particles are of lower conductivity and permittivity than the medium ($\sigma_M > \sigma_p$, $\varepsilon_M > \varepsilon_p$), the resistance at low frequency will be slightly higher (conductivity σ_{LF} and permittivity ε_{LF} lower) than without the particles. At low frequency the ionic conductance dominates the system, and charged particles will be pumped back and forth by electrophoresis. Such a system is representative of the "parallel layer" structure (Figure 7-1B).

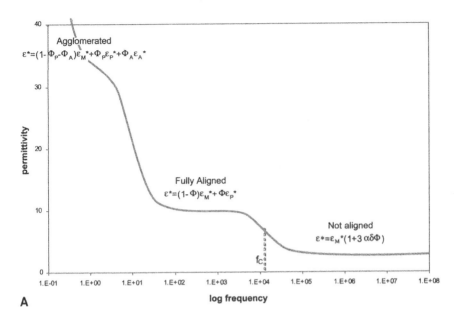

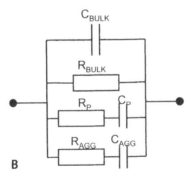

FIGURE 7-6 Colloidal suspension of particles and their agglomerates A. representative permittivity vs. frequency plot; B. circuit model

7.2.   Dielectric analysis of ionic colloidal suspensions

In heterogeneous ionic colloidal suspensions an external AC electric field creates a miniature electrical double layer on each particle, resulting in their polarization [9, p. 29]. The dielectric ionic theory was originally developed for colloidal suspensions of biological cells and globular proteins in conductive ionic environment, where ~ 1 µm cells and proteins could develop ionic cloud. Such systems represent a somewhat unique EIS modeling case. Depending on the composition of biological colloid, at kHz-low MHz frequency range AC current flows through suspended cells. The suspending media solution resistance R_{SOL} ceases to be a sole or even predominant current conductor, and significant capacitive contribution from the current conducted through the cells' membranes and cytoplasm (internal fluid) to the measured impedance and resulting high-frequency dielectric relaxations are observed. This phenomenon will be discussed in more detail in Chapter 11.

Colloidal particles are electrically charged by fixed or adsorbed ions and are surrounded by small counterions. That results in the formation of an electrical double layer of thickness λ_{DEBYE} on the particle, analogous to the electrode process discussed in Section 5-3. The resulting dielectric relaxation is primarily caused by interfacial polarization due to the buildup of charges on boundaries and interfaces between materials with very different electrical properties. Under the influence of an external electric field, the charges are redistributed along the surface of a particle, and the double layer becomes deformed and polarized, leading to the interfacial or "Maxwell-Wagner" polarization and corresponding dielectric relaxation [4]. This relaxation mechanism is different from the Debye relaxations in pure polar liquids, where the molecular dipoles change orientation in the external AC field (Section 1-2). Although dielectric relaxations in heterogeneous systems include both mechanisms, the magnitude of the interfacial polarization is much larger than relaxation due to orientation of polar molecules [11].

Maxwell-Wagner interfacial polarization theory explains relaxations in a double layer of colloidal particles as a result of ionic species' current conductance in parallel with capacitance. As a result of these processes in the double layer of a colloidal particle, the particle interface is charged by conductivity,

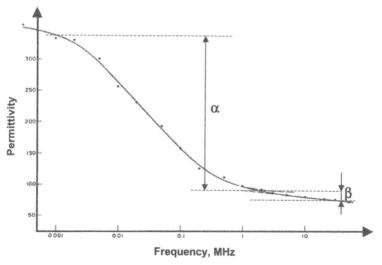

FIGURE 7-7 Dielectric constant ε as function of frequency in suspension of polysterene particles, with α and β relaxation processes indicated

and the resulting relaxation time constant is $\tau = RC$. Experimentally, significant permittivity increments in dielectric studies of several colloidal systems have been observed, most prominently in cellular dispersions. This interfacial single-layer surface charge is not equivalent to the double-layer charge. With liquid interphase formed by solution suspensions of particles and cells, the double-layer effects are additive to the Maxwell-Wagner interfacial polarization effect [9, p. 75]. In wet systems with very small colloidal particles the double-layer effects dominate [9, p. 78].

Schwan [12] and Asami [13] offered expressions describing experimentally observed kHz range incremental increases in permittivity $\Delta\varepsilon_\beta$ due to the Maxwell-Wagner relaxation (also known as "β-relaxation") process for biological cellular colloids:

$$\Delta\varepsilon_\beta = \varepsilon_{LF} - \varepsilon_M = \frac{9}{4\varepsilon_0} \frac{\Phi R_C C_{MBR}}{\left[1 + R_C G_{MBR}(1/\sigma_C + 1/2\sigma_M)\right]^2} \tag{7-10}$$

For typical parameters' values at 25 °C, C_{MBR} = cell membrane capacitance ($\sim 1\ \mu F/cm^2$), G_{MBR} = effective conductance of membrane for "live cell" ($\sim 2*10^{-3}$ S/cm²), R_C = cell radius ($\sim 2\ \mu m$), σ_M = media conductivity ($\sim 10^{-5}$ S/cm), σ_C = cell cytoplasm conductivity ($\sim 5*10^{-3}$ S/cm), Φ = volume fraction taken by the cells, ε_{LF} = permittivity at low frequency, ε_M = media permittivity at high frequency, Eq. 7-10 results in:

$$\Delta\varepsilon_\beta = \frac{9}{4*8.85*10^{-14}} \frac{\Phi*2*10^{-4}cm*10^{-6}F/cm^2}{(1+2*10^{-4}cm*2*10^{-3}S/cm^2*(200cm/S+0.5*10^5cm/S))^2}$$
$$\sim 2500\Phi$$

For 1% cells in the suspension the permittivity increment becomes $\Delta\varepsilon_\beta \sim 25$ (Figure 7-7), a value that is two to five orders of magnitude lower than typical values of permittivity increments observed in experimental dielectric studies of cellular colloidal dispersions [12]. Experimental assessment of the dielectric data in ionic solutions is further complicated by the masking effects of interfacial polarization at the electrode/solution interfaces that often become apparent below ~ 1–10 kHz.

The Schwartz theory [6, 11] offered an alternative mechanism based on diffusion-based relaxation of counterions in the electrical double layer of a large colloidal particle (known as "α-relaxation"). In this model the counterions are electrostatically bound to the surface charges of a large spherical particle but are free to move laterally along its surface. The counterions need to overcome a high potential barrier in order to escape from the surface of a colloidal particle. The movements of the small counterions along the surface of the large particle or macro-ion are much less restricted. When an external electrical field is applied, cation and anion counterions will separate in the particle's double layer without leaving the particle, creating a polarization effect (Figure 7-8). This relaxation is determined by a rate of diffusion of counterions of radius a in the double layer of the colloidal particle of radius A. In this model diffusion and migration govern the motion of ions and suspended colloidal particles (which are behaving like macro-ions in the double layer),

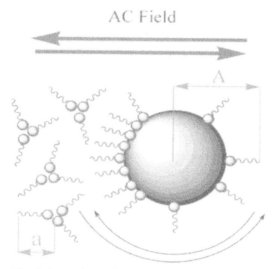

FIGURE 7-8 The interaction of counterions with radius a adsorbed onto the surface of a large colloidal particle, radius A, in an ac electrical field

with the time constants being predominantly dependent on $\tau = l_D^2/D$ instead of Maxwell-Wagner's $\tau = RC$.

The corresponding kHz range frequency "α-relaxation" is characterized by relaxation time:

$$\tau_\alpha = \frac{l_D^2}{D} = \frac{A^2 e_0}{2u_0 K_B T} \tag{7-11}$$

where: u_0 = counterion surface mobility in double layer, $A = \dfrac{K_B T}{6\pi D\eta}$ = radius of the spherical colloidal particle derived from the Stokes–Einstein equation, η = solvent viscosity, K_B ($1.38*10^{-23}$ J/K) = Boltzmann constant, $T(K)$ = absolute temperature, and D = diffusion coefficient of particle in solution.

Counterion mobility in the particle's double layer u_0 should be smaller than its mobility in the solution $u = \dfrac{ze_0}{6\pi\eta a}$, as it is subjected to additional electrostatic activation energy E_A. This energy is related to the height of the potential wall of the double layer, charge of the counterion, minimum distance δ between the counterion and its counter charge on the surface, equivalent to a thickness of a compact portion of the double layer (several angstroms), and dielectric constant of the solvent media ε_M. In general:

$$E_A \sim \frac{e_0^2}{\delta\varepsilon_M} \tag{7-12}$$

with E_A being on the order of ~1kcal/mol in both aqueous and nonaqueous colloidal systems [5]. Therefore:

$$u_0 = ue^{-E_A/K_B T} \tag{7-13}$$

The counterions are bound within the double layer of the spherical particle and restricted to lateral motions, with the Schwartz theory also allowing ions to enter and leave the double layer. The characteristic "α-relaxation" frequency derived from the Schwartz theory [6] becomes:

$$f_\alpha = \frac{1}{2\pi\tau_\alpha} = \frac{u_0 K_B T}{\pi A^2 e} = \frac{z K_B T e^{-\frac{E_A}{K_B T}}}{6\pi^2 \eta A^2 a} \qquad (7\text{-}14)$$

The above theory describing relaxations due to polarization of counterion cloud and processes of adsorption of counterions onto a central colloidal particle can be referred to as a "macromolecule" theory." The permittivity increment from the counterion relaxation theory was offered as [11]:

$$\Delta\varepsilon_\alpha = \varepsilon_{LF} - \varepsilon_M = \frac{9}{4\varepsilon_0} \frac{\Phi}{(1+0.5\Phi)^2} \frac{e_0^2 R_c q_0}{K_B T} \qquad (7\text{-}15)$$

Typical parameter values at 25 °C (T = 298K) are: e_0 = elementary charge $(1.6*10^{-23}$ coulombs), q_0 = average counterion charge density on a typical cell surface obtained from microelectrophoretic measurements $(\sim 10^{13}$ cm$^{-2})$ [11], and other parameters similar to Eq. 7-10, the permittivity increment becomes:

$$\Delta\varepsilon_\alpha = \frac{9}{4*8.85*10^{-14}} \frac{\Phi}{(1+0.5\Phi)^2} \frac{(1.6*10^{-19}C)^2 * 2 * 10^{-4} cm * 10^{13} cm^{-2}}{1.38*10^{-23} J/K * 298K} \sim 3*10^5 \Phi$$

Compared to the Maxwell-Wagner theory, the Schwarz model offers better theoretical estimates for high-permittivity increments that are much closer to the experimentally observed results at ~ 1–100 kHz. The counterion relaxation theory was originally developed for globular proteins in highly conductive ionic environments, where proteins could develop ionic cloud. In a purely mechanistic modeling sense both Schwarz α- and Maxwell-Wagner β-relaxations coexist and can sometimes be separated in kHz-MHz frequency range (Figure 7-7). The magnitude of the β-relaxation, however, is often negligible compared to that of the counterion α-relaxation.

Small proteins and biopolymers of ~ 1 μm sizes also possess the Debye-type orientation relaxation in MHz (for small molecules of several angstroms this relaxation would be in the GHz range) that can overlap with the β-relaxation. Small dipoles and molecules exhibiting rotational orientation, the relaxation mechanism can be approximated as spherical particles of radius a in solvent of viscosity η, where their charge z can often be assumed to be unity [6]. The high-frequency relaxation times τ_{HF} corresponding to this phenomenon can be described in a simplified expression:

$$\tau_{HF}^{-1} = \frac{z K_B T}{4\pi\eta a^3} \qquad (7\text{-}16)$$

$$a = \sqrt[3]{\frac{z K_B T}{8\pi^2 f_{HF} \eta}} \qquad (7\text{-}17)$$

For small nanometer-size molecules, τ_{HF} may reach the order of 10^{-6} sec with corresponding MHz range relaxation.

For rodlike molecules with $a = a_x \gg a_y = a_z$, Eq. 7-16 is modified as:

$$\tau_{HF}^{-1} = \frac{3K_B T (2 \ln \dfrac{2a}{a_Y} - 1)}{8\pi\eta a^3} \tag{7-18}$$

An alternative derivation [1 p. 147] leads to a slightly different equation where the frequency of orientational relaxation is determined by volume of the particle V_p, viscosity of the dispersion medium η, and the particle shape. The formula for relaxation time is given as:

$$\tau_{HF}^{-1} = 2\pi f_{HF} = \frac{K_B T}{3\eta V_p} \tag{7-19}$$

Or, for a spherical particle with $V_p = 4/3\pi a^3$ and charge $z = 1$, the expression becomes essentially identical to Eq. 7-16.

Combining Eqs. 7-11 through 7-19 with mobility-driven bulk conductivity (Eq. 5-5) can be used to model ionic colloidal suspensions, allowing sizes of individual colloidal particles and their agglomerates to be estimated. The theory, originally developed for polar biological suspensions, was expanded to moderately resistive (~1-10 Mohm) ionically conducting nonaqueous colloidal dispersions [5]. The modulus and permittivity data representations are often better able to resolve both media relaxations than the impedance complex plot (Section 7-1). Large colloidal particles and agglomerates often have low characteristic capacitances similar to those of counterions, while the counterions, colloidal particles, and agglomerates typically have very different characteristic conductivities and characteristic resistances. Figure 7-9 illustrates this point for a colloidal system composed of partially agglomerated conductive soot particles (10%) and aminic dispersant counterions (6%) suspended in silicon base fluid. The complex modulus plot reveals all three relaxations—for the counterions at very high frequency, for the large soot particles at medium frequency, and for the agglomerated soot particles at medium-to-low frequency, while the complex impedance plot reveals only the medium-to-low frequency relaxation for the agglomerated soot particles.

Eqs. 7-11 through 7-19 hold true for low concentrations of particles, when their presence has little effect on the apparent viscosity of the fluid. The variable viscosity η in the double layer of colloidal particles "felt" by the counterions may differ from bulk viscosity because of the anomaly of viscous properties of the liquid at the surface of the particles. For nonspherical particles several relaxation times may be present, and corrections allowing for interactions of the particles with dispersion media are often needed [1, p. 147]. Often nonpolar solvents with lower permittivity values produce slightly larger particles, as insulating solvents restrict repulsion and enhance attraction of the particles, leading to slightly higher degree of agglomeration [6, 11]. With a rise in counterion concentration the local field surrounding each colloidal particle may change, and conditions of the particle polarization are altered. As a result, characteristic relaxation time and critical frequency should depend on relative concentrations of counterions and dispersed colloidal particles [1, p. 118]. Loss

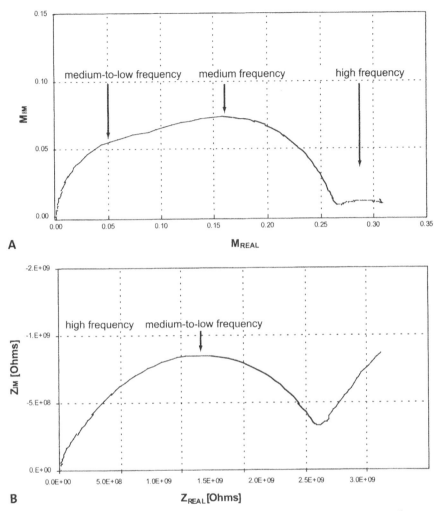

FIGURE 7-9 Complex modulus and Impedance plots for a system of conductive soot particles, their agglomerates, and dispersant counterions suspended in silicon-base fluid at 25 °C (unpublished data)

of counterions and double-layer charge on colloidal particles leads to their agglomeration of [9, p. 31].

A separate aspect of the Schwarz theory [6] covers electrical property changes due to chemical reactions in the bulk solution. Some reactions lead to formation of molecules of very different size and electrical properties than the original reactants, leading to a change in measured permittivity of the system. This effect, however, can be observed only if chemical process occurs faster than rotational diffusion, which is not the case for small molecules but may occur sometimes for macromolecular dipoles. For the chemical reaction process of reactant A producing a product B, with forward rate constant k_1 and reverse constant k_2, when product B of concentration C^* has a permanent dipole moment D_D and reactant A has none, the dielectric increment $\Delta\varepsilon_{RX}$ resulting from the reaction is:

$$\Delta\varepsilon_{RX} = \varpi \frac{N_A D_D^2 C^*}{3\varepsilon_0 K_B T} \left\{ \frac{k_1/k_2}{1+k_1/k_2} \frac{1}{1+i\omega\tau_1} + \frac{\frac{1}{1+k_1/k_2}}{1+i\omega\tau_2} \right\}$$ (7-20)

Where N_A = Avogadro's number, $\varpi = \dfrac{\varepsilon(n^2+2)^2}{(2\varepsilon+n^2)^2} \dfrac{2\varepsilon+1}{3}$ is a geometry factor of the particles, which is ~ 1 for rodlike particles and n = refractive index of solution. Relaxation time is τ_1 from Eq. 7-16, which becomes:

$$\tau_1 = \frac{4\pi\eta a^3}{K_B T}$$ (7-21)

Measured relaxation time τ_2 is related to "relaxation time" (kinetics) of the reaction, where $\dfrac{1}{\tau_{RX}} = k_1 + k_2$, resulting in:

$$\frac{1}{\tau_2} = \frac{1}{\tau_1} + \frac{1}{\tau_{RX}} = \frac{1}{\tau_1} + k_1 + k_2$$ (7-22)

In principle, two relaxation-time constants in the impedance spectrum can always be expected—rotational-diffusion (orientation) relaxation time τ_1 and chemical-reaction relaxation time τ_{RX}. In practice, if rotational diffusion is faster than reaction, the measured time constant τ_2 will be practically equal to τ_1. If the chemical step is faster than rotational diffusion, a distinct second relaxation occurs at higher frequencies than those of rotational diffusion.

7.3.   AC electrokinetics and dielectrophoretic spectroscopy of colloidal suspensions

In the growing research investigating the interaction of AC electric fields with colloidal particles and biological cells, AC electrokinetic methods [13, 14, 15] are increasingly replacing or supplementing classical impedance methods. AC electrokinetic effects, being dependent on dispersion frequency and mobility mismatches between conducting particles and continuous media, can be viewed as a segment of impedance or dielectric spectroscopy. Electrokinetics has higher resolution than EIS for the electrical parameters of single objects due to its different measurement principle and data representation. While impedance methods register the direct response to AC electric field (that is, AC current), AC electrokinetics register force effects that arise from the interaction of the induced polarization charges with the inducing field. Whereas the impedance of a suspension depends, though in a complex way, on the sum of the current contributions through and around the suspended cells, the forces in AC electrokinetics depend on impedance differences between particles and suspending medium. By choosing a medium of appropriate conductivity and permittivity, particles with similar dielectric properties can be efficiently trapped and identified. In general, the particle and medium properties in impedance studies are qualitatively reflected in an integrative manner, whereas AC electrokinetic methods are reflected in a differential manner. Nevertheless, impedance and electrokinetic methods detect the same polarization processes and their dispersions and thus generally yield the same information on the dielectric particle properties. The forces at work behind the functionality of the theories of AC electrokinetics are very general and can be applied to many areas of science.

An electrokinetic phenomenon is composed of a family of several different effects that occur in particle-containing fluids or in porous bodies filled with fluid. Various combinations of the driving force and moving phase determine different electrokinetic effects, such as electrophoresis (EP), electroosmosis (EO), and dielectrophoresis (DEP). At least two phenomena—EO as a flow of fluid through a channeled solid (capillary or a tube) bearing a surface charge and EP as a migration of charged particles through an otherwise quiescent fluid—are associated with the relative motion between the fluid and the solid generated by an externally applied electrostatic force.

AC electrokinetic techniques have been utilized for manipulation, separation, and analysis of various microscale and recently even nanoscale particles, such as blood and pathogenic cells, DNA, proteins, nucleic acids, and latex beads [16, 17, 18, 19]. These methods are appealing because they offer direct (no tags needed), sensitive, and selective detection of extremely low levels of analytes, with the ability to be packaged with electrochemical detectors into a single microfluidic device. A polarized particle may align with and move in an electric field, experiencing EP and drag-force effects. As demonstrated below, additional effects exist, such as induced polarization in a nonuniform AC field (DEP and electrothermal forces), effects of charged surfaces (EO), and Brownian random force and heating effects.

Electrophoresis (EP) is the induced migration (relative to the often static bulk liquid) of charged colloidal particles or molecules suspended in ionic solutions that results from the application of an electric DC field. Particles migrate with a constant velocity that arises when the generated viscous drag forces equal that of the coulombic forces propelling them down their potential gradients. Particles with different EP properties can be separated. For instance, bacteria of different species have different EP mobilities and potentials of zero charge when suspended in common buffers. Such differences in the EP behavior of particles have been extensively used at the macroscale to separate various types of biological cells from one another and from nonbiological material. EP has been used extensively to sort several types of "bioparticles," such as viruses, bacteria, eukaryotic cells, and even subcellular organelles.

The electromigration of these species is classified into two regimes, based on the ratio of the size of a particle a to the Debye length λ_{DEBYE} of the double layer formed on the particle in solution [16] (Figure 7-10). First, consider the

EP of an ionic molecule with characteristic radius much smaller than the Debye length, or conditions $a/\lambda_{DEBYE} \ll 1$. The motion of these charged particles (essentially ions) can be described as a simple balance between the electrostatic force on the ion and the viscous drag associated with its resulting motion. As a result, the EP motilities and velocities of the particles are a function of the ions' effective size a (and therefore of their molecular weight) and are directly proportional to their valence number, as is shown in Eq. 5-6:

$$v_{EP\_ION} = \frac{qV_t}{6\pi\eta a} = \frac{ze_0 V_t}{6\pi\eta a} \tag{7-23}$$

where: q = total charge on the particle, z = valence of the ion (or number of elementary charges $e_0 = 1.6 \ 10^{-19}$ C on the particle), V_t = electric field strength tangential to the ion, and η = media bulk viscosity. Nanometer particles placed in a 10 V electric field in aqueous solutions achieve μm/sec range velocities.

A second limiting situation arises during electrophoresis of relatively large particles, where the particle diameter-to-Debye length ratio is large ($a/\lambda_{DEBYE} \gg 1$). Examples of large particles relevant to microfluidic analysis include 100–10,000 nm diameter polystyrene spheres and ~10 μm diameter cells or single-celled organisms. In this second limiting case the EP velocity is a function of the electrostatic forces on the surface charges of the particles, the electrostatic forces on their charged double layers, and the viscous drag associated with both the motion of the particle as well as the motion of the ionic cloud around the particle. Locally, the ionic cloud near the particle surface can be approximated by the double-layer relations for a flat plate, and the velocity of the particle reduces to Smoluchowski equation:

$$v_{EP\_PARTICLE} = \frac{\varepsilon\varepsilon_0 V_t \zeta}{\eta} \tag{7-24}$$

where: ε = bulk media permittivity, ε_0 = dielectric constant of vacuum, and ζ = zeta-potential of the particle, which is the electrostatic potential drop across the diffuse part of the particle double layer (analogous to the ϕ_D in Figure 5-4 for the electrode double layer).

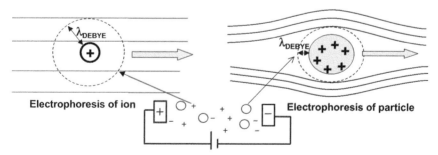

FIGURE 7-10 The two limits for the EP behavior of ions and particles A. EP of an ion ($a/\lambda_{DEBYE} \ll 1$), with electric field lines not appreciably distorted by the presence of the ion B. EP of a micron-sized particle ($a/\lambda_{DEBYE} \gg 1$) with surface/charge- layer interaction similar to the flat-wall EO flow

To define the zeta-potential ζ, it may be recalled that in the Debye-Huckel theory for the counterion cloud around a point charge, the potential profile around the central "particle" of charge $q = ze_0$ is given by:

$$\phi(x) = \frac{q}{4\pi\varepsilon\varepsilon_0 x} e^{-x/\lambda_{DEBYE}} \qquad (7\text{-}25)$$

Now the zeta potential can be redefined as the electrostatic potential at the surface of an ionic sphere ($x = a$):

$$\zeta = \phi(a) = \frac{ze_0}{4\pi\varepsilon\varepsilon_0 a} e^{-a/\lambda_{DEBYE}} \qquad (7\text{-}26)$$

Although Eq. 7-24 is derived for a sphere, this result applies to any shape of particle, provided that the counterion cloud is thin compared to any radius of curvature of the particle. For instance, for condition $a/\lambda_{DEBYE} \ll 1$ $\zeta = \frac{ze_0}{4\pi\varepsilon\varepsilon_0 a}$ and Eq. 7-26 reduces to Eq. 7-23 for small particles. These effects are also valid both for spherical particles in nonuniform fields and for particles with a broken symmetry in a uniform field.

Eqs. 7-23 and 7-24 demonstrate direct proportionality between particle velocity and electric field. The particle mobility (which equals the ratio of velocity/electric field strength, as in Eq. 5-6) only depends on the total particle charge or average zeta-potential. This remarkable result holds for any particle size and shape, even for polarizable particles with a nonuniform charge in the applied field. Dukhin [1] stated that these assumptions are valid only for a constant double layer on the particle and predicted that the EP mobility of particles where charges are induced (u_{EP_IC}) may change as a function of V^2 and velocity (v_{EP_IC}) as V^3. In an AC field that results in nanodielectrophoresis-type EP effects at frequencies below 100 kHz, and Eq. 7-27 for uniform and Eq. 7-28 for nonuniform fields become:

$$u_{EP\_IC} = \frac{\varepsilon\varepsilon_0 aV^2}{\eta(1+\chi)} \qquad (7\text{-}27)$$

$$u_{EP\_IC} = \frac{\varepsilon\varepsilon_0 a^2 \, gradV^2}{\eta(1+\chi)} \qquad (7\text{-}28)$$

Here $\chi = \frac{C_{HELMHOLTZ}}{C_{DIFFUSE}}$ is a parameter controlling how much of the total double-layer voltage ends up across the Helmholtz compact layer—the ratio of the compact layer and the diffuse layer capacitances at the point of zero charge.

Electroosmosis (EO) is the fluid motion induced by the movement of surface charges at the solid–liquid interface under the influence of typically low-frequency AC and DC electric fields [16]. Most wetted material surfaces will acquire fixed surface charges when coming into contact with a fluid that contains ions—either an electrolyte or a dielectric liquid with ionic impurities generated locally via reactions. By electrostatic attraction, the surface

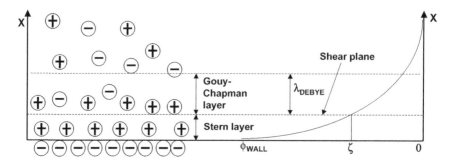

FIGURE 7-11 Schematic of electrochemical double layer model and plot of electric potential vs. distance from the wall

charges attract counterions from the solution and repel co-ions from the solid surface to maintain local charge neutrality. Consequently, excess charge is built up near the solid surface of a channel, thus forming an electrical double layer or a capacitor. The ions in the liquid phase of the double layer are mobile and will migrate under the influence of an electric field tangential to the electrode surface.

The immobile counterions adsorbed to and immediately adjacent to the wall form the compact Stern layer, while the Gouy-Chapman layer comprises the diffuse and mobile counterion layer that is set in motion upon the application of an external electric field. The shear plane separates the Stern and Gouy-Chapman layers and, in simple double-layer models, is the location of the fluid motion's no-slip condition (Figure 7-11). The magnitude of the potential at the wall surface $x = 0$ (ϕ_{WALL}) decays from the wall, and the bulk fluid far from the wall is assumed to be net neutral. The potential at the shear plane, or the electrostatic potential drop across the diffuse part of the double layer, is called the zeta-potential ζ, analogous to the concept of zeta-potential derived above for the ionic particle in EP and to a similar concept for the double layer on the electrode where ϕ_D potential is present at the OHP (Figure 5-4). Analogous to the definitions presented in Section 5-3 and Eq. 5-17, the Debye length λ_{DEBYE} is the length from the shear plane at which the electrochemical double-layer potential has fallen to its e^{-1} value of ζ. The surface charge density q_{SURF} then can be estimated from the Gouy-Chapman equation using surface potential ϕ_{WALL}, as:

$$q_{SURF} = \frac{2\varepsilon\varepsilon_0 K_B T}{z e_0 \lambda_{DEBYE}} \sinh\left(\frac{\phi_{WALL}}{2}\right) = \sqrt{32\pi\varepsilon\varepsilon_0 K_B T C^*} \sinh\left(\frac{\phi_{WALL}}{2}\right) \quad (7\text{-}29)$$

Finally, the expression for the zeta-potential can be represented for a solid surface area A with a double layer charge density q_{DL} [C/cm²] as:

$$\zeta = \frac{q_{DL}\lambda_{DEBYE}}{\varepsilon\varepsilon_0} \quad (7\text{-}30)$$

Smoluchowski analysis considers a microchannel with a cross-section in the y-z plane, which is uniform along the x direction and is subject to a field directed along the x-axis. Thus, the applied electrostatic forces should act per-pendicularly to the electrodes along the x-axis and are expected to have only a tangential component V_t, which is independent of z and y. The velocity of the EO effect—the motion of the bulk liquid along the charged surface in response to an applied electric field in a liquid channel with electric double layers on its wetted surfaces—is then:

$$v_{EO} = -\frac{\varepsilon\varepsilon_0 V_t \zeta}{\eta}\left[1 - \frac{\phi(y,z)}{\zeta}\right] \qquad (7\text{-}31)$$

Numerical solutions to predict EO flows for all relevant length scales in complex geometries with arbitrary cross-sections are difficult for Debye lengths much smaller than the characteristic dimensions of the channels (e.g., the hydraulic diameter). Indeed, this is the case for typical microfluidic elec-trokinetic systems, which have Debye length-to-channel-diameter ratios of order 10^{-4} or less. For these cases, thin double-layer assumptions are often ap-propriate as approximations of EO flow solutions in complex geometries. For the case of thin double layers, the electric potential throughout most of the cross-sectional area of a microchannel is zero, and the previous equation, for the case of zero pressure gradients, reduces to:

$$v_{EO} = -\frac{\varepsilon\varepsilon_0 V_t \zeta}{\eta} \qquad (7\text{-}32)$$

which is the Helmholtz-Smoluchowski relation for EO flow. For a typical value of the ζ potential for aqueous solutions ~ 50mV, a predicted EO velocity is ~3.5 µm/s per V/cm. Thus, except for sign, the EP velocity of a charged sphere through a stagnant fluid is identical to the EO velocity of fluid through a stationary solid membrane. Both EO and EP forces are relatively long-reach-ing forces, scaling as the inverse of distance from the electrode surface and independent of particle size.

Using Eqs. 7-30 through 7-32, the DC EO velocity for a thin double layer becomes:

$$V_{EO} = -\frac{V_t q_{DL} \lambda_{DEBYE}}{\eta} \qquad (7\text{-}33)$$

For these conditions, assuming a characteristic Debye length λ_{DEBYE}~10nm, double-layer charge density q_{DL} can be approximately estimated from an as-sumption that the double-layer voltage is 10% of the total charge across the solution (which is not true at low frequencies, where practically all voltage drop will be across the double layer), reaching in cases of applied voltage 10 V peak-to-peak ~ 50 mC/m². With an electrode gap of 25µm, applied voltage $V_{pp} = 10$V, at $d = 10$µm distance from the electrode edge, total field becomes V~$3.4*10^4$ V/m and tangential field strength $V_t = V^{1/3}$ ~ 50 V/m. For a media viscosity of 10^{-3} Pas, that translates into an average EO velocity ~21 µm/sec.

Practical EO applications include both DC and AC effects. In AC EO, where electrode arrays are energized with time-varying AC potentials, undesirable reactions at the electrodes are suppressed due to interruption and relaxation of the reactions. Similar to DC EO, AC EO is also generated by exerting a force on double-layer charges at the solid–liquid surface by tangential electric fields. In AC EO, the charges in the electrical double layer are induced by electric potentials over electrodes instead of surface charges, as in in DC EO, and the applied potentials provide tangential electric fields to drive the ions. Because the electric fields of symmetric electrode pairs exhibit mirror symmetry and charges in the double layer and electric fields change signs simultaneously, AC EO produces steady counterrotating local vortices above the electrodes [20, 21].

Dielectrophoresis (DEP) is an electrokinetic movement of dielectric particles caused by polarization effects in a nonuniform electric field. A particle suspended in a medium with different dielectric characteristics becomes polarized in a nonuniform electric field (Figure 7-12). The interaction between the nonuniform field and the induced dipole generates a force, which in turn induces some translation of the particle, since the forces are not equal due to the field gradient. If the particle is more conductive than the dielectric medium ($Z_p^* < Z_M^*$), the dipole aligns with the field, causing attraction. If the particle is less conductive than the medium ($Z_p^* > Z_M^*$), then the induced dipole aligns against the field, causing repulsion of the particle. Due to the difference in nonuniform electric field strength on the two sides of the polarizable particle, a DEP force is generated. The nonuniform field is created by a pair of unevenly shaped electrodes. Many nano- and microfabricated interdigitated devices with opposite electrodes of different geometry, shape, and uneven separation are capable of creating nonuniform high-strength electrical fields that can be utilized for dielectrophoretic manipulation of nanoparticles. In the electric field a dielectric particle behaves as a dipole with induced dipole moment proportional to the RMS value of the electric field V and with the constant of proportionality depending on the particle geometry, particle size (radius) a, permittivity of the suspending medium ε_M, and real part of the Clausius-Mossotti factor [β] determined by complex permittivity (impedance) of the medium ε_M^* and complex permittivity (impedance) of the particle ε_p^*. The force on the dipole becomes [22]:

$$F = 2\pi a^3 \varepsilon_M \varepsilon_0 RE\left[\frac{\varepsilon_p^* - \varepsilon_M^*}{\varepsilon_p^* + 2\varepsilon_M^*}\right] grad V_{RMS}^2 = 2\pi a^3 \varepsilon_M \varepsilon_0 RE[\beta] grad V_{RMS}^2 \quad (7\text{-}34)$$

$$F = 2\pi a^3 \varepsilon_M \varepsilon_0 \frac{\varepsilon_0^2(\varepsilon_p - \varepsilon_M)(\varepsilon_p + 2\varepsilon_M) + (\sigma_p - \sigma_M)(\sigma_p + 2\sigma_M)/\omega^2}{\varepsilon_0^2(\varepsilon_p + 2\varepsilon_M)^2 + (\sigma_p + 2\sigma_M)^2/\omega^2} grad V_{RMS}^2$$

Where gradient of the square of the electric field intensity $grad V_{RMS}^2$ (V^2/m^3) is dependent on the position of a particle with respect to the source of the electric field:

$$V_{RMS} = V_{PEAK}/\sqrt{2} = V_{PP}/2^{3/2} \quad (7\text{-}35)$$

The Clausius-Mossotti factor can vary between -0.5 and $+1.0$, depending on relative values of complex permittivity (or impedance) of the particle and of the media. Homogeneous dielectric particles experience a Maxwell-Wagner

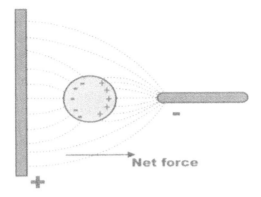

FIGURE 7-12   Dielectrophoretic force diagram showing electrostatic fields and charges

interfacial polarization at a frequency determined by the relationship between complex impedances Z_p^* and Z_M^*. When the factor is positive ($Z_M^* > Z_p^*$ or $\varepsilon_p^* > \varepsilon_M^*$), the particle will move in a "positive" direction towards higher electric field regions (positive or p-DEP). When the factor is negative ($Z_M^* < Z_p^*$ or $\varepsilon_p^* < \varepsilon_M^*$), the particles move towards smaller electrical field regions in a "negative" direction (negative or n-DEP). When using high-frequency DEP, the frequency can always be found where the differences between Z_p^* and Z_M^* are maximized or minimized, depending on the need for separation, contrary to DC methods, where the frequency is always infinitely small and often $Z_p^* \sim Z_M^*$. It is important to note that the direction of the electrical field does not play a role in DEP (unlike in EP with charges and mobile ions). For example, in cases of positive DEP the polarized particles move from weaker electric field to stronger field irrespective of the field's direction.

In an aqueous medium at low frequencies ($\omega < 1$ kHz), a particle's DEP behavior is completely dominated by differences in the particle's and the medium's conductivities, while at high frequencies ($\omega > 100$ MHz) its DEP behavior is dictated by differences in the permittivity values. However, for frequencies that can be practically realized, ~10 kHz to 100 MHz, DEP behavior depends in a complex manner on both of these quantities. It is often advantageous to suspend the particles in a low-conductivity aqueous medium (~100 mS/cm) to develop conditions for p-DEP at a wider frequency range [23-24].

The media and particle complex permittivities ε^* and complex impedances Z^* are highly frequency-dependent, which can be used for practical nano- and micromanipulations based on DEP. The analysis of the relaxation frequency can often be used for the analysis of a given particle. The AC electric field also orients the particle with respect to the electric field by means of the induced moment. This effect is responsible for alignments of many polarizable molecules, including neutral molecules, making AC DEP analysis, which is independent of the charge on a particle, more versatile in manipulating live cells, macromolecules, nanostructures, drugs, DNA, etc., than EP or EO.

Since DEP phenomena rely on field-induced dielectric polarization on particles with variable compositions and properties, the crossover frequency (the particular frequency at which DEP behavior changes from positive to negative) can be estimated from the medium's and particles' electrical properties [23]. For a particle undergoing a single interfacial relaxation process, the characteristic Maxwell-Wagner crossover frequency exists as [25]:

$$f_{MW} = \frac{\sigma_p + 2\sigma_M}{2\pi\varepsilon_0(\varepsilon_p + 2\varepsilon_M)} \tag{7-36}$$

at which the direction of the electrophoretic force alternates (Clausius-Mossotti factor = 0), again depending on frequency dependence of the particle permittivity ε_p^* and the media permittivity ε_M^*. Let us consider an example of a $2a = 10\ \mu m$ diameter latex particle with permittivity $\varepsilon_p = 2.6$, bulk conductivity $\sigma_{PB} = 10^{-7}$ S/m, and surface conductivity $\sigma_{PS} = 10^{-9}$ S suspended in water with permittivity $\varepsilon_M = 80$ and conductivity $\sigma_M = 10^{-6}$ S/m. Surface conductivity of a particle σ_{PS} is typically predominant over the bulk particle conductivity σ_{PB}, and total effective particle conductivity σ_p can be calculated as:

$$\sigma_p = \sigma_{PB} + \frac{2\sigma_{PS}}{a} \tag{7-37}$$

And total particle conductivity can be estimated as $\sigma_p = 2*10^{-4} Sm/m$. The resulting Maxwell-Wagner crossover frequency becomes $f_{MW} = \dfrac{(2*10^{-4}+2*10^{-6})}{6.28*(2.6+2*80)*8.85*10^{-12}} \sim 2*10^4\ Hz$.

Thus, there can exist certain frequencies where the DEP behavior of target cells or particles is different from that of other particles, allowing for their

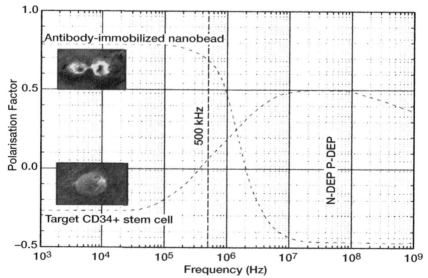

FIGURE 7-13 Frequency dependence of the Clausius–Mossotti factor for the antibody-immobilized nanobead and the target cell [23] (with permission from The Electrochemical Society and the author)

isolation from complex suspensions containing many other species [26, 27, 28, 29]. Figure 7-13 shows the frequency dependence of the Clausius–Mossotti factor for a system containing nonpolar antibody-immobilized nanobead (diameter ~ 800 nm, permittivity $\varepsilon = 2.56$, conductivity 10 mS / m) and polar target stem cell (diameter ~3 μm, permittivity $\varepsilon = 85$, conductivity 2.3 nS / m), in an aqueous medium with conductivity $\sigma = 1$ mS/ m [23]. The plot reveals that at a frequency $f = 500$ kHz the nanobeads and stem cells both experience p-DEP so that the compounds composed of nanobeads adsorbed on the surface of target stem cells move toward the region of higher field strength.

For a planar arrangement of a MEMS IDT device with width of the electrodes d equal to the gap between the electrodes, for a spherical particle of volume πa^3 positioned at a distance h above the plane in a field with voltage V (in volts), the electrophoretic force F_{DEP} changes as a function of distance h above the plane of the electrode and applied voltage V as: $F \sim V^2$ and [26, 27]:

$$F = 8\pi a^3 \varepsilon_M \varepsilon_0 \frac{V^2}{\pi d^3} RE\left[\beta\right] \exp(-\pi h / d) \qquad (7\text{-}38)$$

It should be remembered that DEP is size-sensitive, relatively weak when compared to EP and EO forces, and of rather short range. Therefore, DEP is not entirely effective for manipulating micron and submicron bioparticles (bacteria, viruses) more than 20 μm away from the electrodes. For example, at an applied potential $V_{PP} = 10$ V at a point 10 μm in from the edge of the electrode and at a height of 10 μm, the field gradient becomes $gradV_{RMS}^2 \sim 10^{15}$ $V^2/$ m^3. For a particle with a size of ~ 300 nm and assuming $RE[\beta] = 1$, the DEP force becomes $F_{DEP} \sim 10^{-14}$ N. Using Eq. 7-41, the DEP velocity can be estimated at ~ 5 μm/sec for aqueous media with viscosity 10^{-3} Pa s. However, the field gradient increases rapidly towards the electrode edge (by up to two orders of magnitude), and the particle velocity may reach up to ~ 500 μm/sec [22].

In a special case where a spherical particle is suspended in a liquid, the levitation height H can be found where the DEP force F_{DEP} equals the gravitational force $F_G = \Delta\rho_D g \pi a^3$ as:

$$H = \frac{d}{\pi} \ln\left[-\frac{24\varepsilon_M \varepsilon_0 V^2 RE\left[\beta\right]}{\pi a^3 \Delta\rho_D g} \right] \qquad (7\text{-}39)$$

And for a 1 μm diameter particle suspended in water at the IDT array with $d = 20$μm and $V = 5$V at Clausius-Mossotti factor of –0.5 and particle density being two times higher than water (2000 kg/m^3), the levitation height is $H = 43$ μm. The height does not change significantly with particle density.

The DEP force becomes smaller with decreases in a particle volume, although double-layer charging processes may introduce additional effects that become particularly pronounced at lower AC frequencies. Typically microfabricated irregular "3-dimensional" (3D) electrodes are used. These electrodes offer easy mass production, can be seamlessly integrated with microfluidics, allow reducing the dimensions of the device, and offer analytical electrical field solutions for modeling. Such designs employ aligned electrode pairs fabricated either along the sidewalls or on the top and bottom of channels through which the particle-laden fluid flows. The 3D configuration can generate strong electric fields normal to the flow direction that penetrates across the entire height of the channel with a reasonable voltage. The DEP force was shown to be more effective in 3D structures, showing less decrease of the force profile with distance above the plane, greatly improving the trapping efficiency for the microscale particles and achieving isolation rates of ~300 particles/sec [29, 30, 31]. Through tailoring the orientation of the 3D electrode, the particles can be directed to different streamlines/exits or prevented from flowing past a point in the channel altogether.

Often very small devices and very high voltages are needed, especially for effective nanoscale particle manipulation. Electrokinetics has been shown to be capable of manipulation of particles with sizes down to 20 nm, when small electrode channels (100 nm) and high voltages (~10V) are utilized. Figure 7-14 demonstrates, for example, that for a typical interdigitated transducer with an electrode gap of ~100 μm, DEP manipulation of 1μm particles is possible at ~1 V voltage amplitude levels, while for 100 nm particles a ~100V amplitude is required [19]. Figure 7-15 [23] represents an example of simulation of the electric-field distribution created by strips of microelectrodes with a width of 20 μm and a spacing gap of 20 μm. For this example the p-DEP effect is predominant at the strong field gradients at the edges of electrodes, while n-DEP dominates at low field regions away from the electrodes.

DEP has been traditionally used for manipulation and analysis of micrometer-size molecules and recently of nanometer-size viruses, nanowires, nanotubes, and macromolecules, including precise alignment and placement of nano-objects for the fabrication of nanodevices [28]. Design of appropriate interdigitated transducers for such manipulation includes careful analysis of several forces and effects resulting from applied high electric fields used to influence the manipulated macromolecules and particles, such as DEP, EP, EO forces, drag forces, thermal gradients and heat dissipation, fluid motion buoyancy, Brownian motion, and electrothermal force [22].

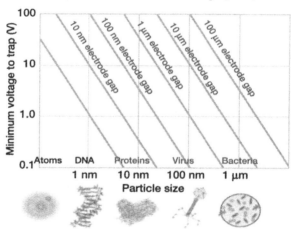

FIGURE 7-14 Applied field requirements for trapping particles of various sizes [19] (with permission from IEEE and the authors)

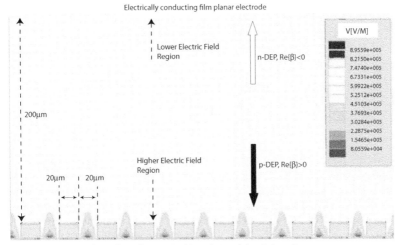

Electrically conducting striped (irregular interdigitated) electrodes

FIGURE 7-15   Simulation of the electric-field distribution created by strips of microelectrodes [23] (with permission from The Electrochemical Society and the author)

The expression for the DEP force was presented above, using Stokes derivation when only the drag and the DEP forces on a spherical particle are considered:

$$F_{DRAG} = 6\pi\eta a(v_{FLUID} - v_{PARTICLE}) \qquad (7\text{-}40)$$

Such a force on a particle in a medium of viscosity η at stationary conditions when $F_{DRAG} = F_{DEP}$ produces velocity:

$$v_{PARTICLE} = v_{FLUID} + \frac{a^2\varepsilon_M\varepsilon_0 RE[\beta]gradV_{RMS}^2}{3\eta} \qquad (7\text{-}41)$$

In microfabricated devices the effects of electrical heating should also be considered. In a device with internal resistance R the electrothermal force generates power V_{RMS}^2/R, and in a medium with electrical conductivity σ the power per volume generation (W/m^3) for applied field V/d becomes:

$$W = \sigma\frac{V^2}{d^2} \qquad (7\text{-}42)$$

For an example of applied voltage 10V for a gap $d = 50\mu m$ and a microfabricated device with a channel length of 2 cm, and electrode-to-electrode gap of $50\mu m$, the volume over which the heat is generated becomes $50\mu m * 50\mu m * 2\ cm = 5*10^{-12}\ m^3$, giving an average power dissipation of 2mW for applied field $2*10^5\ V/m$ and conductivity 0.01 S/m. It can be compared with the power dissipation measured for the electrode, where for a measured resistance of the electrode array R in a medium of 33 kOhm and $V_{pp} = 20V$, the total power dissipation is $V_{pp}^2/R = 20^2/33000 = 1.5mW$. In a medium with thermal conductivity κ the increase in temperature due to electrical heating is:

$$\Delta T = \frac{\sigma V_{RMS}^2}{\kappa} \tag{7-43}$$

and the corresponding adiabatic temperature rise per second $\Delta T / t$ is:

$$\frac{\Delta T}{t} = \frac{\sigma V_{RMS}^2}{C_{HEAT}\rho_D} \tag{7-44}$$

For highly conductive solutions the temperature rise is likely to be excessive (~100 °C) and could lead to denaturation in biological systems. For a system with a typical water thermal conductivity $\kappa = 0.6$ J/(m s K), heat capacity $C_{HEAT} = 4200$ J/(kg K), density $\rho_D = 1000$ kg/m³, peak-to-peak voltage $20V$, and electrical conductivity 1 S/m, heating can reach ~85 °C, while for a conductivity of 0.01 S/m it is only ~1 °C. Data was also presented [26] illustrating that the 3D electrode structure exhibited about 10 times less heating in similar conditions than was shown by a 2D planar electrode structure.

Another aspect of excessive heating is its dependence on the local field, which in case of DEP is nonuniform, creating local changes in medium density, viscosity, permittivity, and conductivity. These differences result in other fluid forces, such as buoyancy force (due to the difference in density, which is typically negligible) and electrothermal force (due to local changes in conductivity and permittivity) [22]. For buffer solutions empirical data showed [23] that $\frac{1}{\sigma}\frac{d\sigma}{dT} = +1\%$ and $\frac{1}{\varepsilon}\frac{d\varepsilon}{dT} = +0.5\%$ per Kelvin, or essentially a change in electrical properties of approximately 1% per every 1 K of temperature change ΔT due to electrical heating. The electrothermal force can be estimated as:

$$F_{ELTHERM} = \frac{V_{RMS}^2}{2\varepsilon^*}\frac{d\varepsilon^*}{dT} \sim V_{RMS}^2 \Delta T \frac{\varepsilon^*}{200} \tag{7-45}$$

The dimensionless frequency- and temperature-dependent factor $M(\omega, T)$ shows the effect of changes in these parameters on the electrical force per volume unit:

$$M(\omega,T) = \frac{\dfrac{T}{\sigma}\dfrac{d\sigma}{dT} - \dfrac{T}{\varepsilon}\dfrac{d\varepsilon}{dT}}{1+(\omega\tau)^2} + \frac{1}{2}\frac{T}{\varepsilon}\frac{d\varepsilon}{dT} \tag{7-46}$$

This factor changes from positive to negative at a frequency approximately equal to f_{MW}. As an example [22], the M factor is positive at low frequencies where the force is dominated by space charges and is moving the particles across the electrode from the edges to the center of the metal. The factor M becomes negative for high frequencies where the force is reversed and is moving the particles away form the electrodes into the middle of the gap between the electrodes. Analysis of Coulomb forces on the particle leads to an equation showing proportionality between the fluid velocity due to electrothermal and

DEP forces, where h is the distance between the electrode surface and the manipulated particle:

$$\frac{v_{ELECTROTHERMAL}}{v_{DEP}} = \frac{0.39}{4\pi} \frac{M(\omega,T)}{RE[\beta]} \frac{\sigma V_{RMS}^2 h^2}{\kappa T a^2} \propto \frac{\sigma V^2 h^2}{a^2} \tag{7-47}$$

The expression shows that high media conductivity σ increases flow due to the electrothermal effect, especially for smaller particles ($a \to 0$), while the DEP force dominates close to the electrode edge ($h \to 0$).

Finally, Brownian random force can be estimated based on the temperature of the fluid and particle of diameter a:

$$F_{BROWN} \approx \frac{K_B T}{2a} \tag{7-48}$$

Thus, the ratio of the DEP force to the thermal force is proportional to the fourth power of the particle radius. If one assumes that the particle is as close to the electrode as its radius a, then it is possible to show that the minimum voltage V_{MIN} required for the DEP force to exceed the thermal force scales as the inverse of the electrode radius r (or principal dimension) and the 1.5th power of the particle radius [27]:

$$V_{MIN} \sim a^{3/2} / r \tag{7-49}$$

AC electrokinetic fields work optimally with microscale particles and devices, integrating seamlessly with microfluidics and miniaturized electrochemical impedance-based sensors [18, 21]. Microfluidic technologies reduce detection times, reduce biological samples (and reagents) required on a single chip, and channel geometries and dimensions that often aid in the isolation and/or detection of the target species concentrations to varying degrees of certainty. EP and DEP in combination with EIS can effectively trap, manipulate, separate, and quantitatively detect particles ranging from viruses to large DNA strands to bacteria and mammalian cells in planar and 3D microfabricated devices, both in a steady pool and in a continuous flow [26]. Electrokinetic delivery methods also find wide applications as microfluidic concentrators serving as sample preparation subsystems in complete microorganism detection systems [32, 33]. A concentrator works by accumulating the particles of interest from a sample and resuspending them in a smaller volume of solution, resulting in an increased local sample concentration. For example, *in vivo* bacteria concentration is typically only 1–30 CFU (colony- forming units/ml). Even for patients who are near death, the load may still be as low as ~1000 CFU/ml and has to be detected in the presence of other cells in concentrations of millions per milliliter. As such, microfluidic concentrators perform two important functions. First, by increasing the concentration of samples presented to the detector, they increase the overall sensitivity of the system. Second, the microconcentrator corrects the volume mismatch between the samples obtained by standard collection methods (~mL) and the volume most microfluidic devices can process in a reasonable amount of time (~μL). For example, bacteria E coli preconcentration capture and detection using AC DEP, EO, and EP-based microfluidic diagnostic systems has been demonstrated [20, 21]. These devices are capable of increasing the concentration of the target pathogen cells (in absolute terms and/or with respect to that of the other cells) and detecting/quantifying their presence. The effectiveness of transdermal drug delivery using electrokinetics for the administration of topical anesthesia and analgesia has been reported [34, 35, s36].

The effect of various electrokinetic forces depends on electrochemical properties of media and particle types, electric field, and characteristic dimensions of electrodes, as demonstrated in Figure 7-16 for the case of 100-nm polystyrene particles ($\varepsilon_p = 2$) suspended in aqueous media ($\varepsilon_M = 80$) when applied

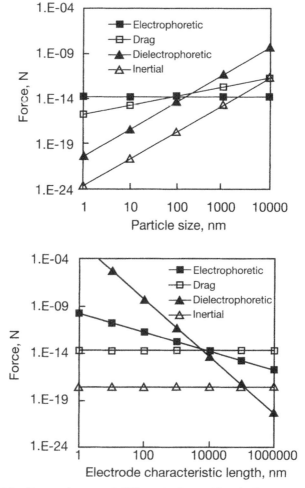

FIGURE 7-16 Dependence of different electrokinetic forces on A. size of the particles; B. electrode characteristic dimensions [37] (with permission from IEEE)

voltage 1 V created a field of 10^5 V/m for electrodes of 10 μm [37]. The magnitude of forces can range widely, depending on particle size, position from electrode, and characteristic length of the electrode. In most cases, inertia force is not important in small scale. For example, the EP and drag forces have strong effects and can be significant even in large scale. On the other hand, DEP is effective for larger particles at short distances, as it is a function of the potential gradient, which decays rapidly away from the electrode edges. The calculated buoyancy force is orders of magnitude smaller than other forces for submicrometer particles.

Capillary EP, where particles are separated in the DC field, is a "microfluidic" technique performed within channels housed in microfluidic chips. The narrow dimension of the microchannels plays a key role in the success of the technique by preventing the development of recirculation vortices that tend to arise due to temperature differences generated by the Joule heating of the ionic current through the system. EP also requires high voltages (kV) to drive the EO flow through the capillaries. This high-voltage requirement not only necessitates the use of bulky (and dangerous) high-voltage supply units but also produces extensive electrolysis and gas bubbles. To avoid these drawbacks inherent in using DC methods such as EP, an AC method such as DEP can be employed. If frequencies higher than ~1 kHz are used, the half-cycle time is shorter than the charge-transfer time at the electrode for most electrochemical reactions, and consequently most Faradaic electrolysis reactions are suppressed.

The biggest DEP shortcoming is that typical particle velocities are low, and forced convection may be required. The DEP force changes as cube function with particle radius and quadratically with the applied voltage. This cubic dependence on particle radius in conjunction with practical limitations on the applied voltage render the particle velocity produced by a DEP force minuscule, typically on the order of 10 μm/sec for bacteria and 1 μm/sec for viruses. Concentration times on the orders of hours are thus required. Also problematic is the fact that the field gradient necessary to drive DEP motion can only be achieved with relatively narrow interdigitated electrodes, whose field penetration depth is limited by the electrode width. As a result, DEP channels are usually less than 50 μm in transverse dimension. For example, for a device 5 cm long employing a 50 μm transverse dimension and a 10 μm/s DEP velocity, pathogen capture can only be ensured if the throughput is less than 25 nL per second. At this flow rate, it would require more than 10 hours to process a 1 mL sample. This low throughput and batch volume can be somewhat alleviated with a massively parallel array for laboratory use, but such arrays cannot be easily designed for a miniature and portable diagnostic kit. Bacteria isolation and concentration can be accelerated by using fluidic convective forces to transport suspended particles such as the target pathogens to a specific localized region and then capturing the targets utilizing the DEP force.

7.4. Specific adsorption and multistep heterogeneous kinetics

The adsorption of reaction intermediates constitutes the most widely known cause of inductive phenomena in low-frequency impedance studies. Corresponding kinetic models represent a total reaction that combines multiple nonequilibrated steps (and, therefore produces intermediates) and does not include "tree" or single electron mechanisms. Both the adsorption process and the "pure" discharge reactions are electrochemical in nature and therefore are potential-dependent. The net result of these reaction rates, displayed as dependencies of current on applied voltage, can be an inductive phase shift in the cell current. A negative capacitance (or inductance) appears when the current decreases with increasing voltage. It should be remembered that inductive behavior can also result from other kinetic and instrumentation effects, such as nonhomogeneous current distribution, high-frequency effects of cables, multi-electrode probes, potentiostat nonidealities, and system instability with impedance decreasing during the time of the experiment.

The appearance of the inductive or capacitive adsorption loop depends on relative rates of adsorption and desorption. These mechanisms often generate a number of different intermediates and surface reaction limiting processes. The inductive loops appear when the rates of adsorption and desorption become approximately equal and their potential derivative changes sign [38, 39, 40, 41, 42, 43]. The result is often a decrease of surface coverage with increasing potential for one of the adsorbed intermediates [44, 45]. Another way to explain this phenomenon is to consider a situation where the kinetic current is a function of the adsorbed intermediate coverage and the response of this intermediate coverage to changes in the potential is sluggish. The delay between the time of the AC potential perturbation and the time when the new coverage value and kinetic current stabilizes leads to a positive (inductive) phase shift.

The impedance characterization approach to modeling heterogeneous systems displaying single and multistep adsorption kinetics was developed by Eppelboin and Keddam [46, 47, 48, 49, 50] and was further expanded by Conway and Harrington [51, 52, 53, 54]. In this model only two reactions take place as species A arrive at the surface, adsorbs, and reacts at the electrode surface, producing species B in two electrochemical reaction steps. The general case can be represented by simple two-step Faradaic process involving reversible adsorption(s) followed by reversible desorption:

$$A \xrightarrow{K1} B_{ADS}; B_{ADS} \xrightarrow{K-1} A \qquad (7\text{-}50)$$

$$B_{ADS} \xrightarrow{K2} C; C \xrightarrow{K-2} B_{ADS} \qquad (7\text{-}51)$$

The model is based on the following assumptions:

1. Only adsorption and charge-transfer processes occur on a two- dimensional interface.

2. Adsorption of the intermediate B obeys a Langmuir isotherm and is characterized by a surface coverage θ with adsorbates forming only a monolayered, fixed number of equivalent adsorption sites, and no energetic interaction between adsorbed species occurs.

3. The reactions are governed by heterogeneous kinetics with diffusion not playing a major role.

4. The reaction rates are exponentially potential-dependent (and obey Tafel's Law) $K_i = k_i e^{b_i \Delta V}$ in mol/cm^2 sec. The k_i may also include, for

convenience, such respective concentration terms as $k_i = k_i^0 C_i$ and Tafel's coefficient $b_i = -\alpha n F / RT$ with $\Delta V = V - V^\circ$ where V_0 is a standard redox potential.

5. The maximum number of sites per surface unit that can be occupied by the absorbate B is characterized by a coefficient Γ_{MAX} (mol/cm²).

The reaction rates can be expressed as:

$$v_1 = K_1(1-\theta) - K_{-1}\theta$$
$$v_2 = K_2\theta - K_{-2}(1-\theta) \tag{7-52}$$

The resulting total Faradaic current through the system becomes:

$$I_F = FA(v_1 + v_2) = FA\left[K_1(1-\theta) + K_2\theta - K_{-1}\theta - K_{-2}(1-\theta)\right] \tag{7-53}$$

The classical macroscopic description gives the mass and charge balances, respectively:

$$\Gamma_{MAX}\frac{d\theta}{dt} = v_1 - v_2 = (K_1 + K_{-2})(1-\theta) - (K_{-1} + K_2)\theta \tag{7-54}$$

For a steady-state condition at equilibrium potential the rates of both reactions are zero, and the changes in surface coverage θ with time are $d\theta / dt = 0$. The equilibrium surface coverage θ_{EQ}, can be estimated as:

$$\theta_{EQ} = \frac{K_1 + K_{-2}}{K_1 + K_{-1} + K_2 + K_{-2}} \tag{7-55}$$

If a small sine-wave perturbation V_A of angular frequency ω is superimposed on the polarization voltage, the sine wave responses $\Delta\theta$ and ΔI_F can be obtained by linearizing Eqs. 7-53 and 7-54:

$$j\omega\Gamma_{MAX}\Delta\theta = [(b_1 K_1 + b_{-2} K_{-2})(1-\theta) - (b_{-1}K_{-1} + b_2 K_2)\theta]V_A - [K_1 + K_{-1} + K_2 + K_{-2}]\Delta\theta \tag{7-56}$$

$$\Delta I_F = FA\Big(\left[(b_1 K_1 - b_2 K_{-2})(1-\theta) + (b_2 K_2 - b_{-1}K_{-1})\theta\right]V_A +$$
$$\left[-b_1 K_1 - b_{-1}K_{-1} + b_2 K_2 + b_{-2}K_{-2}\right]\Delta\theta\Big) \tag{7-57}$$

After introducing the value of surface coverage θ_{EQ}, the Faradaic impedance at equilibrium can be expressed in the following classical form [54]:

$$Z_F = \frac{1}{A + \dfrac{B}{j\omega + G}} = \frac{j\omega + G}{A(j\omega + G) + B} \tag{7-58}$$

where:

$$A = \frac{F^2}{RT}[b_1 K_1(1-\theta) + b_{-1}K_{-1}\theta + b_2 K_2\theta + b_{-2}K_{-2}(1-\theta)] = \frac{1}{R_{CT}} \tag{7-59}$$

$$B = \frac{F^2}{RT\Gamma_{MAX}}(-K_1 - K_{-1} + K_2 + K_{-2})[b_1 K_1(1-\theta) + b_{-1} K_{-1}\theta - b_2 K_2\theta - b_{-2} K_{-2}(1-\theta)]$$

(7-60)

$$G = \frac{1}{\Gamma_{MAX}}(K_1 + K_{-1} + K_2 + K_{-2})$$

(7-61)

Examining Eq. 7-58 shows that at very low frequencies ($\omega \to 0$) total Faradaic impedance approaches the charge-transfer resistance R_{CT}. Depending on the sign of the B term in Eq. 7-58, the low-frequency response of the Faradaic impedance predicts the existence of either a capacitive (negative sign) or inductive (positive sign) loop. Both A- and G-parameters are always positive, and the resulting shape of the Nyquist plot depends on the sign of the parameter B, which can be positive, zero, or negative.

The options for the values of parameter B can be analyzed by examining Eq. 7-60. The analysis shows that only over a relatively narrow potential range, where the kinetic parameters change (due to their different potential dependence) from the values corresponding to the conditions $(K_1 + K_{-1}) > (K_2 + K_{-2})$ to those of $(K_1 + K_{-1}) < (K_2 + K_{-2})$ or vice versa, does the inductive behavior arise. The relative values for the Tafel parameters b_i and the relative surface coverage θ also play a role. In a simplified system with two irreversible processes (K_1 and K_2) and similar Tafel parameters values $b_1 \sim b_2$, the parameter B is positive when there is a transition from small predominance in adsorption (K_1) to desorption (K_2) processes or vice versa, with $K_2 > K_1, b_2 K_2 > b_1 K_1$ and $\theta \to 0$, or when $K_1 > K_2, b_1 K_1 > b_2 K_2$, and $\theta \to 1$, which is the predominant desorption at low coverage or predominant adsorption at high coverage, with the system being thermodynamically "off balance."

The example in [55] presents two more universal model cases. The values were chosen as rate constants $K_1 = 0.05$, $K_{-1} = 0.05$, $K_2 = 1.5$, $K_{-2} = 0$ and the Tafel coefficients as $b_1 = b_{-1} = 0.5$, $b_2 = 0.01$, with voltage amplitude $V_A = 0.05V$. At any value of surface coverage θ the resulting relationships become $b_1 K_1(1-\theta) + b_{-1} K_{-1}\theta - b_2 K_2\theta - b_{-2} K_{-2}(1-\theta) > 0$; $-K_1 - K_{-1} + K_2 + K_{-2} > 0$, and $B > 0$, resulting in an inductive loop. In the case of $K_1 = 0.5$, $K_{-1} = 1.5$, $K_2 = 0.5$, $K_{-2} = 0$, $b_1 = b_2 = b_1 = 0.5$, $V_A = 0.05V$ at any value of surface coverage θ the relationships becomes $-K_1 - K_{-1} + K_2 + K_{-2} < 0$ and $b_1 K_1(1-\theta) + b_{-1} K_{-1}\theta - b_2 K_2\theta - b_{-2} K_{-2}(1-\theta) > 0$, resulting in B < 0 and capacitive loop.

When parameter $B = 0$, the Faradaic impedance is real and equals R_{CT}, with a resulting single semicircle. That may occur, for instance, when the system is in total equilibrium at a given potential so that $(K_1 + K_{-1}) = (K_2 + K_{-2})$.

When the parameter $B > 0$, the closed form for the Faradaic impedance in Eq. 7-58 can be represented by an equivalent circuit composed of a parallel combination of the charge-transfer resistance $R_{CT} = 1/A$, a series combination of resistance $R_{ADS} = G/B$, and an adsorption inductor $L = 1/B$. The double-layer capacitance and the solution resistance can be added to complete the circuit (Figure 7-17A). There are two semicircles in the complex impedance plane— a high-frequency capacitive and a low-frequency inductive (Figure 7-17B):

$$Z_F = \frac{1}{A + \dfrac{1}{j\omega/B + G/B}} = \left[\frac{1}{R_{CT}} + \frac{1}{j\omega L + R_{ADS}}\right]^{-1}$$

(7-62)

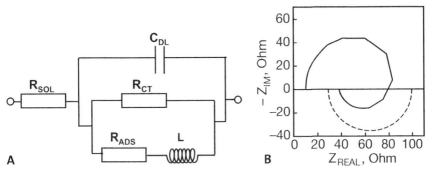

FIGURE 7-17   Equivalent circuit diagram (A) and representative data (B) for adsorbed species where B > 0

Continuous line = total impedance, dashed line = Faradaic impedance, parameters used: R_{CT} = 100 ohm, R_{ADS} = 40 ohm, L = 0.2 H, C_{DL} = 2*10^{-5} F, R_{SOL} = 10 ohm

When the parameter $B < 0$, as happens in a majority of cases when $\alpha \sim 0.5$, $\theta \sim 0.5$, and one of the processes is very predominant—either $K_1 \gg K_2$ or $K_2 \gg K_1$—the closed form for the Faradaic impedance in Eq. 7-58 can be represented by:

$$Z_F = \frac{1}{A - \frac{|B|}{j\omega + G}} = \frac{R_{CT}(j\omega + G) - R_{CT}^2\,|B| + R_{CT}^2\,|B|}{j\omega + G - R_{CT}\,|B|} = R_{CT} + \frac{R_{CT}^2\,|B|}{j\omega + G - R_{CT}\,|B|} \quad (7\text{-}63)$$

The last equation can be further rearranged as:

$$Z_F = R_{CT} + \frac{1}{j\omega / R_{CT}^2\,|B| + (G - R_{CT}\,|B|)/R_{CT}^2\,|B|} = R_{CT} + \frac{1}{j\omega C_{ADS} + \frac{1}{R_{ADS}}} \quad (7\text{-}64)$$

The Faradaic circuit can be described as a series of the charge-transfer resistance and a parallel combination of the adsorption capacitance $C_{ADS} = \dfrac{1}{R_{CT}^2\,|B|}$ and the adsorption resistance $R_{ADS} = \dfrac{R_{CT}^2\,|B|}{G - R_{CT}\,|B|}$ (Figure 7-18A). At very high frequencies ($\omega \to \infty$) $Z_F \to R_{CT}$. At lower frequencies further options have to be examined. If $G - R_{CT}\,|B| = 0$, then $Z_F = R_{CT} - \dfrac{j}{\omega C_{ADS}}$, represented by a series connection of R_{CT} and C_{ADS} with a limiting capacitive behavior (vertical - 90° line) at the low frequencies (Figure 7-18B). If $G - R_{CT}\,|B| < 0$, then $R_{ADS} < 0$, and there are two semicircles, the second one with negative real resistance values (Figure 7-18C). If $G - R_{CT}\,|B| > 0$, then all elements are positive, and Faradaic impedance depends on the relative values of double-layer capacitance C_{DL} and C_{ADS}. There may be one ($C_{ADS} \ll C_{DL}$) or two ($C_{ADS} \gg C_{DL}$) semicircles (Figure 7-18D).

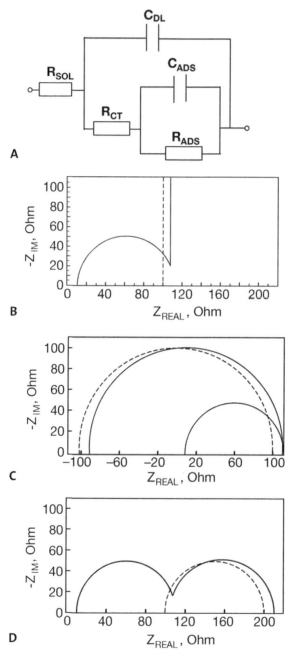

FIGURE 7-18 Equivalent circuit diagram (A) and representative data (B, C, D) for adsorbed species and B < 0.

Continuous line = total impedance, dashed line = Faradaic impedance, parameters used: R_{CT} = 100 ohm, R_{ADS} = 40 ohm, L = 0.2 H, C_{DL} = 2*10⁻⁵ F, R_{SOL} = 10 ohm

A more general descriptive condition for an inductive response phenomenon was offered by Lasia [56, 57] and Conway [41]. They presented a situation where the potential V is near the maximum or the minimum of $\theta = f(V)$ plot, and surface coverage change $\Delta\theta$ decreases with changing electrochemical potential $d\theta/dV < 0$. This condition occurs, for instance, in the case of "specific adsorption / chemisorption in IHP," which is driven by chemical attraction of the adsorbing species to the electrode and may not be based on the charges of the electrode or the species. In the case of a single adsorbed intermediate the condition $d\theta_1/dV < 0$ results in one inductive loop at low frequency. For the case of two and more adsorbed intermediates the rule of $d\theta/dV < 0$ still applies, and in the condition $d\theta_1/dV < 0$ and $d\theta_2/dV < 0$ leads to two separate low-frequency inductive loops.

A great number of reaction mechanisms, among them some very complicated ones, have been analyzed by employing this method. In [58, 59] various kinetics-limited pathway processes of oxygen reduction and hydrogen evolution were simulated and investigated. When surface-related kinetics was a limiting step, an inductive impedance behavior was observed. Several specific models with regard to passivation of metals, electrodeposition, and corrosion processes have been developed by Harrington [51, 52, 53, 54], Keddam [55], Lasia [56, 57], Gabrielli [60], and Tribollet and Orazem [61]. For instance, Lasia [57] offered an expanded analysis of various kinetic options of hydrogen reaction processes such as evolution (discharge), diffusion, adsorption, desorption, absorption, and underpotential deposition on metal electrodes. A simple adsorption in the absence of diffusion complications $R_{ADS}|C_{ADS}$ circuit model represented by Figure 7-18A may evolve into additional parallel conduction interfacial impedance structures (Figure 7-19). The Nyquist plots reveal:

1. Two semicircles appear in the case of hydrogen adsorption or desorption limitations coupled with evolution only (without absorption or diffusion limitations), which are related to two time constants, charge transfer $R_{CT}|C_{DL}$, and adsorption $R_{ADS}|C_{ADS}$ or desorption impedance Z_{DES}.

2. Three semicircles occur in the presence of hydrogen absorption, corresponding to the charge-transfer $R_{CT}|C_{DL}$, adsorption $R_{ADS}|C_{ADS}$, and absorption resistance together with hydrogen diffusion effects $R_{ABS} - Z_{DIFF}$ (portions of a straight line at $-45°$ may be observable). When the absorption reaction is very fast, the semicircle corresponding to hydrogen absorption disappears.

3. Two semicircles are followed by a finite diffusion line in the case of hydrogen underpotential deposition (UPD), coupled or not with hydrogen adsorption. In the case of underpotential deposition the adsorption resistance is very large ($R_{ADS} \to \infty$), and $R_{ABS} + Z_{DIFF}$ are in parallel with C_{ADS}. The features of the Nyquist plots are related to two remaining time constants, charge transfer $R_{CT}|C_{DL}$, and the UPD represented by absorption/adsorption impedance $R_{ABS}|C_{ADS}$, followed by Z_{DIFF} (finite-length diffusion—i.e., a line at $-45°$ and capacitive $-90°$ or semicircular features). In the absence of the absorption reaction the semicircle connected with R_{CT} is followed by a vertical $-90°$ pseudocapacitance line.

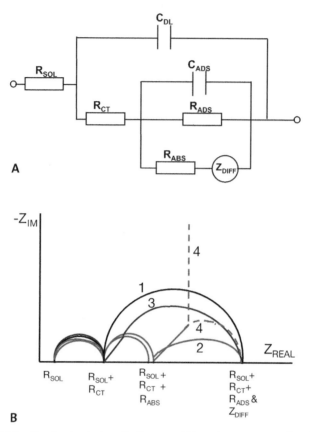

FIGURE 7-19   Equivalent circuit diagram (A) and representative data (B) for hydrogen evolution with effects of UPD, adsorption, absorption, and diffusion reactions

(1-adsorption and evolution; 2-evolution, adsorption, absorption with diffusion effects; 3-evolution, adsorption, fast absorption with diffusion effects; 4-evolution, UPD deposition, and finite diffusion)

7.5.   Impedance kinetics studies on porous electrodes

Several important technological applications, such as battery devices and electrocatalysis, need a very large effective surface of contact between the electrode and the electrolyte. This expanded surface can be developed on porous electrode surfaces. The complexity of the random structure of the porous electrode and various experimental situations related to mass-transport impedance in the pores, coupled with interfacial kinetics inside the pores, led investigators initially to investigate simple single-pore models. Of the possible shapes modeled, the cylindrical pore with a length l and a radius r has been most thoroughly investigated, both by analogy with the distributed electrical line and by direct integration of the equations describing the pore [56, 57, 62, 63, 64, 65]. In order to take into account the random nature of a porous electrode, a size-distribution function of the pores should be considered.

As the first approximation, impedance of a porous electrode can always be considered as a series combination of two processes—a mass-transport resistance inside the pores and impedance of electrochemical reactions inside the pores. De Levie was the first to develop a transmission line model to describe the frequency dispersion in porous electrodes in the absence of internal diffusion limitations [66]. De Levie's model is based on the assumption that the pores are cylindrical, of uniform diameter $2r$ and semi-infinite length l, not interconnected, and homogeneously filled with electrolyte. The electrode material is assumed to have no resistance. Under these conditions, a pore behaves like a uniform RC transmission line. If a sinusoidal excitation is applied, the transmission line behavior causes the amplitude of the signal to decrease with the distance from the opening of the pore, and concentration and potential gradients may develop inside the pore. These assumptions imply that only a fraction of the pore is effectively taking part in the double-layer charging process. The R_{PORE} [ohm] resistance to current in a porous electrode structure with number of pores n, filled with solution with resistivity ρ, is:

$$R_{PORE} = \rho l \,/\, n\pi r^2 \tag{7-65}$$

For a typical situation with $\rho = 10$ ohm cm, $l = 0.05$ cm, $r = 10^{-4}$ cm, n $= 10^6$, and $R_{PORE} \sim 16$ ohm, the pore impedance Z_{PORE} becomes a combination of the electrolyte resistance in the pore R_{PORE} and interfacial impedance $Z_{INTERFACIAL}$. According to De Levie [66]:

$$Z_{PORE} = \sqrt{R_{PORE} Z_{INTERFACIAL}}\; e^{j\phi/2} = \sqrt{R_{PORE} Z_{INTERFACIAL}}\; \coth\sqrt{\frac{R_{PORE}}{Z_{INTERFACIAL}}} \tag{7-66}$$

It is important to note that total impedance varies as $Z^{1/2}$, with Z being a corresponding impedance of a plane electrode. In Eq. 7-66 it is assumed that $Z = Z_{INTERFACIAL}$. Depending on the number and geometry of the pores, more or less significant distortions to the interfacial-impedance feature (ideal semicircle or diffusion-limited profile) are expected. The phase-angle value for the impedance of the pore is also two times smaller than for a plane electrode, leading, for example, to a situation where a porous electrode with purely capacitive walls yields a phase angle of $-45°$ instead of $-90°$ and can be easily mistaken for a Warburg diffusion to a plane surface. A semi-infinite diffusion in pores exhibits a $-22.5°$ line at low frequencies in the impedance diagram in cases of Warburg infinite diffusion in a pore instead of a conventional $-45°$. Impedance of charging of a porous electrode is mathematically analogous to the frequency-dependent Faradaic charging of an ion/electron-conducting polymer layer on the electrode surface [67] (Chapter 9).

The $Z_{INTERFACIAL}$ component can be broadly defined as impedance that results from porosity, percolation, Warburg diffusion, accumulation of corrosion products in pores of paint (which can be finite or infinite diffusion), and other "disorder" processes in the electrolyte or solid. $Z_{INTERFACIAL}$ is typically composed of the Faradaic impedance Z_F (resistance to charge transport or low frequency diffusion impedance of the $-45°$ Warburg region), and the double-layer capacitance is represented by C_{DL} (or CPE_{DL}):

$$Z_{INTERFACIAL} = \frac{1}{\dfrac{1}{Z_F} + (j\omega)^\alpha Q_{DL}}$$ (7-67)

When Faradaic charge transfer is absent and the interfacial impedance is represented only by a low-frequency (differential) capacitance C_{DL}, the impedance in pores becomes:

$$Z_{PORE} = \sqrt{\frac{R_{PORE}}{j\omega C_{DL}}} \coth \sqrt{R_{PORE} j\omega C_{DL}}$$ (7-68)

A more complicated expression for pore impedance develops when Faradaic charge transfer is present [68], such as that for the impedance of an electrode consisting of cylindrical pores in the absence of DC current (that is, in the absence of diffusion) and represented by a parallel combination of charge-transfer resistance R_{CT} and CPE_{DL} [64]:

$$Z_{PORE}(\omega) = \sqrt{\frac{R_{PORE}}{\dfrac{1}{R_{CT}} + (j\omega)^\alpha Q_{DL}}} \coth \sqrt{R_{PORE}\left(\frac{1 + (j\omega)^\alpha R_{CT} Q_{DL}}{R_{CT}}\right)}$$ (7-69)

The behavior of porous electrodes depends on the details of their geometry, which is frequently represented by penetration depth λ_p of the AC current in the pore and the ratio l/λ_p. A penetration depth λ_p of the signal can be defined as a square root of the ratio of interfacial impedance for a flat electrode (Eq. 7-67) to the resistance of a pore (Eq. 7-65). For a simple interfacial impedance case without charge transfer, as presented in Eq. 7-68, the penetration depth becomes inversely proportional to the square root of frequency ω and directly proportional to the square root of the radius of the pores r and solution conductivity σ:

$$\lambda_p = \sqrt{\frac{Z_{INTERFACIAL}}{R_{PORE}}} \sim \frac{1}{2}\sqrt{\frac{\sigma r}{C_{DL}\omega}}$$ (7-70)

When the relative porosity effect is large, the penetration depth is small compared to the depth of the pores. De Levie [66] reported that for $l \geq 3\lambda_p$ the pore behaves as semi-infinite ("deep pore"), and for $l \leq 0.2\lambda_p$ the impedance behavior is similar to that of a flat electrode ("shallow pore"). For a simplified case where the concentration of electroactive species is high and the concentration gradient in pores is absent but there is a potential gradient, the details of the impedance representation for the porous electrode depend on the pore's length l, penetration depth λ_p, and applied frequency ω. The analysis of Eqs. 7-66 through 7-69 [56] results in three cases (Figure 7-20).

1. The simplest case is a "shallow pores" condition or $l << \lambda_p$. For this case the "real" surface area available for the electrochemical reactions exceeds the geometrical (or visible) area for a perfectly flat electrode due to the expanded surface area in the shallow pores. The AC signal can penetrate to the bottom of the shallow pores even at high frequencies. The contribution of the resistance in pores R_{PORE} is relatively small due to the small depth of the pores l, and the high-frequency impedance feature due to porosity effects is very small or absent. The penetration depth, and therefore the influence of the pores becomes progressively more important since the inner surface of the electrode is often very large compared to the outer surface. As the ratio of $R_{PORE}/Z_{INTERFACIAL} \rightarrow 0$,

   $\coth \sqrt{\dfrac{R_{PORE}}{Z_{INTERFACIAL}}} \rightarrow \sqrt{\dfrac{Z_{INTERFACIAL}}{R_{PORE}}}$, and Eq. 7-66 transform into a simple

   expression for a flat electrode as $Z_{PORE}(\omega) = Z_{INTERFACIAL}$. As $\omega \rightarrow 0$, in the presence of charge transfer the impedance in pores becomes $Z_{PORE}(\omega) \rightarrow R_{CT}$ and $Z(\omega) \rightarrow R_{SOL} + R_{CT}$. The interfacial impedance $Z_{INTER-FACIAL}$ in that case is typically smaller than that for a true "flat electrode" by a factor that equals to the ratio between the "real" surface area of the electrode due to porosity and the "geometrical" or "visible" area of the flat nonporous electrode. Depending on the exact limiting process controlling the behavior of the flat electrode interfacial impedance at a given frequency, such as the nature of mass transport and charge transfer, $-45°$ Warburg line (semi-infinite diffusion), $-90°$ capacitive line (finite diffusion with absorption boundary), a distorted semicircle (finite diffusion with transmission boundary), an ideal or depressed charge-transfer semicircle (Faradaic case), or deformed variations of the semicircle can be observed (example A).

2. The case of "deep pores" emerges when the penetration depth is much smaller than the pore length l, creating $l >> \lambda_p$ condition. In that case the penetration depth is negligible compared to the depth of the pores. The large R_{PORE} creates significant impedance at high frequencies. As the expanded surface area inside the pores is not readily accessible at the high AC frequencies, sometimes only the capacitance (or interfacial impedance) effects of the flat external segments of the electrode surface are measured. For these conditions at higher frequencies $\coth \sqrt{R_{PORE} / Z_{INTERFACIAL}} \rightarrow 1$, and $Z_{PORE}(\omega) = \sqrt{R_{PORE} Z_{INTERFACIAL}}$, with pores behaving as if they were semi-infinitely deep and with Z_{PORE} impedance showing the $\omega^{-1/2}$ dependency. At sufficiently low frequencies $\omega \rightarrow 0$ the AC signal can penetrate through the pore, and in the presence of charge transfer the impedance in pores becomes $Z_{PORE}(\omega) = \sqrt{R_{PORE} R_{CT}}$ and $Z(\omega) = R_{SOL} + \sqrt{R_{PORE} R_{CT}}$. Often the high-frequency Warburg-like $-45°$ line corresponding to the transport inside the pores is followed by a low-frequency interfacial impedance-

distorted semicircle (Example B). At very low frequencies a condition for full AC signal penetration becomes valid. In principle, depending on a surface inside the pores (walls and bottom) being available to discharge the electroactive material, the low-frequency (DC) impedance limit for the case of deep pores can be larger or smaller than that for the case of shallow pores. In the case shown in Figure 7-20 it is assumed that deep pores present a much larger effective surface area than shallow pores and that the DC impedance limit for deep pores is smaller than that for shallow pores.

3. The general case follows Eqs. 7-66 and 7-69. Eq. 7-66 predicts a straight line at $-45°$ at high frequencies characteristic of mass-transport limitation in the pores, which is equivalent to a semi-infinite Warburg impedance or a CPE with $\alpha = 0.5$. At low frequencies ($\omega \to 0$) it is followed by a semicircle or capacitive line, depending on the characteristics of the interfacial impedance (Example C). If the model is extended to pores with finite and uniform length, the penetration depth will eventually reach the bottom of the pores when the frequency is sufficiently low. If charge transfer R_{CT} is present, all impedance becomes real at DC, and the Eq. 7-66 transforms into $Z_{PORE}(\omega) = \sqrt{R_{PORE}R_{CT}} \coth \sqrt{R_{PORE}/R_{CT}}$. If charge-transfer resistance is infinitely large, the interfacial impedance is purely capacitive (Eq. 7-68). The low-frequency impedance in the Nyquist plot becomes represented by a capacitive line.

Experimental results usually match predictions of de Levie's model at high frequencies. At low frequencies, however, deviations from ideal capacitive behavior are frequently observed, which is often attributed to pore distribution and a distribution of the penetration depth even at a single frequency.

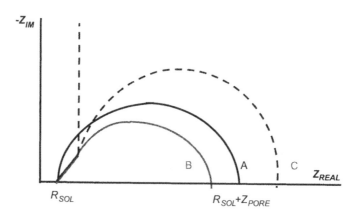

FIGURE 7-20 Complex plane plots for a porous electrode according to de Levie's model

A. limiting case for shallow pores; B. limiting case for very deep pores; C. general case (Eq. 7-66)

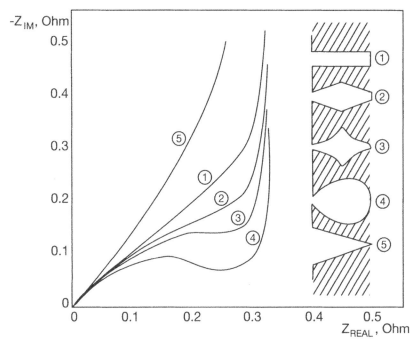

FIGURE 7-21 Calculated impedance for various shapes of a single pore-blocking electrode (Figure reproduced by permission of Solartron Analytical, a part of AMETEK Advanced Measurement Technology [60])

Use of a "CPE-CPE model" representation with two parallel $R \,|\, CPE$ elements connected in series as $R_{SOL} - R_{PORE} \,|\, CPE_{PORE} - R_{CT} \,|\, CPE_{DL}$ typically results in a good fit with both experimental results and de Levie's model [65]. The higher-frequency semicircle is dependent on solution resistance in pores and the factors that influence R_{PORE}, such as porosity, surface geometry, temperature, and viscosity. R_{PORE} essentially describes resistance to transport of electroactive species inside a pore. The lower-frequency feature depends solely on the electrode kinetics ($Z_{INTERFACIAL}$) and discharge reactions of the electroactive species [63]. In experimental systems with lower concentrations of electroactive species a concentration gradient and related diffusion impedance are often present. In such a case, according to Eq. 7-66, the high-frequency $-45°$ line may develop into a very depressed semicircular shape with the phase angle approaching $-22.5°$, and/or two semicircles are observed.

The variation of the impedance with frequency has been examined for various geometries of a single pore, and the results are summarized in Figure 7-21 for cases with no Faradaic charge transfer, where all interfacial impedance is determined by the double-layer capacitance (Eq. 7-68) [69]. It can be shown that the more occluded the shape of a pore, the more the impedance exhibits a pseudo-transfer resistance. The RC transmission line model was also successfully applied to the impedance analysis of more complicated electrodes representing fractal structures composed of variable sizes of double and triple pores [70].

References

1. S. S. Dukhin, *Dielectric properties of disperse systems*, Surface and Colloid Science, E. Miti-jevic (Ed.), Wiley Science, New York, 1971, pp. 83–165.

2. D. G. Han, G. M. Choi, *Simulation of impedance for 2-D composite*, Electrochim. Acta, 1999, 44, pp. 4155–4161.

3. M. Siekierski, K. Nadara, *Modeling of conductivity in composites with random resistor networks*, Electrochim. Acta, 2005, 50, pp. 3796–3804.

4. E. Barsukov, J. R. MacDonald, *Impedance spectroscopy*, J. Wiley & Sons, Hoboken, New Jersey, 2005.

5. M. F. Smiechowski, V. F. Lvovich, *Characterization of carbon black colloidal nanoparticles by electrochemical impedance spectroscopy*, J. Electroanal. Chem., 2005, 577, 1, pp. 67–78.

6. G. Schwarz, *Dielectic relaxation of biopolymers in solution*, Adv. Mol. Relaxation Processes. 1972, 3, pp. 281–295.

7. T. Hanai, *Electrical properties of emulsions*, Emulsion Science, P. Sherman (Ed.), Academic Press, London, 1968, pp. 353–478.

8. K. Asami, *Evaluation of colloids by dielectric spectroscopy*, HP Application Note 380-3, 1995, pp. 1–20.

9. S. Grimnes, O. G. Martinsen, *Bioimpedance and bioelectricity basics*, Academic Press, 2000.

10. B. J. Ingram, T. O. Mason, *Powder-solution-composite technique for measuring electrical conductivity of ceramic powders*, J. Electrochem. Soc., 2003, 150, 8, pp. E396–E402.

11. G. Schwarz, *A theory of the low-frequency dielectic dispersion of colloidal particles in electrolyte solution*, J. Phys. Chem., 1962, 66, pp. 2636–2642.

12. H. P. Schwan, S. Takashima, V. K. Miyamoto, W. Stoeckenius, *Electrical properties of phospholipid vesicles*, Biophys. J., 1970, 10, pp. 1102–1119.

13. K. Asami, T. Yonezawa, H. Wakamatsu, N. Koyanagi, *Dielectric spectroscopy of biological cells*, Bioelectrochem. Bioelectr., 1996, 40, pp. 141–145.

14. T. Jones, *Basic theory of dielectrophoresis and electrorotation*, IEEE Eng. Med. And Biol., 2003, 6, pp. 33–42.

15. J. Gimsa, *Characterization of particles and biological cells by AC electrokinetics*, Interfac. Electrokinetics and electrophoresis, 2001, 13, pp. 369–400.

16. S. Devasenathipathy, J. G. Santiago, *Electrokinetic flow diagnostics*, Micro- and Nano-Scale Diagnostic Techniques, K. Brewer (Ed.), New York, Springer Verlag, 2003, pp. 1–19.

17. P. J. Burke, *Nanodielectrophoresis: electronic nanotweezers*, Encyclopedia of Nanoscience and Nanotechnology, H. S. Nalva (Ed.), 2003, 6, pp. 623–641.

18. R. Diaz, S. Payen, *Biological cell separation using dielectrophoresis in a microfluidic device*, robotics.eecs.berkeley.edu/~pister/245/project/DiazPayen.pdf.

19. L. Zheng, S. Li, P. J. Burke, J. P. Brody, *Towards single molecule manipulation with dielectrophoresis using nanoelectrodes*, Proceedings of the Third IEEE Conference on Nanotechnology, 2003, 1, pp. 437–440.

20. J. Wu, Y. Ben, H.-C. Chang, *Particle detection by electrical impedance spectroscopy with asymmetric polarization AC electroosmotic trapping*, Microfluid, Nanofluid., 2005, 1 , pp. 161–167.

21. R. Zhou, P. Wang, H.-C. Chang, *Bacteria capture, concentration and detection by alternating current dielectrophoresis and self-assembly of dispersed single-wall carbon nanotubes*, Electrophoresis, 2006, 27, pp. 1376–1385.

22. A. Ramos, H. Morgan, N. Green, A. Castellanos, *AC electrokinesis: a review of forces in microelectrode structure*, J. Phys. D: Appl. Phys. (1998) 31, pp. 2338–2353.

23. M. W. Wang, *Using dielectrophoresis to trap nanobead/stem cell compounds in continuous flow*, J. Electrochem. Soc., 2009, 156, 8, pp. G97–G102.

24. V. Lvovich, S. Srikanthan, R. L. Silverstein, *A novel broadband impedance method for detection of cell-derived microparticles*, Biosens. Bioelectr., 2010, 26, pp. 444–451.

25. S. K. Ravula, D. W. Branch, C. D. James, R. J. Townsend, M. Hill, G. Kaduchak, M. Ward, I. Brener, *A microfluidic system combining acoustic and dielectrophoretic particle preconcentration and focusing,*. Sens. Act. B, 2008, 130, pp. 645–652.

26. F. Tay, L. Yu, A. J. Pang, C. Iliescu, *Electrical and thermal characterization of a dielectroiphoretic chip with 3D electrodes for cells manipulation*, Electrochim. Acta, 2007, 52, pp. 2862-2868.

27. H. Morgan, A. Izquierdo, D. Bakewell, N. Green, A. Ramos, The dielctrophoretic and traveling wave forces generated by interdigitated electrode arrays: analytical solution using Fourier series, J. Phys. D: Appl. Phys., 2001, 34, pp. 1553–1561.

28. S. Kumar, S. Rajaraman, R. A. Gerhardt, Z. L. Wang, P. J. Hesketh, *Tin oxide nanosensor fabrication using AC dielectrophoretic manipulation of nanobelts*, Electrochim. Acta, 2005, 51, pp. 943–951.

29. S. Sengupta, J. E. Gordon, H.-C. Chang, *Microfluidic diagnostic systems for the rapid detection and quantification of pathogens*, Microfluidics for Biological Applications, W.–C. Tien, E. Finehout (Eds.), Springer, 2008.

30. K. H. Kang, D. Li, *Force acting on a dielectric particle in a concentration gradient by ionic concentration polarization under an externally applied DC electric field*, J. Coll. Inter. Sci., 2005), 286, pp. 792–806.

31. B. G. Hawkins, A. E. Smith, Y. A. Syed, Brian J. Kirby, *Continuous-flowpParticle separation by 3D insulative dielectrophoresis using coherently shaped, dc-biased, AC electric ields*, Anal. Chem., 2007, 79, pp. 7291–7300.

32. H.-Y. Lee, J. Voldman, *Optimizing micromixer design for enhancing dielectrophoretic microconcentrator performance*, Anal. Chem., 2007, 79, 5, pp. 1833–1839.

33. N. Gadish, J. Voldman, *High-throughput positivedDielectrophoretic bioparticle microconcentrator*, Anal. Chem., 2006, 78, pp. 7870–7876.

34. B. A. Berkowitz, J. H. Asling, S. M. Shnider, *Relationship of pentazocine plasma levels to pharmacological activity in man*, Clin. Pharmacol. Ther., 1969, 10, pp. 320–328.

35. M. Ehrnebo, L. O. Boreus, U. Lonroth, *Bioavailability and first-pass metabolism of oral pentazocine in man*. Clin. Pharmacol. Ther., 1977, 22, pp. 888–892.

36. T. Kinoshita, T. Shibaji, M. Umino, *Transdermal delivery of lidocaine in vitro by alternating current*, J. Med. Dent. Sci., 2003, 50, pp. 71–77.

37. P. K. Wong, T.-H. Wang, J. H. Deval, C.-M. Ho, *Electrokinetics in micro devices for biotechnology applications*, IEEE/ASME Transactions on Mechatronics, 2004, 9, 2, pp. 366–376.

38. M. Ciureanu, H. Wang., *Electrochemical impedance study of electrode-membrane assemblies in PEM fuel cells. 1– Electrooxidation of H_2 and H_2/CO mixtures on Pt-based gas-diffusion electrodes*, J. Electrochem. Soc., 1999, 146, 11, pp. 4031–4040.

39. X. Wang, I. - M. Hsing, *Kinetics investigation of H_2/COelectro-oxidation on carbon supported Pt and its alloys using impedance based models*, J. Electroanal. Chem., 2003, 556, pp. 117–126.

40. N. Wagner, M. Schulze, *Change of electrochemical impedance spectra during CO poisoning of the Pt and Pt-Ru anodes in a membrane fuel cell (PEFC)*, Electrochim. Acta, 2003, 48, pp. 3899–3907.

41. L. Bai, B. Conway, *Three-dimensional impedance spectroscopy diagrams for processes involving electrosorbed intermediates, introducing the third electrode-potential variable—examination of conditions leading to pseudo-inductive behavior*, Electrochim. Acta, 1993, 38, 14, pp. 1803–1815.

42. I. Eppelboin, M. Keddam, *Faradaic impedances: diffusion impedance and reaction impedance*, J. Electrochem. Soc., 1970, 117, pp 1052–1056.

43. T. Thomberg, J. Nerut, K. Lust, R. Jjager, E. Lust, *Impedance spectroscopy data for $S_2O_8^{2-}$ anion electroreduction at Bi(111) plane*, Electrochim. Acta, 2008, 53, pp. 3337–3349.

44. M. V. ten Kortenaar, C. Tessont, Z. I. Kolar, H. van der Weijde, *Anodic oxidation of formaldehyde on gold studied by electrochemical impedance spectroscopy: an equivalent circuit approach*, J. Electrochem. Soc., 1999, 146, 6, pp. 2146–2155.

45. V. D. Jovic, B. M. Jovic, *EIS and differential capacitance measurements onto single crystal faces in different solutions*, J. Electroanal. Chem., 2003, 541, pp. 1–21.

46. I. Epelboin, M. Keddam, *Faradaic impedances: diffusion impedances and reaction impedances*, J. Electrochem. Soc., 1970, 117, pp. 1052–1056.

47. I. Epelboin, R. Wiart, *Mechanism of the electrocrystallization of nickel and cobalt in acidic solution*, J. Electrochem. Soc., 1971, 118, pp. 1577–1582.

48. I. Epelboin, M. Keddam, J. C. Lestrade, *Faradaic impedances and intermediates in electrochemical reactions*, Faraday Disc. of the Chem. Soc., 1973, 56, pp. 265–275.

49. I. Epelboin, C. Gabrielli, M. Keddam, H. Takenouti, *A model of the anodic behaviour of iron in sulphuric acid medium* Electrochim. Acta, 1975, 20, pp 913–916.

50. C. Gabrielli, M. Keddam, F. Minouflet-Laurent, K. Ogle, H. Perrot, *Investigation of zinc chromatation. Part II- electrochemical impedance spectroscopy*, Electrochim. Acta, 2003, 48, pp. 1483–1490.

51. F. Seland, R. Tunold, D. A. Harrington, *Impedance study of methanol oxidation on platinum electrodes*, Electrochim. Acta, 2006, 51, pp. 3827–3840.

52. F. Seland, R. Tunold, D. A. Harrington, *Impedance study of formic acid oxidation on platinum electrodes*, Electrochim. Acta, 2008, 53, pp. 6851–6864

53. D. A. Harrington, *Electrochemical impedance of multistep mechanisms: mechanisms with diffusing species*, J. Electroanal. Chem., 1996, 403, pp. 11–24.

54. D. A. Harrington, B. E. Conway, *AC Impedance of Faradaic reactions involving electrosorbed intermediates—I. kinetic theory*, Electrochim. Acta, 1997, 32, pp. 1703–1712.

55. P. Cordoba-Torres, M. Keddam, R. P. Nogueira, *On the intrinsic electrochemical nature of the inductance in EIS. a Monte Carlo simulation of the two-consequitive-step mechanism: the flat surface 2D case*, Electrochim. Acta, 2008, 54, pp. 518–523.

56. A. Lasia, *Electrochemical impedance spectroscopy and its applications, in modern aspects of electrochemistry*, B. E. Conway, J. Bockris, R. White (Eds.), vol. 32, Kluwer Academic/ Plenum Publishers, New York, 1999, pp. 143–248.

57. A. Lasia, *Applications of the electrochemical impedance spectroscopy to hydrogen adsorption, evolution and absorption into metals*, Modern Aspects of Electrochemistry, B. E. Conway, R. White (Eds.), vol. 35, Kluwer Academic/ Plenum Publishers, New York, 2002, pp. 1–49.

58. M. Itagaki, H. Hasegawa, K. Watanabe, T. Hachiya, *Electroreduction mechanism of oxygen investigated by electrochemical impedance spectroscopy*, J. Electroanal. Chem., 2003, 557, pp. 59–73.

59. C. Gabrielli, P. P. Grand, A. Lasia, H. Perrot, *Investigation of hydrogen adsorption-absorption into thin palladium films*, J. Electrochem. Soc., 2004, 151, 11, pp. A1925–1949.

60. C. Gabrielli, *Identification of electrochemical processes by frequency response analysis*, Solartron Analytical Technical Report 004/83, 1998, pp. 1–119.

61. S. –L. Wu, M. E. Orazem, B. Tribollet, V. Vivier, *Impedance of a disc electrode with reactions involving an adsorbed intermediate: local and global analysis*, J. Electrochem. Soc., 2009, 156, 1, pp. C28–C38.

62. A. Lasia, *Nature of the two semi-circles observed on the complex plane of porous electrodes in the presence of a concentration gradient*, J. Electroanal. Chem., 2001 500, pp. 30–35.

63. C. Hitz, A. Lasia, *Experimental study and modeling of impedance of the her (hydrogen evolution reaction) on porous Ni electrodes*, J. Electroanal. Chem., 2001, 500, pp. 213–222.

64. R. Jurczakowski, C. Hitz, A. Lasia, *Impedance of porous Au based electrodes*, J. Electroanal. Chem., 2004, 572, 2, pp. 355–366.

65. P. Los, A. Lasia, H. Ménard, *Impedance studies of porous lanthanum-phosphate-bonded nickel electrodes in concentrated sodium hydroxide solution*, J. Electroanal. Chem., 1993, 360, pp. 101–118.

66. R. De Levie, *Contribution to the sizing of supercapacitors and their applications*, P. Delahay (Ed.), Advances in Electrochemistry and Electrochemical Engineering, Vol. 6, Wiley, New York, 1967, pp. 329–397.

67. M. G. Sullivan, R. Kotz, O. Haas, *Thick active layers of electrochemically modified glassy carbon. Electrochemical impedance studies*, J. Electrochem. Soc., 2000, 147, 1, pp. 308–317.

68. M. Tomkiewicz, *Impedance of composite media*, Electrochim. Acta, 1993, 38, 14, pp. 1923–1928.

69. H. Keiser, K.D. Beccu, and M.A. Gutjahr, *Abschätzung der porenstruktur poröser elektroden aus impedanzmessungen*, Electrochim. Acta, 1976, 21, pp. 539–543.

70. M. Itagaki, Y. Hatada, I. Shitanda, K. Watanabe, *Complex impedance spectra of porous electrode with fractal structure*, Electrochim. Acta, 2010, 55, pp. 6255–6262.

CHAPTER 8

Impedance Instrumentation, Testing, and Data Validation

8.1. Impedance test equipment

An impedance measurement system usually integrates an AC measurement unit such as a frequency-response analyzer (FRA), a potentiostat, or a galvanostat of suitably high bandwidth, and an electrochemical cell composed of two, three, or four electrodes in contact with an investigated sample (Figure 8-1). The analyzed electrochemical interface is located between the sample and the working electrode (WE). A counter-electrode (CE) is used to supply a current through the cell. Where there is a need to control the potential difference across the interface, one or two reference electrodes (RE1 and RE2) with a constant and reproducible potential are used.

In the case of EIS measurements, the potentiostat is not only responsible for maintaining a defined DC potential level but also for applying predetermined AC voltage to the analyzed system. The DC potential and the AC perturbation are added together (V_{IN}) and applied to the electrochemical cell at the counter-electrode. The voltage difference between the reference and the working electrodes (ΔRE) is measured and fed back to the control loop, which corrects the voltage V_{IN} applied to the counter-electrode and the current flowing through the working electrode until the required potential difference between the working and the reference electrodes is achieved. The voltage measured between the working and the reference electrodes (V_{OUT}) and the current measured at the working electrode (I_{OUT}) are amplified by the potentiostat and fed

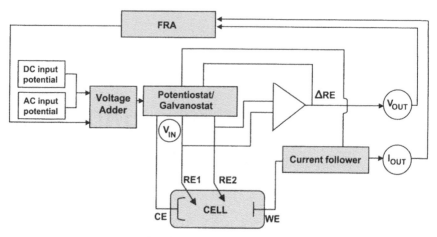

FIGURE 8-1   Example of impedance measurement setup with a frequency-response analyzer and a four-electrode cell

into the FRA as voltage signals. The voltage and current amplifiers should have a high common mode rejection and a wide bandwidth. In addition, they need to have an offset capability in order to eliminate the DC component of the analyzed signal. The potentiostat can provide DC rejection as well as compensate for the solution resistance (IR compensation) if required. The summing amplifier is used with a gain of 0.01 for the analyzing signal. This gain is necessary because digital generators have a low signal-to-noise ratio at very low levels, and it is better to use a 1V signal reduced by a factor of 100 than a 0.01V signal [1].

For many dielectric applications such as analysis of polymers, composite materials, and biomedical tissues, where it is not essential to maintain a DC-voltage control during the impedance measurement, a potentiostat is not required. In this case the AC measurement (dielectric) unit can be used more efficiently by itself, since it usually allows more accurate measurements and the use of higher frequencies. Potentiostats, however, are useful additions to the impedance test equipment when the system under investigation has low and often also high impedance. Potentiostats have high input impedance, rejecting any current flow through the reference electrode, and good current sensitivity.

In the present state-of-the-art equipment it is possible to measure and plot the electrochemical impedance automatically. The generator can be programmed to sweep from a maximum to a minimum frequency in a number of required frequency steps. Thus the measurement frequency is changed automatically, and the total measurement time for one experiment can be predetermined. For example, for a measurement using five samples per every decade of frequency, it takes ~ 31 seconds to take a sweep from 50 kHz to 0.1 Hz, 5 minutes 30 seconds for a sweep from 50 kHz to 0.01 Hz, and almost 54 minutes for a sweep from 50 kHz to 0.001 Hz. Automatic plotting of the experimental data can be performed by computer software in different graphical representations—either as Nyquist or Bode plots for impedance, modulus or permittivity notations, which also allow subsequent digital processing, and circuit and mechanistic parameters identification. The commonly used modern equipment in AC impedance measurement techniques can be subdivided into two main groups—single-sine (lock-in amplification and frequency response analysis) and multiple-sine techniques such as fast Fourier transforms (FFT).

8.2.   Single-sine equipment—lock-in amplifier and frequency-response analyzer

In single-sine techniques, a small amplitude sinusoidal signal with a fixed frequency is applied to the test cell. The response signal is then analyzed to extract the two components of the impedance (real and imaginary parts or magnitude and phase). This experiment is then repeated at a series of different test frequencies, usually starting at the highest frequency and finishing at the lowest in order to minimize sample perturbation. The main advantages of single sine techniques are [2]:

- high-quality data
- simple instrumentation since the sine wave can be produced with a frequency generator
- fast measurements at very high frequencies

A schematic diagram of a lock-in amplifier is shown in Figure 8-2. A lock-in amplifier is based on multiplication of two sine wave signals—one being the signal carrying the amplitude-modulated information of interest and the other a reference signal with the chosen frequency and phase. An oscillator produces a sinusoidal waveform, which is simultaneously applied to the electrochemical cell and fed into the reference input. The measured system output signal is mixed with the reference square-wave signal of the same frequency, and the resulting amplified signal goes through a low-pass filter. The filtering step produces an average of all components and reduces the bandwidth of the noise [3].

The phase shifter allows for precise adjustment of the reference-signal phase in order to zero the phase difference ϕ between the reference and the input signals. The in-phase and out-of-phase components of the cell response to the perturbation signal can be successively analyzed by various techniques. For instance, in two-phase combined lock-in amplifiers the measured signal is mixed with the reference signal to obtain the in-phase component and additionally with the reference signal shifted by $\pi/2$ to resolve the imaginary component. The cell response is successively compared to the reference signals using a phase-sensitive detector, with the perturbation signal shifted by $\Delta\phi$ and $\Delta\phi + \pi/2$. The impedance is therefore measured in polar and not Cartesian coordinates. A lock-in amplifier is often interfaced with a potentiostat that provides a DC output that is proportional to the AC input signal. The rectifier, which performs this AC to DC conversion, is called a phase-sensitive detector (PSD) [1].

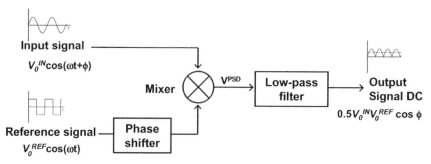

FIGURE 8-2   A schematic diagram of a lock-in amplifier

If a reference signal (typically a square waveform) is a voltage of frequency ω according to:

$$V^{REF} = V_0^{REF} \cos \omega t \tag{8-1}$$

and the input signal a sinusoidal voltage of the same frequency ω proportional to the AC current flowing through the system under investigation according to:

$$V^{IN} = V_0^{IN} \cos(\omega t + \phi) \tag{8-2}$$

then the product of these two signals would be a sinusoid at twice the reference frequency ω:

$$V^{PSD} = \frac{1}{2} V_0^{IN} V_0^{REF} [\cos(2\omega t + \phi) + \cos \phi] \tag{8-3}$$

The resulting signal contains a one-time independent component, depending on the phase difference of two signals ϕ, and is proportional to the amplitude of the measured AC signal. It reaches a maximum when the phase difference of the two mixed signals is zero. The output signal is subsequently applied to a low-pass filter, which averages the signal components with frequencies above the filter cutoff frequency. It produces a DC signal proportional to the amplitude.

The mean level of V^{PSD}, which is the DC component of the detector signal, can be calculated as:

$$V_{mean}^{PSD} = \frac{1}{2} V_0^{IN} V_0^{REF} \cos \phi \tag{8-4}$$

Since there is no phase shift between input and reference signals ($\phi = 0$) after passing through the phase shifter, the mean level of the detector signal assumes its maximum possible value of $\frac{1}{2} V_0^{IN} V_0^{REF}$. If the reference signal is fixed at a constant level, the input signal and therefore the impedance can be calculated from the mean level of the detector signal. The DC component of the detector signal is isolated from the signal by using a low-pass filter and by measuring the filtered output with a DC voltmeter. Noise associated with the input signal would also be multiplied with the reference signal. However, since noise does not have a fixed frequency or phase relationship to the reference signal, it does not cause a change of the mean DC level, as lock-in amplifiers are highly frequency-selective.

Measurements using a lock-in amplifier provide high-quality data on relatively inexpensive equipment but are usually more time-consuming than those carried out with a frequency-response analyzer. Modern digital lock-in amplifiers are designed for signal detection in a high-noise background and are therefore capable of very accurate and sensitive readings. The analysis results depend very much on the utilized phase-detector technique, but it is not normally possible to measure impedances with a perturbation frequency below 1Hz. Older analog multipliers (where the input signal is multiplied by the reference signal via an electronic circuit) often show poor noise rejection, as it is difficult to guarantee linear operation of analog multipliers in the presence of large noise. Digital switching multipliers are linear, but they do not only detect signals at the frequency of the reference signal but also at odd harmonics of the reference frequency.

The disadvantage of the lock-in technique is that it retains sizable contributions of the harmonic frequencies $(2n + 1)\omega$ of the reference square-wave signal and harmonics and related noise present in the input signal, although their influence with increasing n is attenuated by $1/3$, $1/5$, $1/7$, etc. The digital–multiplier multiplied the initially digitized and amplified signal with a digitized version of the reference signal using a digital signal processor (DSP). This technology provides perfect multiplication and excludes odd harmonics of the reference frequency. A problem associated with instruments using DSPs is the dynamic range, which is inherent to all digital instruments.

Lock-in amplifiers typically operate in a fairly limited frequency range from 1 Hz to ~100kHz with a precision of 0.1 to 0.2 percent. But the measurement can be slow, as it is done on the basis of a frequency-by-frequency scan and may not be fast enough for unstable or dynamic systems. The upper frequency is typically limited at ~ 100 kHz due to the conversion time of digital converters, but radio-frequency lock-in amplifiers capable of reaching 200 MHz with relative phase error at ~ 2 percent in the upper MHz range are also known. Modern lock-in amplifiers are controlled by a microprocessor and permit automated measurements with automatic range selection.

Frequency-response analysis [1] is the most widely used technique for impedance testing. Similar to the lock-in technique, it can extract a small signal from a very high background of noise, automatically rejecting DC and harmonic responses. The difference is that a frequency-response analyzer (FRA) correlates the input signal with the reference sine waves. To achieve faster measurements, FRAs are usually equipped with separate analyzers for each input channel.

Frequency-response analyzers have a sine wave generator, which outputs a small amplitude voltage signal V^{IN} to the investigated system. The response signals, usually the voltage measured between two reference points in the electrochemical cell and a voltage signal proportional to the current flowing through the cell, are fed into the input channels, digitalized, and then integrated over several cycles (N) in order to reject noise. FRA therefore correlates cell response with two reference signals—one in phase with sine-wave perturbation and the other shifted by 90°.

In addition to the waveform at the frequency of interest, the measured signals usually contain a DC component, harmonics, and noise. The process used to reject these spurious components is analogous to the digital multiplication of the input signal with a reference waveform performed by a phase-sensitive detector. However, FRA does not use a phase shifter to compensate for the phase shift between input and reference signals; rather, the integrated waveforms are immediately correlated by multiplying them by sine and cosine reference waveforms, and the resulting signals are integrated (Figure 8-3).

For a pure sinusoidal input signal a measured signal V^{IN} with amplitude V_0^{IN} is proportional to the AC current from the potentiostat as:

$$V^{IN} = V_0^{IN} \sin(\omega t + \phi) = a \sin \omega t + b \cos \omega t \qquad (8\text{-}5)$$

This measured signal is multiplied by a sine and cosine of the reference signal at the same frequency and then is integrated during one ($N = 1$) or more ($N > 1$) wave-periods $T_t = 2\pi / \omega$ of the perturbation signal. The results of the integrations would be:

$$Z_{REAL} = \frac{a}{2} = \frac{1}{T_t N} \int_0^{2\pi/\omega} (V_0^{IN} \sin(\omega t + \phi) \sin \omega t) dt = \frac{V_0^{IN}}{2} \cos \phi \qquad (8\text{-}6)$$

$$Z_{IM} = \frac{b}{2} = \frac{1}{T_t N} \int_0^{2\pi/\omega} (V_0^{IN} \sin(\omega t + \phi) \cos \omega t) dt = \frac{V_0^{IN}}{2} \sin \phi \qquad (8\text{-}7)$$

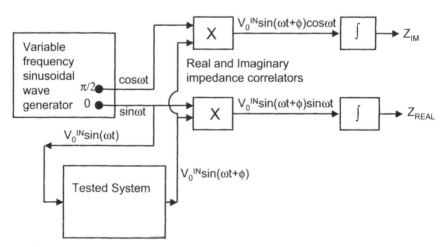

FIGURE 8-3 Frequency-response analyzer working principle

These two quantities, proportional to the real and to the imaginary part of the input signal V^{IN}, are obtained, and hence the impedance of the analyzed system is known. Once real and imaginary parts of the input signals have been determined by these correlations, the complex impedance of the test object can be calculated. It can be mathematically proven that all the spurious components (such as harmonics) are removed by this technique of correlation, provided that a sufficiently large number of cycles have been used for the integration. Signal-to-noise ratio is enhanced as a function of $N^{1/2}$.

Digital FRAs are also capable of analyzing harmonics and very low-frequency impedance response [1]. This could be achieved by multiplying the input signal by a waveform of the appropriate frequency or by carrying out a fast Fourier transform (FFT) analysis on the input signal and the digital Fourier integral method. The technique of digital correlation can be applied to a large range of frequencies. However, to allow easier analysis at high frequencies, the measured signal is shifted to lower frequencies using a technique called heterodyning. The input signal is multiplied by a reference signal of slightly different frequency. The resulting signal has two components, one at the sum of both frequencies and one low-frequency component at the difference of both frequencies. The high-frequency signal can easily be filtered out, and the low-frequency signal is then analyzed by the above correlations. Modern digital frequency-response analyzers cover the range between 10 μHz and 32 MHz with resolution of 16 bits and basic accuracy ~ 0.1 percent for magnitude and 0.1 percent for phase measurements. FRA gives good accuracy for stationary systems and can be performed more rapidly than PSD over a wider frequency range. FRA instruments are costlier than PSD, have higher noise and lower sensitivity [2], and typically perform well only for stable electrochemical systems where data can be safely integrated over several wave periods.

8.3.   Multiple-sine impedance equipment

Impedance spectroscopy is frequently used to characterize unstable systems that change with time. One way to reduce the effects of system alteration during the measurement is to reduce the total measurement time by using a multisine technique, which is also frequently called time-domain or FFT (fast Fourier transform) technique [2].

Signal-processing theory refers to data domains, where the same data can be represented in different domains. EIS uses two of these domains, the time domain and the frequency domain. In the time domain the impedance-signal amplitudes are represented as a function of time (Figure 8-4A), and in the frequency domain these signal amplitudes are displayed as functions of frequency (Figure 8-4B), shown here for a signal consisting of two superimposed sine waves. In modern EIS systems, lower-frequency data are usually measured in the time domain.

In the case of multisine techniques, a measurement can be carried out at several frequencies simultaneously when the waveforms of typically 15 to 20 different frequencies and equal amplitudes are superimposed. The phases of the superimposed signals are randomized to minimize the amplitude of the

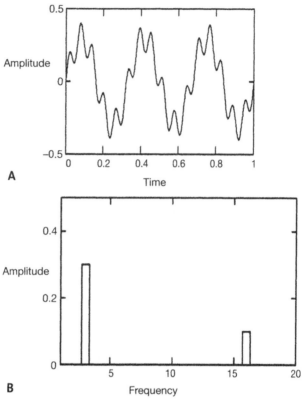

FIGURE 8-4 Two superimposed sine waves in A. the time domain; B. the frequency domain

composite signal. The voltage across the system under investigation and the current, which contains information about the response of the system to each of the frequencies contained in the perturbation signal, are measured in the time domain. The time-domain signals are digitized and transferred into the frequency domain by carrying out FFT. Taking the Fourier transform of the perturbation signal $V(t)$ and that of the resulting current signal $I(t)$ allows determination of the transfer function, and the system AC impedance may be obtained from:

$$Z(j\omega) = \frac{F[V(t)]}{F[I(t)]} = \frac{V(j\omega)}{I(j\omega)} \tag{8-8}$$

where F denotes the Fourier transform. The resulting data for each discrete frequency can be treated the same way as the impedance data obtained with a single-sine technique. The fast Fourier transform (FFT) provides a fast and efficient algorithm of computation of the Fourier transform. Repeated application of the waveform and averaging of the signal before FFT is applied can improve the signal-to-noise ratio of the multisine technique, although it also increases the required measurement time.

In contrast to single-sine techniques, multisine techniques do not require waiting for a full cycle to be completed for each of the frequencies used. Using a multisine technique, a full experiment can be carried out during a full cycle of the lowest frequency, resulting in reduced measurement time. The main advantage of the FFT technique is time saving over single-frequency methods. The FFT impedances method may be more appropriately applied to study unstable systems evolving with time. (Of course, measured impedance must still be considered constant during the time of measurement.) This advantage can be utilized to investigate electrochemical processes in real time, provided that time required for data acquisition is shorter than the time constants of the system under investigation.

It is worth considering the effect of a change in the system during the experiment. Due to the fact that the perturbation signal is a superposition of different frequencies, data are acquired for each frequency throughout the entire experiment. The result for one frequency is then obtained by averaging all the data sampled for this particular frequency. If the system under investigation changes during the measurement, the data averaged for each frequency would vary noticeably, resulting in an impedance plot with considerable scatter. This scatter can be taken as a clear indication of unreliable data, but it also makes the entire data set unusable. The effect of a change in the system during the measurement can frequently be eliminated by discarding the low-frequency data. For a multisine technique, this means that the experiment has to be repeated with a higher low-frequency limit in order to further reduce the measuring time. Otherwise, in order to obtain a good resolution in the lowest part of the frequency range, additional redundant high-frequency measurements should be obtained with additional samples, using a computer with a large memory. When the analyzed system is steady-state over a reasonably long time, then FRA analysis is preferred, since it is more accurate and convenient. When the system is unstable (time-varying) but linear and the demand for measurement accuracy is not too stringent, FFT analysis may be used [1].

Electrochemical systems can also show a nonlinear response—that is, the current response of an electrochemical system can be composed of a response at the excitation frequency and responses at harmonics of the excitation frequency. Therefore, the frequencies superimposed for a multisine experiment have to be chosen very carefully. To prevent faulty results, the frequencies chosen are usually odd harmonics of the lowest frequency in order to eliminate the second harmonic components, which may be caused by a nonlinear response of the system. Nonlinear behavior of the system would cause additional frequencies to appear in the response signal. The response at these additional frequencies would appear in the time-domain signal at places not occupied by the frequencies of the excitation waveform and at places already occupied by the excitation waveform. Hence, not only nonstationary effects but also nonlinearity of the response appear in the FFT spectra as additional scatter.

The weakness of the FFT technique is that the response to individual frequencies is usually weaker than that when a single frequency is applied. Due to the inherent noise encountered in signal measurement, FFT analysis has to be repeated a number of times to average out the effect of noise on the impedance measurement. The Fourier transform defined by Eq. 8-8 involves integration to infinity; while in practice only limited-length data are transformed, causing broadening of the computed frequency spectrum. Another problem, called aliasing, is connected with the presence of frequencies larger than one-half of the time-domain sampling frequency. This problem may be easily eliminated by assuring that the sampling frequency is greater than (or at least equal to) twice the highest frequency present in the measured signal. In some cases the highest frequencies may be filtered out by a low-pass filter. This minimum sampling frequency, necessary to get information about the existing signal, is called the Nyquist sampling rate [2].

8.4. Electrochemical cells

In an experimental electrochemical-cell setup, there is usually a choice between two, three, or four electrodes (Figure 8-5). These options determine which part of the electrochemical cell is characterized by impedance measurements. The sample is located between the working electrode (WE) and the counter electrode (CE), and the electrochemical interface is adjacent to the working electrode. When mass transport needs to be controlled, a well-controlled external convection with a rotating disk working electrode is often employed. A counter (auxiliary or secondary) electrode (CE) is used to allow a current to flow through the cell between the WE and CE. A CE is typically a large surface area (a long wire or mesh), often a "noble metal" current conductor that does not produce electroactive species by electrolysis that may discharge in interfering reactions at WE. A large surface-area CE possesses high capacitance and very low impedance, so all the measured impedance of the cell is related to that of the working electrode (Section 6-2). One or two constant-potential "non-polarizable" or reversible reference electrodes (RE) (Section 6-1) are used for careful control of the electrochemical DC potential at the WE.

A two-electrode cell is the simplest form of electrochemical device. In the two-electrode cell configuration, current-generating electrodes also serve as voltage-measuring sensors with a sample bounded by the two electrodes (Figure 8-5A). One electrode serves as a combined working and reference electrode (WE), and another electrode serves as a combined counter electrode and second reference electrode (CE). The impedance is always measured between two reference electrodes—that is, for the two-electrode probe the total mea-

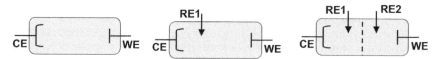

FIGURE 8-5 Experimental electrochemical cells with: A. two electrodes; B. three electrodes; C. four electrodes WE = working electrode; RE = reference electrode; CE = counterelectrode

sured impedance includes that of the counter electrode, the sample, and the working electrode. If the properties of a single WE are to be investigated using the two-electrode arrangement, the impedance of the CE can be minimized by giving it a considerably larger surface area than that of the WE.

The two-electrode arrangement is frequently used if it is impossible to reliably position a RE in the electrochemical cell—for example, in the characterization of many solid-state materials (fuel cells, batteries, etc.). The two-electrode probes are convenient and accurate for dielectric analysis of bulk-impedance response of highly resistive materials where control of DC potential is not needed and it is important to maintain a homogeneous current distribution. It is generally acknowledged that accurate measurements can be made in a "parallel plate" or "symmetrical" two-electrode configuration with homogeneous current distribution and shielded lead wires, where the influence of stray capacitance and lead inductance can be practically eliminated. Another important characteristic of two-electrode cells is the ability to define the frequency range that will correspond solely to bulk-media impedance properties, according to Eqs. 6-7 through 6-9.

The disadvantage of the two-electrode probe lies in the presence of the low-frequency interfacial polarization effects at the WE that serves simultaneously as a voltage and current probe and progressively higher contribution of bulk-material impedance ("IR drop") into total measured voltage. In conductive solutions the actual measured potential at WE may become very different from the total applied voltage between the WE and CE, resulting in loss of actual control over the electrochemical potential-dependent reactions at the WE. In low-conductivity media under an AC field a significant number of ions reach the electrode before reversal of polarity. In the absence of the electrochemical charge-transfer processes at the experimental potential, a charge accumulation at the interfaces occurs, with corresponding interfacial polarization effects dominating the low-frequency impedance response. Polarization of electrodes is a problem in low-conductivity media where there are only a few oxidizable or reducible polar species, which can eliminate polarization through charge transport or electrostatic adsorption on the electrodes, and migration in the bulk occurs under the influence of voltage gradient $\Delta V / \Delta x$ that is nonzero across the whole sample (Section 5-3). In ionic low-conductivity fluids the interfacial polarization results in formation of layers of counterions at each electrode, neutralizing the excess charge on the electrodes. The resulting double layers cause a larger voltage drop at the interfaces and diminish the effective electric field $\Delta V / \Delta x$ in the bulk of the sample. The charged particles in the bulk-media zone far from the electrodes start to move more slowly because of the diminished effective electric field in the bulk and appear to have lower electrophoretic mobility than they actually do [4]. The measured bulk-media resistance (R_{BULK}) increases as a consequence of the electric current reduction due to slower motion of the ions [5].

Another popular and widely used example of practical two-electrode symmetrical cells is a planar interdigitated electrode (Figure 8-6). Such arrays have become prominent as miniaturized sensors, transducers, and actuators. Typically these structures are composed of two symmetrical comb-shaped low micrometer-sized interdigitated metallic electrodes prepared on nonconductive

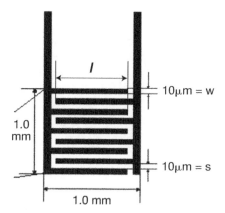

FIGURE 8-6   Interdigitated two-electrode probe

supporting substrates (such as alumina and silicon). Such devices can be placed in the sample, remain innocuous to the environment, and be used over a wide range of applications. An important advantage of these structures is a simple and inexpensive microfabrication-based mass-production process that delivers inexpensive and easily packaged miniature structures with well-defined geometry, compactness, and sensitivity. Interdigitated sensors have been used for the detection of capacitance, dielectric constants, and bulk conductivity in gases or fluids.

Several studies have been conducted regarding the electrostatic properties of interdigitated electrodes, leading to the development of mathematical equations for their capacitance and cell constant values [6, 7, 8, 9, 10]. The cell constant c_K [1/cm] is defined as the proportionality between the measured media resistance and its specific resistivity:

$$c_K = R_{BULK} / \rho = \varepsilon_o \varepsilon / C_{BULK} = d / A \qquad (8\text{-}9)$$

where R_{BULK} = measured bulk media resistance, ρ = specific resistance, ε = permittivity (dielectric constant) of investigated media, ε_0 = permittivity of vacuum constant (8.85 $\cdot 10^{-14}$ F/cm), and C_{BULK} = bulk media capacitance. Devices with lower cell constants are more effective at measuring high-resistivity materials. The cell constant is determined by the geometric parameters of the measurement device—effective distance between plate d and surface area A. These parameters in the case of planar interdigitated electrodes include the number of fingers, their width, and the space between them.

A simple method of calculating the cell constant for interdigitated electrodes, based on the meander length of the cell, was developed by Zaretsky [7]. For interdigitated arrays with number of fingers N and length of finger overlap l, where the line and space dimensions are equal (Figure 8-6) it was shown that:

$$c_K = 2/(N \cdot l) \qquad (8\text{-}10)$$

Alternatively, the cell constant for an interdigitated system can be determined through calculation of the capacitance of the device. The capacitance between any two planar electrodes can be determined through Laplace's equation. When dealing with an interdigitated array, this becomes slightly more complex; however, after several integral calculations [6], the capacitance C_{IDT} of an interdigitated array can be expressed as:

$$C_{IDT} = (N-1)l \cdot \frac{\varepsilon_o \varepsilon}{2} \frac{K\left[(1-k^2)^{1/2}\right]}{K(k)} \qquad (8\text{-}11)$$

In systems where N is greater than 2, k is defined as:

$$k = \cos\left(\frac{\pi}{2}\frac{w}{s+w}\right) \qquad (8\text{-}12)$$

where w = width of the finger and s = width of the space between fingers (Figure 8-6). $K(k)$ is a first-order elliptic integral defined as:

$$K(k) = \int_{t=0}^{1} \frac{dt}{\left[(1-t^2)(1-k^2t^2)\right]^{1/2}} \qquad (8\text{-}13)$$

The resulting cell constant for the interdigitated arrays can be determined by substituting the capacitance value C_{IDT} from Eq. 8-11 into Eq. 8-9. This approach usually results in reported experimental values for the cell constant being smaller than the theoretical ones [6, 9]. This systematic discrepancy was explained by the error introduced by the fringing effects at the electrode-solution interface, causing apparent changes in the electrode width. An alternative equation to approximate the capacitance of an interdigitated electrode was developed through a simplification of Engan's equations for integer N [8]:

$$C_{IDT} = NI \frac{4\varepsilon_o \varepsilon}{\pi} \sum_{N=1}^{\infty} \frac{1}{2N-1} J_0^2\left(\frac{(2N-1)\pi s}{2(s+w)}\right) \qquad (8\text{-}14)$$

where J_0 is a zero-order Bessel function. The resulting cell constant usually compares well with Zaretzky's Eq. 8-10 and experimental data, showing a loss in accuracy of only under 2 percent [8, 10]. It was noted, however, for small values of $s/(s+w)$ there would be poor convergence even for simple systems. This is partially due to the assumptions made in the derivation of the equation, such as the thickness of the fingers being negligible.

A three-electrode cell is a standard electrochemical analysis tool for characterization of electrode kinetic processes requiring accurate control of electrochemical potential at the WE (Figure 8-5B). The potential at the WE is monitored relative to nonpolarizable RE that is placed close to the WE to keep measured media resistance (impedance) R between the two electrodes as small as possible. The RE has a constant and reproducible potential when no current flows through it. The saturated calomel electrode (SCE) and the silver-silver chloride (Ag/AgCl) are the most common types of RE in aqueous media. Depending on the electrolyte, other types of reference electrode, such as metal/metal oxide pseudo-reference electrodes, can be used.

The potentiostat device used to measure the potential difference between the RE and the WE has very high internal input impedance Z_{INPUT}, so a negligible current is drawn through the RE. Since the voltage difference in the three-electrode cell is measured between the stable RE and WE and current is supplied by the CE, the measured impedance will only be influenced by the properties of the WE interface and the properties of the analyzed media resistance R between the WE and the RE. The influence of the media resistance between the WE and the CE on the voltage measurement is eliminated.

In spite of the wide popularity and convenience of the three-electrode configuration for electrochemical analysis and control of the WE electrochemical potential, one has to keep in mind a possibility of measurement artifacts. For instance, the finite input impedance of the reference electrode Z_{INPUT} can cause artifacts in high-frequency impedance measurements and instability in the feedback loop of the regulating device. The electrodes often have sizeable resistance (up to several kohm), and stray capacitances of ~ 1nF are not uncommon. Better understanding and elimination of impedance measurement artifacts during examination of the three-electrode electrochemical cell under potentiostatic control requires careful selection of the model of the working electrode in real measuring conditions. Various attempts were made to account for this complexity and to eliminate artifacts [11, 12, 13, 14, 15, 16].

One complication of such an approach results from the fact that the electrode impedance is represented by a two-terminal network, while the real three-electrode electrochemical cell connected to the potentiostat involves a three-terminal network. Fletcher [16] developed a representation for a three-terminal electrochemical cell (Figure 8-7A) by its two-terminal electrical equivalent for the purposes of analysis of its electrical responses, identification of sources of the artifacts, and optimal experimental conditions for their minimization. In this analysis all impedances of the three electrodes were approximated by simple resistances, and the coupling between the lines by capacitances as R_W, R_R, R_C—resistances of working, reference, and counter electrodes—and C_{WR}, C_{RC}, C_{WC}—capacitance between working and reference, working and counter, and reference and counter electrodes—respectively:

$$L^* = R_R R_C C_{RC} \qquad R^* = \frac{R_R R_C}{R_R + R_C} \qquad C^* = C_{WR} + C_{WC} + \frac{C_{WR} C_{WC}}{C_{RC}}$$

$$R^{**} = \frac{C_{RC}(R_W R_R + R_W R_C + R_R R_C)}{R_W C^{**}} \qquad C^{**} = \frac{R_R C_{WR}}{R_W} + \frac{R_C C_{WC}}{R_W} - \frac{C_{WR} C_{WC}}{C_{RC}} \qquad (8\text{-}15)$$

This analysis (Eq. 8-15) demonstrated that inductive and capacitive artifacts may contribute to the data at all sampling frequencies. The criterion for near artifact-free measurements of the sample resistance $R_1 = R_W$ at the working electrode was derived as [16]:

$$\omega R_R R_C C_{RC} \ll R_W \ll \frac{1}{\omega C_{WR}(1 + \frac{R_R}{R_W}) + \omega C_{WC}(1 + \frac{R_C}{R_W})} \qquad (8\text{-}16)$$

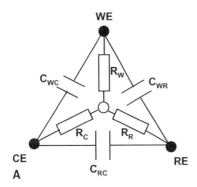

FIGURE 8-7 A. Three-terminal representation of the electrochemical cell with minimum number of components; B. its equivalent two-electrode circuit; C. simplified equivalent circuit (personal communication with the authors [16, 20])

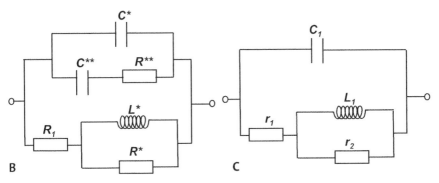

The strategy for minimizing measurement artifacts in the three-electrode cell is to keep very small all "parasitic components"—the resistances associated with the reference and the counter electrodes, including both the sample and interfacial impedances (R_R and R_C), and all coupling capacitances between working, reference, and counterelectrodes (C_{WR}, C_{WC} and C_{RC}). That can be achieved by using large reversible metal electrodes and highly conducting media. Many of these requirements can be difficult to implement in lower-conductivity samples.

Fletcher's model was expanded by Horvat-Radosevic [17, 18, 19], and Sadkowski and Diard [20]. The equivalent circuit diagram on Figure 8-7B was simplified and replaced by a circuit (Figure 8-7C) where the branch containing C^{**} and R^{**} was eliminated. The modified expressions for the circuit become:

$$r_1 = R_1 = R_W \qquad C_1 = C^* = C_{WR} + C_{WC} + \frac{C_{WR}C_{WC}}{C_{RC}}$$

$$L_1 = \frac{(R_R C_{RC} - R_W C_{WC})(R_C C_{RC} - R_W C_{WR})}{C_{RC}}$$

$$(8\text{-}17)$$

$$r_2 = \frac{(C_{WC}R_W - C_{RC}R_R)(C_{WR}R_W - C_{RC}R_C)}{C_{WR}(C_{RC}R_R - C_{WC}R_W) + C_{RC}(C_{WC}R_C + C_{RC}(R_R + R_C))}$$

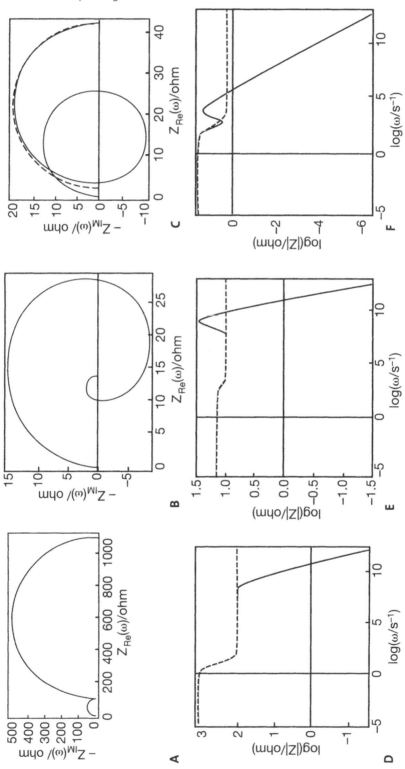

FIGURE 8-8 Nyquist (A, B, C) and Bode (D, E, F) diagrams of the working electrode ($R_{SOL} - C_{DL}|R_{CT}$) - dashed line, and equivalent three-terminal cell - solid line A., D. $R_{SOL} = 100$ ohm, $R_{CT} = 1000$ ohm, $C_{DL} = 400\mu F$, $R_R = R_C = 100$ ohm, $C_{WR} = C_{RC} = C_{WC} = 10^{-11}$ F; B., E. $R_{SOL} = 10$ ohm, $R_{CT} = 4$ ohm, $C_{DL} = 400\mu F$, $R_R = R_C = 100$ ohm,

This approach significantly assisted in elucidating inductive and capacitive artifacts of the electrochemical impedance measurements. The analysis of a model system with the working electrode interface (originally simplified as R_W) replaced by a more realistic "Randles expression" as $Z^* = R_{SOL} - C_{DL} | R_{CT}$ revealed that this approach is largely valid (Figure 8-8). The influence of "parasitic" elements R_R, R_C, C_{WR}, C_{RC}, and C_{WC} with resulting inductance and capacitance artifacts should be in most cases limited to the high-frequency part of the spectrum. In case of well separated time constants of the WE and internal time constants of parasitic elements, the capacitive or inductive deformation is located outside of the experimentally accessible frequency range at $f > 10^6$ Hz. The Fletcher general recommendations remain intact, as significant distortions were observed when the capacitive parasitic parameters were changed to higher values, such as $C_{WR} = C_{WC} = C_{RC} = 10^{-6}$ F, resulting in the recommendation to minimize the parasitic time constants $R_i C_i$ in order to shift the distorted part of the spectrum to higher frequencies outside of the measurement range. Horvat-Radosevic [17, 18, 19] identified two important "parasitic" time constants as $\tau_1 = R_{SOL} C_1$ and $\tau_2 = \sqrt{L_1 C_1}$. Increases in solution resistance R_{SOL}, parasitic capacitance C_1 and parasitic inductance L_1 always shift the artifact-induced distortions to lower frequencies. One must also keep in mind that realistic impedances of WE, RE, and CE may be represented by more complex functions than simple resistance. That may lead to frequency dependency for the reference and counterelectrodes' impedances and for the coupling capacitances. The analysis of the artifacts becomes infinitely more complex, as will be shown below for the case of a four-electrode cell.

UBEIS applications include characterization of "bulk" material properties, such as ion-exchangeable solid-state devices [14, 15], biological membranes [21], fuel cells [22, 23], and nonpolar colloidal dispersions [24, 25]. For this kind of application, four-electrode cells often provide the best results. The logic behind the four-electrode cell is somewhat different than that for the three-electrode cell. In a four-electrode cell two external auxiliary electrodes (still noted in Figure 8-5C as WE and CE) supply current to the cell, and two additional voltage reference electrodes (RE1 and RE2), connected through a high-impedance device and placed between two current leads, are used to measure the potential difference directly in the cell. The analyzed sample, such as a membrane, is contained in the volume bounded by the two reference electrodes (Figure 8-9).

The four-electrode arrangement offers an advantage of excluding combined contact interfacial polarization and cable impedances (Z_C) of the current-generating electrodes from the overall measurement. However, achieving "minimal error measurement" in the four-electrode cell configuration often depends on several experimental and instrumentation parameters that point in opposite directions and therefore are difficult to achieve. The sample impedance, type of experimental media, geometry and positioning of voltage electrodes, internal impedance of the analyzer, and capacitive coupling between the signal lines introduce impedance measurement artifacts that may appear as additional inductive and capacitive features at medium and low AC frequencies [25] (Figure 8-10).

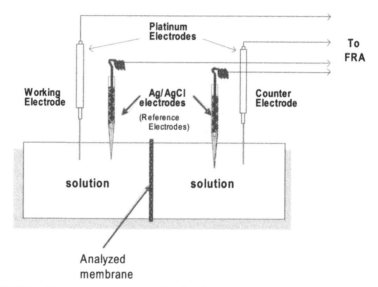

FIGURE 8-9 Experimental setup for the frequency-response analysis (FRA) of the membrane

The goal of the "ideal" four-electrode measurement is to minimize these additional artifacts to keep the measured impedance $Z_{MEASURED}$ as close as possible to the sample impedance Z_{SAMPLE}, which becomes:

$$Z_{MEASURED} = V_{SAMPLE} / I_{SAMPLE} \qquad (8\text{-}18)$$

As Figure 8-10 illustrates, two voltage-measuring reference electrodes are connected through a high-impedance input device that in principle should reject all current flow through these electrodes. This input impedance Z_{INPUT} consists of the capacitance C_{INPUT} with a typical value of ~10 pF in parallel with resistance R_{INPUT} with a value of ~10^{12} ohms. Except for very high impedance samples, R_{INPUT} can usually be neglected, and the input impedance at a frequency of 100 Hz becomes:

$$Z_{INPUT} = \frac{R_{INPUT}}{1 + j\omega R_{INPUT} C_{INPUT}} \sim \frac{1}{j\omega C_{INPUT}} = \frac{1}{2\pi * 100Hz * 10^{-11}F} \sim 160MOhm \quad (8\text{-}19)$$

As the voltage reference points V_{HIGH} and V_{LOW} are nonzero, a finite Z_{INPUT} load on both voltage electrodes may still allow parasitic current $I_{PARASITIC}$ to flow. This parasitic current flows out from the sample and may create an error in the total sample current measurement that is negligible as long as $Z_{INPUT} \gg Z_{SAMPLE}$. Measured sample current becomes:

$$I_{SAMPLE} = \frac{V_{HIGH} - V_{LOW}}{Z_{SAMPLE}} - I_{PARASITIC} \qquad (8\text{-}20)$$

The contact impedances of the voltage electrodes (Z_R) may be represented by a parallel combination of R_R and C_R and can be estimated as $Z_R = \dfrac{R_R}{1 + j\omega R_R C_R}$. Contact impedance of the voltage electrodes Z_R increases with a decrease in sampling frequency ω. A combination of measurable parasitic current $I_{PARASITIC}$ and contact impedances Z_R creates a voltage drop across the contact impedance $V_R = I_{PARASITIC} Z_R$ which further causes an error in the total measured output voltage V_{SAMPLE} of the system with decrease in frequency as:

$$V_{SAMPLE} = V_{HIGH} - V_{LOW} - V_R = V_O - I_{PARAZITIC}\frac{R_R}{1 + j\omega R_R C_R} \qquad (8\text{-}21)$$

The errors in voltage V_{SAMPLE} and current I_{SAMPLE} measurements (Eqs. 8-20 and 8-21) result in a frequency-dependent error in measurement of the sample impedance $Z_{MEASURED}$ (Eq. 8-18). The voltage divider effect [15] takes place, creating a situation where the measured impedance $Z_{MEASURED}$ decreases with

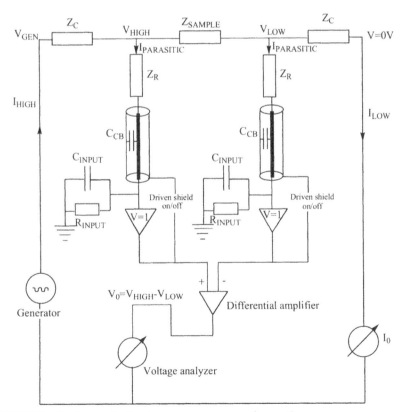

FIGURE 8-10   Four-electrode measurement schematic

decreasing frequency ω, resulting in what appears as an inductance at the low frequencies. This measurement artifact, which has no physical meaning, can be modeled by a parallel combination of ideal resistor R_{LOAD}, inductance L_{LOAD}, and capacitance C_{LOAD} [25]. R_{LOAD} and C_{LOAD} values are in fact positive, resulting in a positive corresponding time constant $\tau = R_{LOAD}C_{LOAD}$ and hence are not in violation of the Kramers-Kronig transformation rules [12]. The inductive artifact reflects the property of the sample cell with the type of four-electrode measurement and is not an error introduced by the instrumentation. The measured sample impedance between points V_{HIGH} and V_{LOW} from the circuit (Figure 8-10), considering input Z_{INPUT} and contact impedances of the voltage electrodes Z_R, is:

$$Z_{MEASURED} = Z_{SAMPLE} \frac{2Z_{INPUT} + 2Z_R}{2Z_{INPUT} + 2Z_R + Z_{SAMPLE}} \qquad (8\text{-}22)$$

It becomes obvious from Eq. 8-22 that for the optimal measurement $Z_{MEASURED} \approx Z_{SAMPLE}$ the condition should exist so $Z_R \ll Z_{SAMPLE} \ll Z_{INPUT}$. Therefore, in addition to the need to suppress parasitic current $I_{PARASITIC}$ by maintaining $Z_{SAMPLE} \ll Z_{INPUT}$ within the analyzed frequency range, one has to make sure that V_R is very small compared to measured total voltage difference $V_0 = V_{HIGH} - V_{LOW}$. For this purpose a voltage electrode configuration with $Z_R \ll Z_{SAMPLE}$ should be chosen. For certain analyzed samples, such as biological membranes with resistance even as high as ~ 100 kohm, that can be achieved by choosing reversible nonpolarizable reference electrodes with extremely low interfacial impedance Z_R. These reference electrodes are placed in well-conducting media on both sides of the membrane [21, 26] (Figure 8-9), resulting in no noticeable voltage divider effect and accurate impedance measurements for samples with resistance under 1 Mohm [14].

Unfortunately, for samples with bulk resistance in excess of 1 Mohm accurate impedance characterization becomes difficult and requires a careful analysis of several contributions. Traditional "aqueous type" reversible reference electrodes such as Ag/AgCl cannot be used in nonaqueous environments, and metal "pseudoreference" electrodes have to be employed [25, 27]. That leads to experimental complications, represented in Figure 8-11. The frequency analysis of the input, reference, and sample impedances demonstrated that the difference between Z_{INPUT} and Z_{SAMPLE} is minimal at high frequencies, leading to a relatively high $I_{PARASITIC}$ with large loss in the measured current I_{SAMPLE}. There is a relatively small loss in the measured voltage V_{SAMPLE} and the measured impedance becomes $Z_{MEASURED} > Z_{SAMPLE}$, with a substantial parasitic high-frequency capacitive effect. At lower frequencies where $Z_{INPUT} \gg Z_{SAMPLE}$ the parasitic current decreases and the measured current error becomes minimal. However, the reference-electrode resistance Z_R becomes larger with the frequency decrease, leading to an increase in V_R, large loss (error) in V_{SAMPLE}, and decrease in the measured impedance $Z_{MEASURED} < Z_{SAMPLE}$, which manifested itself in a pronounced inductive artifact [24, 25].

This analysis is also applicable to determination of impedance characteristics of membranes in conductive solutions that typically do not develop sizable impedance measurement artifacts [21]. For the setup in Figure 8-9, a

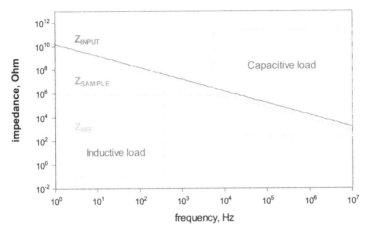

FIGURE 8-11 Dependence of Z_{INPUT}, Z_{SAMPLE}, and Z_R vs. frequency for a highly resistive sample with $R_{BULK} \sim 1\text{Mohm}$

biological membrane with thickness of ~ 0.01 cm, surface area 5 cm², permittivity $\varepsilon = 5$, and resistivity $\rho = 10^8$ ohm cm is placed in the electric field between two large metal counterelectrodes with cross-sectional surface area of 5 cm² and two Ag/AgCl reference electrodes in highly conductive buffer solution with negligible resistance. If one assumes the membrane to be homogeneous without parallel conduction processes in pores, etc., then the equivalent circuit for the membrane will be represented by capacitance $C_{MBR} \sim 200$ pF and resistance $R_{MBR} \sim 200$ kohm. The input impedance of the analyzer Z_{INPUT} consists of the capacitance $C_{INPUT} \sim 10$ pF in parallel with resistance $R_{INPUT} \sim 10^{12}$ ohms. At high frequencies (100MHz to 10kHz) the capacitive impedance of the membrane will be only ~20 times less than that of the analyzer, and that will result in ~5 percent error in the current reading with parasitic current flowing through the reference electrodes and the analyzer input. The capacitive error on the order of 5 percent will take place, which may be important if investigation of multiple conduction processes through the membrane at the high-frequency range is a subject of study. Reversible RE with low resistance prevents the voltage losses at low frequencies and the appearance of inductive effects. Although the distortions of the impedance data under these realistic circumstances are relatively minor and are confined to the high-frequency region, the recommendations developed in the sections on three-electrode and four-electrode probes remain valid. Selecting an extremely high-input impedance analyzer, keeping impedances of the CE low with a large surface area, employing reversible low-impedance RE, analyzing thin membranes with large surface areas, and making the cables short and shielded to minimize coupling capacitance all remain essential experimental tasks.

In complex cases of four-electrode measurement in highly resistive media with metal "pseudo-reference" electrodes the contact impedance may become significant, resulting in confounding results and equivalent circuits consisting of inductive, capacitive, and resistive elements related to an instrumental artifact and not to a physical condition of a studied system [15]. One therefore has to make a careful assessment of the effects of contact impedances, capacitive coupling between signal lines and signal lines and ground, nonuniform current distributions, voltage drop, cable shielding, and finite input resistance of the amplifier in considering the cells' geometric arrangements. It is advisable to create an electrode configuration with relatively small sample impedance by increasing the current and voltage electrode active areas, minimizing the spacing between two voltage electrodes and between the voltage and current electrodes, and keeping $Z_{SAMPLE} << Z_{INPUT}$ [13]. This can be accomplished by the use of two flat metal mesh "pseudoreference" electrodes mounted in between the two current electrodes and placing each of the voltage electrodes as close to the two outer current electrodes as possible.

Depending on the resulting electrode configuration, this solution may introduce additional errors due to the creation of a strongly nonhomogeneous electric field as a result of the insertion of large voltage electrode barriers to the current flow. Minimization of the interference of the voltage electrodes to the homogeneity of the current flow can be realized by using voltage electrodes with a small surface area such as needles or rings [11, 12], increasing the gap between the voltage electrodes, and decreasing the gap between the voltage and the current electrodes. However, this approach is exactly opposite to the configuration with larger areas for the electrodes and results in higher interfacial impedance of the voltage electrodes Z_R and significant voltage divider effect [13, 14]. For instance, both studies of a Nafion^R membrane with bulk resistance on the order of 1 kohm [11] and industrial colloids with bulk resistance on the order of 1 Mohm [25] revealed a substantial medium-frequency inductive loading at ~ 100 Hz for four-electrode cells with small surface-area voltage electrodes (Figure 8-12). Larger surface-area RE showed less of the inductive interference but produced an additional parallel resistance-capacitance load at low frequency (~0.1 Hz), which has no physical meaning

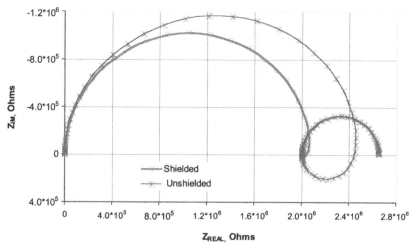

FIGURE 8-12 Complex impedance Nyquist and Bode plots of shielded vs. nonshielded cables *(continued on next page)*

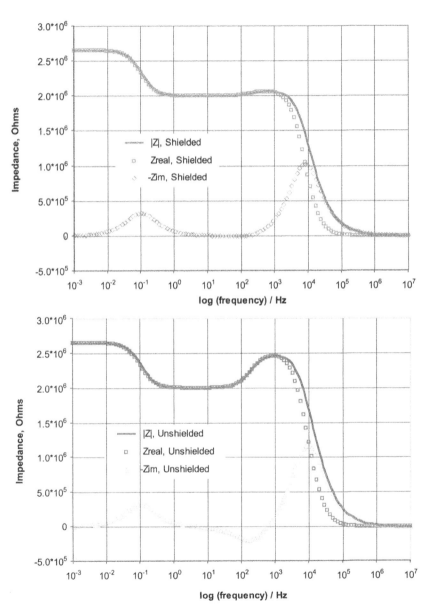

FIGURE 8-12   *(Continued)*

but is due to the geometry of the sampling cell, namely the relative thickness and positioning of the voltage reference electrodes [4, 13] (Figure 8-12). The depression in the low-frequency semicircle was explained by a transmission-line effect along the electrode strip width due to the differences in the local electric-field distribution at high and low frequencies caused by the capacitive blocking nature of the solid-state voltage electrodes. Since the thickness of the

RE is not zero, a part of the sample is shorted by the metal of well-conducting RE behaving as bipolar electrodes. At low frequencies the voltage is measured between the centerlines, while at high frequencies the voltage is measured between the inner edges of the voltage electrodes. The resulting maximum error for an electrode with a rectangular cross-section equals the ratio of the non-zero thickness of the reference electrodes and the distance between the centerlines of the electrodes. Therefore, experimental development of a higher-resistance media sample calls for a very delicate balance between the two conflicting approaches.

Another source of error in impedance measurements are loads of connecting cables (Section 5-1). Minimizing the cables' loading impedances using shorter cables remains the simplest physical solution to reducing measurement errors. This recommendation is true for any cell, not just the four-electrode cell. The voltage electrode cable capacity and the leakage resistance from the signal line to the grounded shield are loads for signal sources with large source impedance. As can be seen from Figure 8-10, the capacity of the voltage electrode cables C_{CB} are in parallel to the input capacitance C_{INPUT} if the outer cable shield is connected to ground ("driven shields off"). Therefore, the input impedance Z_{INPUT} (Eq. 8-22) can be more broadly defined as a combination of the cable capacitance and insulation resistance with the input capacitance and resistance of the amplifier circuit. The capacity of a standard BNC cable is about 100 pF m^{-1} and therefore can significantly increase the total voltage input capacity as $C_{CB} = l_{CB} *100$ pF m^{-1}, where l_{CB} = the length of the cable. In that case the total input impedance becomes smaller as:

$$Z_{INPUT} = \frac{R_{INPUT}}{1 + \omega R_{INPUT}(C_{INPUT} + C_{CB})} \tag{8-23}$$

That introduces a change in Eq. 8-22 and also increases the parasitic current $I_{PARASITIC}$ determined from the total input impedance. The result leads to larger values for the contact reference-electrode voltage drop V_R and therefore increases the inductive artifact due to the voltage divider effect (Figure 8-12).

Unfortunately, circumstances do not always permit changing the physical properties of the electrochemical cell; therefore, active driven shielding must be used. Active shielding is accomplished by surrounding the cables with a screen that is at the same potential as the signal lead. By connecting the shield to the input amplifier voltage output, the shield is kept at the same potential as the electrode. In principle the cable capacity is still present using this setup, but active shielding rejects the capacitive current [4] and minimizes the effects of the amplifier open-circuit parasitic capacitance by an order of magnitude [13, 14]. Experimentally, driven shields reduced the combined effect of the cable and amplifier open circuit capacitance and resulting inductive and total medium-frequency impedance loading by a factor of 5 to 10 at frequencies below 10 kHz (Figure 8-13). With the shielding engaged the total input parasitic current is reduced to the current related to high impedance of the input amplifier Z_{INPUT}. Thus, the sole source of interfacial loads will be the input capacitance of the amplifier circuit. In practice, this concept works only partially [24]. It is necessary to extend the active shield on the leads as close as possible

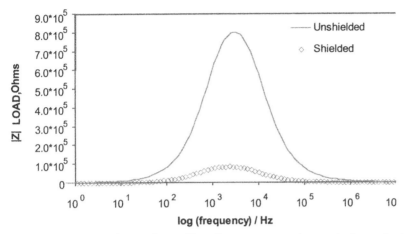

FIGURE 8-13 Total impedance "load" measurement error in four-electrode cells using shielded and no-shielded cables

to the sample. This prevents the leads to the current electrodes from affecting other signal lines due to stray capacitance.

Four-electrode cell impedance measurements are reliable if the sample impedance Z_{SAMPLE} is sufficiently small with respect to the input impedance Z_{INPUT}, which requires a reasonably high conductivity sample or a medium where the reversible reference electrodes can operate and the selection of an instrument with very high input impedance. Additional recommended improvements to the experimental setup include reduction of the influence of the cable impedance through shorter leads and active shielding. When metal reference electrodes have to be used, choosing finer-grade meshes is recommended in order to expand the available surface area; however, the tradeoff will be increased electric-field inhomogeneity and larger low-frequency transmission-line geometric artifact. The interface impedances of current electrodes Z_C and of voltage electrodes Z_R should be very low to minimize the voltage divider effect and to prevent a substantial interruption in the current flow from the outside current electrodes while the thickness of the voltage electrodes should be kept to a minimum with respect to the distance between these electrodes. These conditions lead to complicated requirements for shape and surface area properties of the voltage electrodes. Additionally, a careful and difficult analysis of several contributions, such as sample resistance, positioning of the electrodes, geometry of the electrodes, and active shielding, becomes necessary. However, even with perfect instrumentation and experimental arrangements, measurements will not be completely artifact-free.

8.5. Linearity, causality, stability, consistency, and error analysis of impedance measurements

Impedance is only properly defined as a transfer function when the system under investigation fulfils the conditions of causality, linearity, stability, and consistency during the measurement [28, 29]. Deviation from causality can arise when the response is not caused by the input but rather, for example, by a concentration, current, or potential relaxation upon departure of the system from equilibrium. Causality can also be disturbed as result of instrument artifacts or noise. Impedance analysis at very high frequencies where low impedances and low capacitances are measured often introduces errors due to instrumental artifacts, such as the cable inductive and capacitive effects described above.

In general, linearity of a given electrochemical system (that is, a constant linear relationship between the applied voltage and the measured current) can be studied as a function of sampling frequency, AC voltage amplitude, applied electrochemical (DC) potential, and other factors. Electrical-circuit theory distinguishes between linear and nonlinear systems. As Figure 8-14A shows, electrochemical systems can be pseudolinear, when current becomes not proportional to the higher applied AC and DC voltage. As the current-voltage characteristic of any electrode shows, electrochemical systems fundamentally are non-linear. In the case of AC voltage perturbation, the problem of nonlinearity can be generally overcome by making the amplitude of the perturbation signal small enough to approach quasilinear conditions. In normal EIS practice, a small (1 – 10 mV) AC voltage amplitude is applied to the cell. A small enough portion of a cell's current vs. voltage curve seems to be linear. The signal is small enough to be confined to a pseudolinear segment of the cell's current vs. voltage curve; however, this condition is detrimental to the signal-to-noise ratio, in particular for higher impedance samples. In polymer electrolytes it is convenient to apply high-voltage amplitude between the electrodes to measure higher currents, with typical input voltages in volts to kvolts. Under these conditions, when the voltage amplitude is higher than the thermal voltage of the electron ($V_t = K_B T/e \sim 20 mV$), the system tends to be nonlinear.

Applied electrochemical (DC) potential also causes nonlinearity. This nonlinearity in electrochemical systems typically results from the potential dependence of Faradaic processes at low frequencies. For example, charge-transfer reaction kinetics are governed by Volmer-Butler (Eq. 5-23), where the reaction rate (current) shows exponential dependence on the interfacial potential (voltage). For a potentiostated electrochemical cell, the input is the potential and the output is the current. Doubling the voltage will not necessarily double the current. DC studies should be done with minimum overpotential to keep linearity, especially at low frequencies where the model may or may not be linear, and the values of the model's components may vary with the DC voltage [30, p. 45].

In electrochemical systems, nonlinearity mainly affects the low-frequency part of the spectrum, and the electrode usually operates in the linear region at high and the no-linear region at low frequencies [30, pp. 45–48]. When the system operates in a significantly nonlinear region, it is often necessary to repeat the measurement several times to verify that results are repeatable and consistent [31]. If the system is non-linear, the current response will contain harmonics of the excitation frequency. Some researchers have made use of this phenomenon, studying higher harmonic responses in the EIS spectrum and

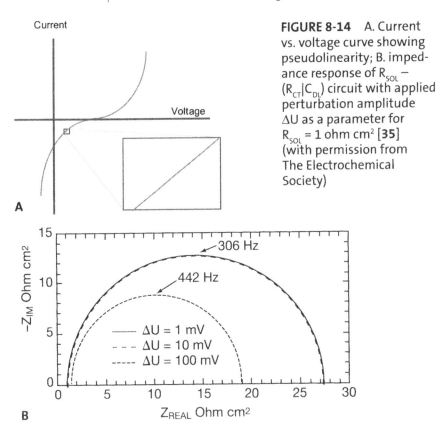

FIGURE 8-14 A. Current vs. voltage curve showing pseudolinearity; B. impedance response of R_{SOL} − ($R_{CT}|C_{DL}$) circuit with applied perturbation amplitude ΔU as a parameter for R_{SOL} = 1 ohm cm² [35] (with permission from The Electrochemical Society)

developing "nonlinear" EIS data. Linear systems should not generate harmonics, so the presence or absence of significant harmonic response allows one to determine the system's linearity.

The stability of a system is usually not guaranteed, especially when changing systems (such as corrosion or batteries) are investigated [3] over a long measurement cycle, in particular when low AC frequency analysis is employed [29, p. 213]. In that respect, rapidly corroding parts and charging/discharging batteries are examples of unstable systems. A common cause of problems in EIS measurements and their analysis is drift in the analyzed system. A system is considered to be stable if it returns to its original state after removal of perturbation, which is difficult to achieve. For that reason no system is absolutely stationary, and in practice a nonstationary system can be best approximated as stationary by minimizing the measurement time.

Nearly ideal experimental data can be obtained by making the perturbation signal small to achieve linearity, rerunning the experiment and comparing results to verify for stability, and using Kramers-Kronig calculations to validate the data consistency. Kramers-Kronig calculations, presented in Eq. 8-24, comprise a series of integral equations that govern the relationship between the real and imaginary parts of complex impedance for systems fulfilling the conditions of linearity, causality, stability, and "finiteness" of impedance magnitude

throughout the whole frequency range and comparing experimental and theoretical results:

$$Z_{IM}(\omega) = \frac{2\omega}{\pi} \int_0^\infty \frac{Z_{REAL}(x) - Z_{REAL}(\omega)}{x^2 - \omega^2} dx$$

$$Z_{REAL}(\omega) = Z_{REAL}(\infty) + \frac{2}{\pi} \int_0^\infty \frac{xZ_{IM}(x) - \omega Z_{IM}(\omega)}{x^2 - \omega^2} dx \qquad (8\text{-}24)$$

where x is an independent variable representing the complex frequency.
The Kramers-Kronig equations have been used by a large number of scientists
as a diagnostic tool for the validation of impedance data to establish which
frequency ranges contain consistent data. To assess whether experimental
data fulfill the Kramers-Kronig relations, one part of the impedance is calculated from the other part of the impedance, which has been experimentally
determined. For example, the imaginary part can be calculated from the experimentally measured real part of the impedance using Eq. 8-24. The calculated imaginary part of the impedance can then be compared with the experimentally measured imaginary part of the impedance. Any difference between
calculated and measured values would indicate a deviation from one of the
conditions of linearity, causality, or stability. This method, however, requires
that at least one part of the impedance is known between the frequency limits
of zero and infinity. An alternative method of data validation calls for the comparison of the response amplitude spectrum with the perturbation spectrum
(often a random-phase multisine input signal), where emergence of nonexcit-
ed frequencies with significant power indicate data distortion [32].

In most cases the frequency range is not sufficiently large to integrate over
the frequency limits zero and infinity. To overcome this problem, various
models for extrapolating the experimental data sets to the frequency limits
have been suggested, such as extrapolating impedance data to low frequencies (useful only for systems with a single time constant) and calculating impedance data below the lowest measured frequency by using Eq. 8-24, simultaneously forcing the measured data set to satisfy the Kramers-Kronig relations
[33]. Any discontinuity between the measured and the calculated data set
would then indicate erroneous impedance data. A method that does not require extrapolating data to frequency ranges not accessible by experiment is
the fit of the data with an equivalent circuit.

Equivalent circuits consisting of ideal circuit elements satisfy Kramers-
Kronig relationships, except for particular cases where deviations were reported, such as "negative capacitance" ("inductance") and "negative resistance" phenomena due to adsorption or dissolution-passivation [34]. Any
successful fit of such an equivalent circuit to a data set would confirm that the
data is consistent with the Kramers-Kronig relations. Failure to satisfy Kramers-Kronig relationships at high frequency can usually be attributed to instrumental artifacts (causality) and at low frequency to nonstationary system behavior. It has been commonly thought that Kramers-Kronig equations can be
used only to test the stability and causality conditions of the system, and for
potentiostatic conditions the Kramers-Kronig transformations should be used

only in admittance and for galvanostatic conditions only in impedance form, respectively. Failure to follow these recommendations can lead to apparent noncompliance with Kramers-Kronig relations, which is often claimed unjustifiably [34].

However, the Kramers-Kronig relationships are generally considered to be insensitive to nonlinearity of the system. As was discussed above, varying the AC voltage amplitude may assist in detecting nonlinear phenomenon (Figure 8-14B). As was demonstrated by Orazem [35], the sensitivity of the Kramers-Kronig relationships to the nonlinearity of an electrochemical system depends in fact on the magnitude of the AC potential perturbation, the magnitude of the bulk material or solution impedance, and the employed frequency range. For a simple model circuit represented by $R_{SOL} - (R_{CT} | C_{DL})$ it was shown that a "transition frequency" f_T exists above which the Kramers-Kronig transformations are not satisfied for higher-amplitude AC voltage perturbations. For the analyzed simple system, the Kramers-Kronig relationship's sensitivity to the system nonlinearity depends on both the AC voltage amplitude and the ratio of charge transfer to the bulk-solution resistances:

$$f_T = \frac{1}{2\pi R_{CT} C_{DL}} \left[1 + \frac{R_{CT}}{R_{SOL}} \right] \qquad (8\text{-}25)$$

As can be seen from Eq. 8-25, this transition frequency value depends on the relative magnitudes of the interfacial impedance (in this case the charge transfer resistance R_{CT}) and the bulk impedance (R_{SOL}). In highly conductive systems the bulk impedance is very small ($R_{SOL} \to 0$ ohm), the transition frequency is infinitely high, and Kramers-Kronig relationships are satisfied in all analyzed frequency ranges, even for higher applied voltage amplitudes causing nonlinear behavior. The Kramers-Kronig relationships are also satisfied, of course, when the applied voltage amplitude is sufficiently small and the system is effectively linear. The Kramers-Kronig relationships are violated at high frequencies when the system is nonlinear and the ratio of R_{CT}/R_{SOL} is small resulting in the frequency f_T falling within the experimental analyzed frequency range. Higher bulk impedance shifts effective interfacial impedance measurements to lower frequencies [35], making the system's nonlinearity more apparent for higher applied voltages and violating the Kramers-Kronig relationships.

The Voigt model (Section 3-3) is consistent with Kramers-Kronig relationships, and if the data can be fit into the Voigt model, it means that the data is consistent, stable, and acceptable [3]. The lack of fit of the Voigt model could be due to causes other than inconsistency with the Kramers-Kronig relationship, such as a limited number of measured frequencies to allow regression with a sufficient number of Voigt parameters, a large level of instrumentation noise, or a poor initial guess for the nonlinear regression. While in principle a complex fit of the measurement model could be used to access the consistency of impedance data, sequential regression to either Z_{REAL} or Z_{IM} provides greater sensitivity to lack of consistency, with Z_{IM} being usually much more sensitive to minor data details [36]. A refined approach was developed to resolve the ambiguity that exists when the model does not provide a good fit to the data and causes bias error. The optimal approach is to fit the model to the component that contains the greatest amount of information. Typically, more Voigt line shapes can be resolved when fitting to the imaginary part of the impedance than can be resolved when fitting to the real part. The solution resistance R_{SOL}, for example, cannot be obtained by fitting the measurement model to the imaginary part of the impedance and therefore is treated as an arbitrarily adjustable parameter in this case [37].

It should be noted that some instrument and experiment-related artifacts may be evident over a broad frequency range and that some may result in impedance features that are consistent with the Kramers-Kronig relations. Thus, the inability to identify inconsistent data does not prove that the measured spectrum reflects only the properties of the electrochemical system under study. The measurement model procedure provides a statistical basis by which one can decide whether data satisfy the Kramers-Kronig relations. The sensitivity of the method can be improved by decreasing the noise level of the measurement and by increasing the number and range of sampled frequencies. Overall, while the Kramers-Kronig relations are useful for validation of impedance data, it should be noted that these relations provide a necessary only, not a necessary-and-sufficient, condition. Failure to satisfy the Kramers–Kronig relations provides unequivocal evidence of non-causal, nonlinear, or nonstationary influences, yet satisfaction of the Kramers-Kronig relations does not prove that the data are representative of a specific electrochemical cell. For example, some instrument artifacts may appear as Kramers-Kronig-consistent features in impedance spectra, as was shown in the example of inductive data in the 4-electrode cell caused by the voltage divider effect (Section 8-4).

A systematic error analysis demonstrates that, in spite of differences between sequential impedance scans and sometimes the appearance of inductive and incomplete capacitive loops, the individual data sets represented a pseudostationary system and could be interpreted in terms of a stationary model [37]. An example of such a system can be impedance analysis of corrosion, which usually includes both capacitive and inductive elements. The residual errors $\psi_{RES}(\omega)$ in an impedance measurement can be expressed in terms of the difference between the observed value $Z(\omega)$ and a model value $Z_M(\omega)$ as:

$$\psi_{RES}(\omega) = Z(\omega) - Z_M(\omega) = \psi_{FIT}(\omega) + \psi_{BIAS}(\omega) + \psi_{STOCH}(\omega) \qquad (8\text{-}26)$$

where ψ_{FIT} = the systematic error that can be attributed to inadequacies of the proposed impedance model, ψ_{BIAS} = the systematic experimental bias error that cannot be attributed to model inadequacies (and leads to inconsistency with Kramers-Kronig relationships primarily due to lack of stationarity and existence of instrumental artifacts), and ψ_{STOCH} = the stochastic error that arises from the integration of time-domain signals that contain noise from the electrochemical equipment, thermal fluctuations of resistivity, thermal fluctuations of concentration of species and the rates of electrochemical reactions, and macroscopic events such as pitting and bubble formation [38].

To reduce the bias error ψ_{BIAS}, it is necessary to reduce the total measurement time (to avoid the system nonstationary; however, that leads to higher stochastic errors as integration becomes restricted), introduce delay or quiet time, avoid harmonics, and select an appropriate modulation technique. Deletion of data that are strongly influenced by bias errors ψ_{BIAS} increases the amount of information that can be extracted from the data. Removal of the biased data results in a better conditioned data set that enables reliable identification of a larger set of parameters. As mentioned above, inconsistencies at low frequencies may be attributed to nonstationary behavior, and inconsistencies at high frequencies may be attributed to instrument artifacts. Inconsistencies in the middle of a frequency range can arise due to switching of current measurement circuitry and/or switching between impedance measurement techniques (i.e., multisine Fourier transform to phase-sensitive detection). Regression of measurement models is sensitive to such discontinuities [19]. A combination of inductive and capacitive loops could be associated with failure to be consistent with the Kramers-Kronig relations.

An iterative approach in which the data identified as being inconsistent with the Kramers-Kronig relations are removed from the data set is used to obtain the stochastic error ψ_{STOCH} structure estimate. The estimate obtained using the measurement model approach will be more accurate than would be obtained by direct calculation of the standard deviation. Thus, the variance of the real and imaginary residual errors provides a good estimate for the frequency-dependent variance of the stochastic noise in the measurement. The standard deviation of the stochastic errors can be identified from the standard deviation of the residual errors for the real and imaginary parts of the impedance. The stochastic error ψ_{STOCH} is a strong function of frequency, therefore one must choose the experimental frequency range carefully, increase the integration time/cycles, introduce a delay and ignore the measured data at the very first frequency (as it is often corrupted by a startup transient), and avoid external fields/harmonics or characteristic frequencies such as 60Hz in the US and 50Hz in Europe.

The sequence of experimental measurements often indicates a substantial change from one measurement to another and therefore the presence of the systematic error ψ_{FIT}. This lack of reproducibility in itself raises the question of whether each individual measurement was corrupted by nonstationary phenomena or can be attributed to shortcomings of the experimental data modeling. Additional experiments are needed to demonstrate that impedance measurements were taken after the system reached a steady state.

The "modulus weighting" proposed by Boukamp, where error ψ or standard deviation of impedance measurement is proportional to the absolute impedance $|Z|$, is accepted as an initial strategy in regression of impedance data obtained by FRA and PSD algorithms. An equation correcting for these small departures is proposed as [38]:

$$\psi = \alpha \mid Z_{IM} \mid + \beta \mid (Z_{REAL} - R_{SOL}) \mid + \gamma \frac{\mid Z \mid^2}{R_M} + \delta \qquad (8\text{-}27)$$

where R_{SOL} = solution resistance, R_M = resistance of the current measuring circuit, and α, β, γ, δ = constants specific for a given potentiostat and the analyzed electrochemical system. However, the details of impedance measurement error analysis are different for FRA and PSD. Variances of real and imaginary parts of impedance in the time domain are equal and uncorrelated for FRA, but they are nonequal and correlated in PSD even with absolute impedance and phase-angle values being uncorrelated. The errors in the frequency domain are normally distributed for FRA and not normally distributed for PSD. In general PSD shows a slightly larger error (~0.3 percent) than the FRA algorithm (~0.07 percent) [39].

The FFT multisine method shows results very similar to the single-sine FRA algorithm [40]. The highest frequency used for FFT EIS without introducing "aliasing" is the Nyquist frequency of the instrument, which is equal to the sampling frequency of the electronics hardware used to record voltage and current data [41]. That makes a very high-frequency end of the spectrum not very useful for quantitative analysis. Usually special filters are introduced to either correct for errors or reject very high frequencies. At low frequencies a noise related to high measured impedance is observed, which also can be corrected for. DC offset voltage can be applied to the system; however, reliable measurements with the offset voltage can be performed only after ~1sec when large current variations are somewhat settled [41].

Early investigators questioned the validity of EIS measurements on the basis of system stability and linearity requirements over the time of the measurement. The FFT method applies the resulting voltage signal from mixed AC waves of several dozen selected frequencies to a system on top of DC voltage bias. The current signal obtained in the time domain is converted back into AC signals in the frequency domain by the Fourier transformation [42, 43]. Recent results showed that the FFT impedance technique indeed records instantaneous impedance data, whereas traditional single-frequency impedance techniques present time-averaged data. Fast impedance methods are indeed fast enough for the electrochemical reaction timescale [42, 43], making it available for sensors and in-situ measurements.

Of different ways of representing and analyzing the impedance data, $|Z|$ and Z_{REAL} vs. frequency usually have poor sensitivity to errors and variability to differences in various impedance models and equivalent circuits. Z_{IM} vs. Z_{REAL}, Z_{IM} vs. frequency, phase angle vs. frequency, and log Z_{IM} vs. log frequency representations have good sensitivity to errors, and a residual-errors plot has the best sensitivity to errors. It is important to look at several impedance representations (for instance, the complex modulus M^* plot for the bulk material and impedance Z^* plots for the interfacial impedance) to approximate the mechanism of the process and initiate the fitting procedure.

8.6. Complex nonlinear least squares (CNLLS) regression fitting

Modern EIS analysis uses computer calculations and commercially available impedance equipment programs to find and evaluate the model parameters that cause the best agreement between a model's impedance spectrum and a measured spectrum. Complex nonlinear least-squares regression techniques for EIS were developed in the late 1960s . Typically a complex nonlinear least-squares fitting (CNLLS) regression algorithm is used, consistent with the expectation that real and imaginary components of impedance data satisfy the constraints of the Kramers-Kronig relations [37]. The regression of a complex function Z^* to a complex data Z can be expressed in a least-squares (χ^2) sense as a minimization of the sum of squares. Evaluation of the kinetic parameters and fitting of the models to the experimental data can be performed in Mathematica® or other commercially available software packages included with impedance analyzer hardware, utilizing a built-in, numerical, nonlinear fitting function based on the Nelder-Mead simplex algorithm [28]. The regression procedure involves minimization of a complex function with the weighted, least-squares fit parameter χ^2 used for the modeling efforts utilizing the impedance data (Z, ϕ, Z_{REAL}, Z_{IM}) as shown below:

$$\chi^2(Z,\phi) = \sum_1^N \frac{\left(\phi_{EXP}(f) - \phi_{CALC}(f)\right)^2}{\phi^2_{STANDARDDEVIATION}(f)} + \frac{\left(|Z(f)|_{EXP} - |Z(f)|_{CALC}\right)^2}{|Z|^2_{STANDARDDEVIATION}(f)} \tag{8-28}$$

$$\chi^2(Z_{REAL}, Z_{IM}) = \sum_1^N \frac{\left(Z_{REAL,EXP}(f) - Z_{REAL,CALC}(f)\right)^2}{Z_{REAL\,STANDARDDEVIATION}^2(f)} + \frac{\left(Z_{IM,EXP}(f) - Z_{IM,CALC}(f)\right)^2}{Z_{IM\,STANDARDDEVIATION}^2(f)} \tag{8-29}$$

The Σ in the above equations indicates that the fit value between the laboratory-based experimental (*EXP*) and model-based calculated (*CALC*) impedance data was determined by the sum of the individual fit parameters for each frequency (*N* total) utilized in the experimental testing. The fit values at each frequency were then summed to provide the overall fit value corresponding to the minimum χ^2 values.

CNLLS starts with initial estimates for all the model's parameters. The fit values at each frequency then are summed to provide the overall fit value for a given set of model parameters. Following the initial estimate of the model parameters, the CNNLS algorithm makes changes in several or all of the parameter values and evaluates the resulting fit. If the change improves the fit (decreases χ^2 value), the new parameter value is accepted. If the change worsens the fit (increases χ^2 value), the old parameter value is retained. Next a different parameter value is changed and the test is repeated. Each trial with new values is called iteration. Iterations continue until the accuracy of fit exceeds an acceptance criterion or until the number of iterations reaches a limit. When several models are compared, the best minimization χ^2 values from each set of starting model values and the sum of all best minimization parameters for a single model across the experimental conditions can be used in comparisons between the models, with the smallest fit parameter combination being determined as the better fit.

CNLLS algorithms are not perfect. In some cases they do not converge on a useful fit. This can be the result of several factors including:

- an incorrect model for the data set being fitted
- poor estimates for the initial values
- excessive noise in the data

In addition, the fit from a CNLLS algorithm can look graphically poor while providing relatively low values for the χ^2 fit parameter when the fit's spectrum is superimposed on the data spectrum. It may appear that the fit completely ignores a region in the data, especially if a small component is missing from the model. The CNLLS algorithm optimizes the fit over the entire spectrum and sometimes overlooks poor graphical fit over a small section of the spectrum.

8.7.   Practical approach to experimental impedance data collection and analysis

The aim of EIS data analysis is to elucidate the electrode process or the material properties and to derive their characteristic parameters. It should be stressed here that EIS is a very sensitive technique, but it does not provide a direct measure of physical phenomena. Other electrochemical (DC potential, transients) and broader analytical (optical, magnetic, rheological) experiments should also be implemented for system characterization. Good physical, chemical, electrical, and mechanistic understanding of the system, such as knowledge of analyzed media, working electrode surface composition, thickness, porosity, presence of various layers, hydrodynamic conditions, etc., is also required. When employing the impedance method to a sample analysis, the following general sequence is advisable:

- develop initial concept of general type and anticipated electrical, mechanical, physical and spatial characteristics of analyzed material

- organize experimental setup—select electrodes, cell geometry, temperature, AC and DC voltage parameters

- collect the impedance data over a wide frequency range to ensure that all essential impedance-affecting processes are identified

- improve the measurement conditions—use optimal current measurement range, increase integration time or number of cycles, introduce quiet time, ignore the first measured frequency (it is often corrupted by a startup transient), repeat the measurement several times to check for repeatability, stability, and consistency using Kramers-Kronig relationships

- collect experimental impedance data from several sets of experiments at different controlled conditions

- combine the data with either a plausible physicochemical model describing the physics and chemistry of processes in the system or a completely empirical equivalent circuit based on physical intuition

- follow with a complex nonlinear least-squares (CNLLS) regression and independent assessment of experimental error structure

EIS is not a technique that can be applied without prior knowledge of an analyzed system's chemical, physical, mechanical, and electrical characteristics. One has to have at least a preliminary idea of what the system "is expected to look like" to analyze the data and make initial correct assumptions. The anticipated experimental data must often be interpreted within the context of the developed model even before the data is acquired. A recommended first step is to determine if the investigated material can be roughly classified as highly conductive (such as an aqueous ionic system) or as dielectric (composite solids, nonpolar liquids with high resistance). The analyzed sample type largely affects the choice of equipment and experimental conditions such as the applied AC frequency and amplitude range, temperature, electrodes, and sample geometry. Highly conductive media respond to DC polarization, and multifrequency EIS analysis can be conducted at selected DC voltages in a three-electrode arrangement. In dielectric samples the analysis may be initiated with a two-electrode probe at rest potential.

Afterwards, EIS measurements should be carried out over a wide frequency range (usually with a sampling rate of 10 frequencies per decade) in order to identify all time constants in the circuit. The highest-frequency selection depends on the type of impedance equipment used, giving consideration to possible equipment-induced artifacts such as high frequency phase shift due to stray capacitance and inductance of cables and electrodes. In modern EIS analyzers the high-frequency range typically is at ~100 kHz, although it may be as high as ~100 MHz in higher-tier equipment with shielded cables and compensation for the cables' impedance.

A typical lowest practical frequency limit is 1 μHz, selected primarily due to the system's potential lack of stability during time-consuming low-frequency measurements and resulting changes in the process kinetics. At this frequency measurement averaged over five wave periods can take more than 1 hour. The complete frequency sweep over the whole range between high (~10 MHz) and low (1 μHz) frequency limits takes a much longer time, sometimes two to four hours depending on selected sampling rate. An attempt to save some time by reducing the sampling rate can lead to erroneous results when the frequency is swept "too fast." Rapid change of frequency may lead to a transient regime. If the measurements are performed during this transient, an error is introduced into the results, especially when FRA is used, and results show dependency on the initial phase of the sinusoidal excitation. It can be neglected when 10 cycles of signal integration and at least five samples per frequency decade are used in measurements. This error becomes negligible at higher frequencies (typically over 10 Hz) because of the internal delay of the measurements at each of the higher frequencies.

In general, it is relatively easy to get precise measurements for impedances between 1 ohm and 100 kohm at frequencies below 50 kHz. For very low (molten salts) and very high (polymeric coatings) materials additional impedance distortions may be observed. The errors in high-impedance measurements originate from the finite potentiostat input impedance, as was discussed in Section 8-4. The internal resistance should be at least 100 times larger than the measured impedance at all sampling frequencies. Even in relatively conducting systems bulk impedance on the order of 100 ohms affects the applicability of Kramers-Kronig transformations for detecting the validity of the Faradaic reaction data for nonlinear systems (Section 8-5) and complicates interpretation of the impedance plots (Section 2-1). It is difficult to obtain accurate results of the Faradaic processes when this ohmic drop is large compared to the interfacial impedance. The Luggin capillary technique could be used to diminish the ohmic drop but not to cancel it completely, as a certain distance between the working electrode and the Luggin capillary always exists. For high frequencies ~100 kHz further consideration must be given to the general structure of the experimental arrangement (such as the use of active shields and short coaxial cables) because of inductive or capacitive leakage and cable loads. For very low impedance measurements (<1 ohm), the same care must be taken where frequencies are lower than ~100 kHz, as an inductance appears in series with the working electrode impedance.

A major aspect of impedance-data interpretation starts with an understanding of the measuring principle of the method and the types of data it generates. The impedance parameter at a given sampling AC frequency is an integrated value. This value is composed of contributions of both circuit components present at the sampling frequency and all components that load the circuit at frequencies higher than this sampling frequency. This is particularly important to remember when dealing with low-frequency impedance features, often representing the electrode-material interface. The interfacial-impedance portion of the circuit appears at low frequencies, but its graphical representation is affected by the higher-frequency loading component of the circuit even when this component is a simple single resistor with a relatively low associated impedance value. It is useful to remember that the critical frequency is determined for a single $R\,|\,C$ type process, while the experimental impedance data shows "total" measured impedance, which includes many potential contributions of higher-frequency (such as the bulk solution, rapid charge transfer, etc.) processes. Therefore, these higher-frequency processes and related impedance data should be analyzed and taken into consideration before the "pure" impedance characteristic of the studied process can be singled out using the Nyquist or Bode graphical interpretations (Section 2-1). In complex cases involving more "preceding" or higher-frequency processes or several "distributed processes," even the highest point in a Z_{IM} plot is not always a correct indication of the critical frequency of a given process, as the analytical expression for the imaginary and real impedance values become affected by the impedance contributions of the higher-frequency and "distributed" processes.

When analyzing the impedance data, it is essential to start the analysis at the highest available frequency and determine all loading components sequentially as the frequency decreases. The low-frequency impedance data will contain the contributions from all components of the circuit appearing at the higher frequencies. The particular lower-frequency impedance component of interest can be properly analyzed only after all higher-frequency contributions are carefully subtracted from the total measured impedance.

After the cable loads contributions are assessed, the bulk-media impedance (Z_{BULK}) evaluation should be taken as the next analysis step. EIS analysis should be initiated by looking at the data at the highest frequency (impedance Z_{BULK} and phase angle at 50 kHz – 10 MHz), taking particular notice of the high-frequency phase-angle values to determine the bulk resistance and ca-pacitance characteristics of the sample. A purely resistive response (with phase angle 0°) and small total measured impedance indicate highly conductive sample with likely interfacial impedance at low frequencies. A capacitive re-sponse at high frequency (with phase angle −90°) indicates a dielectric sample that has a significant capacitive bulk component C_{BULK}. The high-frequency impedance limit is generally equal to R_{SOL} in an aqueous medium or is in-versely proportional to the bulk solution capacitance C_{BULK} in a dielectric me-dium. In a capacitive high-frequency response, at lower frequencies the phase angle may completely transition to 0° in a simple $R|C$ model or may reach ~−60 to −70° and then shift back to −90° in a Debye-type dielectric dispersion structure.

In order to separately study only slow processes, such as low-frequency Faradaic kinetics, frequency-dependent overcompensation or mathematical subtractions (Section 2-1) can be used to eliminate the Z_{BULK} contributions. Various circuits based on a positive feedback principle have been proposed to compensate for the bulk-material impedance effects. With them, however, in-stability problems arise due to exact compensation or overcompensation. This problem can be avoided by correcting the ohmic drop out of the feedback loop of the potentiostat. An analog quantity, proportional to the current flowing through the cell, is subtracted from the potential difference between the refer-ence and working electrodes, and a corrected voltage is obtained. Hence the electrolyte resistance can be corrected exactly, overcorrected, or partially cor-rected without any influence on the stability of the potentiostatic regulation. This real-part control of the impedance allows the impedance to be measured with better accuracy, which is especially important for the phase angle, as it shows significant sensitivity to noise due to instability and general measure-ment error.

Such initial experimental and data-assessment procedures should be sup-ported by a series of measurements at different potentials, temperatures, con-centrations, and convections, with the data to be combined with the error analysis. After the data is acquired, it can be initially represented by an equiv-alent circuit, physical, or continuum level model that is consistent with physi-cal and chemical information and is comparable to previously published EIS and other analytical results on identical or at least similar systems. The pre-liminary selection of the data representation, such as complex impedance, modulus, and phase- angle notations, is often helpful, as quite often some of these graphic notations are more informative than others.

Data interpretation is the most challenging aspect of impedance analysis and is often misunderstood and incorrectly performed. Interpretation of im-pedance data requires use of an appropriate model. Development of a proper model requires knowledge of the chemistry and physics of the system, some prior information about it, and a good understanding of the characteristics of the measured values. This is a difficult task that must be carried out very care-fully. The most accurate model will combine all recent knowledge about relax-ation processes, frequency dispersion, mass-transport, fractals, and redox pro-cesses and will have the same frequency response as the analyzed substance, taking into account all chemical and electrical considerations. In reality such an idealized model does not exist [30, p. 200], therefore an intelligent guess must be taken. As an initial step, an investigator may attempt to find a model that produces a calculated impedance plot that matches the measured EIS data [3]. Such descriptive models are the only ones that can differentiate among different reaction layers, solution components, etc., inside the structure and have therefore at least a prescreening clinical value [30, p. 200]. The descriptive models can be considered as intermediate stages in the development of better explanatory models [30, p. 201]. There are two main approaches to system descriptive modeling [29, p. 85]:

1. Equivalent circuit (formal or mathematical) modeling presents the system in hypothetical electrical circuits consisting of well-defined ideal and sometimes nonideal electrical elements. Measurement modeling explains the experimental impedances in terms of mathematical functions in order to obtain a good fit between the calculated and experimental impedances. In the latter case the parameters obtained do not necessarily have clear physicochemical significance. Such a model describes the system's response to various possible electrical input signals.

2. Physicochemical (molecular, process, or structural) modeling develops equations linking experimentally measured impedances with physicochemical parameters of the process (kinetic rate parameters, concentrations, diffusion coefficients, sample geometry, hydrodynamic conditions, etc.) and attempts to provide a description of the motions of individual charge-carrying particles. The bulk regions of the electrodes and the sample material are presented as continuous media, and mass and charge transport processes can be represented by differential equations

The ambiguity in data interpretation is common to model identification for all electrochemical measurements and presents the greatest challenge for the analysis of impedance data. Ideally, at first equivalent circuit modeling should be carried out, identifying the number and the nature of ideal electrical components (resistors, capacitors, inductors, CPE elements, etc.). Then physicochemical modeling based on a deeper understanding of underlying impedance-affecting processes can be implemented.

Equivalent circuit (EC) analysis is relatively simple for a circuit containing ideal elements R, C, and L. It may also be carried out for circuits containing distributed elements that can be described by a closed-form equation, such as CPE, semi-infinite, finite length, or spherical diffusion. Many "ideal" resistances and capacitances chosen to represent a real physicochemical system are really nonideal as any resistor has a capacitive component and vice versa. However, for the broad frequency range utilized by UBEIS it is usually adequate to incorporate "ideal" resistors, capacitors, and inductances [29, p. 87]. The type of electrical components in the model and their interconnections control the shape of the model's impedance spectrum. The EC model's parameters (i.e., the resistance value of a resistor) control the size of each feature in the spectrum. Both the component's type and its parameter affect the degree to which the model's impedance spectrum matches a measured EIS spectrum. EC analysis is convenient and graphically appealing, and the fit to the experimental data is performed using CNLLS algorithms and commercially available impedance fitting software.

Physicochemical impedance data modeling relies on development of several equations describing the process kinetics, as shown, for example, in Section 7-4. In a physicochemical model each of the model's components is postulated to come from a physical process in the investigated system. The choice of which physicochemical model applies to a given cell is made from knowledge of physical, mechanical, and chemical characteristics of the system and its components. Very often physicochemical impedance-data modeling goes astray because of errors in the experimental data. However, even for excellent quality data many different conditions arise from the numerical calculations—e.g., for porous electrodes, nonlinear or nonuniform diffusion, or nonhomogeneous materials. In such cases a priori model predictions are difficult or impossible. A physicochemical model deduced from the reaction mechanism may have too many adjustable parameters, while the experimental impedance spectrum is simple. On the other hand, one can always get a good-looking fit by adding lots of circuit elements to a model. Unfortunately, these elements may have little relevance to the investigated cell processes. Drawing conclusions based on changes in these elements is especially dangerous. With the availability of distributed elements such as CPE, typically an excellent data fit can be obtained with little physical meaning supporting the interpretation. For example, a system with one adsorbed species may produce two semicircles in the complex plane plots, but experimentally only one semicircle is often identified. In such a case approximation to a full model introduces too many free parameters, and a simpler model containing one time constant should be used. Empirical models should therefore initially use the fewest elements possible. Therefore, first the number and nature of parameters should be determined, and then the process model can be constructed consistent with the found parameters and the physicochemical model can be established. The ultimate objective is not to provide the easiest fit but to gain a valuable model to learn about the physics and chemistry of the system.

Another problem of data modeling is due to the fact that the same data may often be represented by very different equivalent circuits. The model is chosen to give the best possible match between the calculated and the measured impedance data. However, there is not a unique equivalent circuit that describes the spectrum, and one cannot assume that an equivalent circuit that produces a good fit to a data set represents an accurate physical model of the cell. The exclusive use of equivalent circuit models for the analysis of largely unknown materials and systems is fraught with potential sources of error.

Models can also be partially or completely empirical. An empirical model can be constructed by successively subtracting component impedances from a spectrum. If the subtraction of an impedance simplifies the spectrum, the component is added to the model, and the next component impedance is subtracted from the simplified spectrum. This process ends when the spectrum is completely obliterated ($Z=0$). Physicochemical models are generally preferable to empirical equivalent circuit models, but even the physicochemical models are suspect, as some of the actual processes may be unaccounted for. Whenever possible, the equivalent circuit or physicochemical models should be verified. One way to verify the model is to alter a single cell component (for example, a paint layer thickness, temperature, electrode separation, or surface area) and see if the expected changes in the impedance spectrum are produced.

Many experimental systems analyzed by impedance spectroscopy involve many-body problems currently not solvable at the microscopic level. The limitations of impedance spectroscopy and, in particular EC modeling do not always appear to be widely appreciated by those using these techniques. Obviously, a suitably wide frequency range must be used and a sufficiently large number of data points obtained if meaningful interpretation of the data is to be possible. Care must be taken when graphically presenting the measured data so that the chosen plot is optimal for the given application and that vital components of the data are not discarded. The use of suitably chosen three-dimensional or multiple parameter representation plots would appear to be the most rigorous approach, and this should clearly demonstrate the suitability or otherwise of a given mathematical formula or equivalent circuit model. It must be borne in mind that, even the use of a purely mathematical approach, such as the impedance equations, involves approximations that may significantly affect the accuracy of the work, especially when one is interested in the electrical behavior of the system over limited frequency ranges or in the possible presence of several "relaxations."

References

1. C. Gabrielli, *Identification of electrochemical processes by frequency response analysis*, Solartron Analytical Technical Report 004/83, 1998, pp. 1–119.

2. A. Lasia, *Modern aspects of electrochemistry*, B. E. Conway, J. Bockris, R. White (Eds.), vol. 32, Kluwer Academic/ Plenum Publishers, New York, 1999, pp. 143–248.

3. S. Krause, *Impedance methods*, in Encyclopedia of Electrochemistry, A. J. Bard (Ed.), Wiley-VCH, Vol. 3, 2001.

4. G. Fafilek, *The use of voltage probes in impedance spectroscopy*, Solid State Ionics, 2005, 176, 25-28, pp. 2023–2029.

5. S. Seitz, A. Manzin, H. D. Jensen, P. T. Jakobsen, P. Spitzer, *Traceability of electrolytic conductivity measurements to the international system of units in the sub ms m^{-1} region and review of models of electrolytic conductivity cells*, Electrochim. Acta , 2010, 55, pp. 6323–6331.

6. B. H. Timmer, W. Sparreboom, W. Olthuis, P. Bergveld, A. van den Berg, *Planar interdigitated conductivity sensors for low electrolyte concentrations*, Proceedings of SeSens 2001, (2001) pp. 878–883, Veldhoven, the Netherlands.

7. M.C. Zaretsky, L. Mouyad, J.R. Melcher, *Continuum properties from interdigitated electrode dielectrometry*, IEEE Trans. Electr. Insul.,1988, 23, pp. 897–906.

8.   M. W. den Otter, *Approximate expressions for the capacitance and electrostatic potential of interdigitated electrodes*, Sens. Act. A, 2002, 96, pp. 140–144.

9.   W. Olthuis, W. Steekstra, P. Bergveld, *Theoretical and experimental determination of cell constants of planar-interdigitated electrolyte conductivity sensors*, Sens. Act. B, 1995, 24–25, pp. 252–256.

10.   V. F. Lvovich, C. C. Liu, M. F. Smiechowski, *Optimization and fabrication of planar Interdigitated Impedance sensors for highly resistive non-aqueous industrial fluids*, Sens. Act. B, 2007, 119, 2, pp. 490–496.

11.   G. Fafilek, M. W. Breiter, *Instrumentation for ac four-probe measurements of large impedances*, J. Electroanal. Chem., 1997, 430, pp. 269–278.

12.   P. Vanýsek, *Artifacts and their manifestation in impedance measurements*, presentation at 207th Meeting of the Electrochemical Society, Quebec City, Canada, 2005.

13.   B. Boukamp, *Interpretation of an 'inductive loop' in the impedance of an oxygen ion conducting electrolyte-metal electrode system*, Solid State Ionics, 2001, 143, 1, pp. 47–55.

14.   D. D. Edwards, J. –H. Hwang, S. J. Ford, T. O. Mason, *Experimental limitations in impedance spectroscopy: Part V. Apparatus contributions and corrections*, Solid State Ionics, 1997, 99, pp. 85–93.

15.   G. Hsieh, S. J. Ford, T. O. Mason, L. R. Pederson, *Experimental limitations in impedance spectroscopy: Part I. simulation of reference electrode artifacts in three-point measurements*, Solid State Ionics, 1996, 91, pp. 191–210.

16.   S. Fletcher, *The two-terminal equivalent network of a three-terminal electrochemical cell*, Electrochem. Comm., 2001, 3, pp. 692–696.

17.   V. Horvat-Radosevic, K. Kvastek, *Three-electrode cell set-up electrical equivalent circuit applied to impedance analysis of thin polyaniline film modified electrodes*, J. Electroanal. Chem., 2009, 631, pp. 10–21.

18.   V. Horvat-Radosevic, K. Kvastek, *Analysis of high-frequency distortions in impedance spectra of conducting polyaniline film modified Pt-electrode measured with different cell configurations*, Electrochim. Acta, 2007, 52, pp. 5377–5391.

19.   V. Horvat-Radosevic, K. Kvastek, *Quantitative evaluation of experimental artifacts in impedance spectra of conducting polyaniline thin-films measured using pseudo-reference electrode*, J. Electroanal. Chem., 2008, 613, pp. 139–150.

20.   A. Sadkowski, J.-P. Diard, *On the fletcher's two-terminal equivalent network of a three-terminal electrochemical cell*, Electrochim. Acta, 2010, 55, pp. 1907–1911.

21.   Y. Yoon, A. C. Mount, K. M. Hansen, D. C. Hansen, *Electrolyte conductivity through the shell of the eastern oyster using a four-electrode measurement*, J. Electrochem. Soc.,2009, 156, 2, pp. 169–176.

22.   Z. Xie, C. Song, B. Andreaus, T. Navessin, Z. Shi, J. Zhang, S. Holdcroft, *Discrepance in the measurements of ionic conductivity of PEMs using two and four-probe AC impedance spectroscopy*, J. Electrochem. Soc., 2006, 153, 10, pp. E173–E178.

23.   B. D. Cahan, J. S. Wainright, *AC impedance investigations of proton conduction in nafion*, J. Electrochem. Soc., 1993, 140, 12, pp. L185–L186.

24.   V. F. Lvovich, M. Smiechowski, *AC impedance characterization of highly resistive media using 4-electrode electrochemical cells*, Impedance Techniques: Diagnostics and Sensing Applications, Vol. 25, V. F. Lvovich, P. Vanysek, M. Orazem, B. Tribollet, D. Hansen (Eds.), The Electrochemical Soc., Inc., Pennington, NJ, 2010.

25.   V. Lvovich, M. Smiechowski, *AC Impedance investigation of conductivity of industrial lubricants using 2- and 4-electrode electrochemical cells*, J. Appl. Electrochem., 2009, 39, 12, pp. 2439–2452.

26. N. M. Kocherginsky, V. F. Lvovich, *Biomimetic membranes with nano channels but without proteins: impedance of impregnated nitrocellulose filters*, Langmuir, 2010, 26 (23), pp. 18209–18218.

27. M. F. Smiechowski, V. F. Lvovich, *Iridium oxide sensors for acidity and basicity detection in industrial lubricants*, Sens. Actuators B, 2003, 96, pp. 261-267.

28. M. E. Orazem, B. Tribollet, Electrochemical Impedance Spectroscopy, J. Wiley and Sons, Hoboken, New Jersey, 2008.

29. E. Barsukov, J. R. MacDonald, Impedance Spectroscopy, John Wiley & Sons, Hoboken, New Jersey, 2005.

30. S. Grimnes, O.G. Martinsen, Bioimpedance and Bioelectricity Basics, Academic Press, 2000.

31. M.J. Yang, Y. Li, N. Camaioni, G. Casalbore-Miceli, A. Martinelli, G. Ridolfi, *Polymer electrolytes as humidity sensors: progress in improving an impedance device*, Sensors and Actuators B, 2002, 86, pp. 229–234.

32. E. Van Gheem, R. Pintelon, J. Vereecken, J. Schoukens, A. Hubin, P. Verboven, O. Blajiev, *Electrochemical impedance spectroscopy in the presence of non-linear distortions and non-stationary behavior. Part I. theory and validation*, Electrochim. Acta, 2004, 49, pp. 4753–4762.

33. J. M. Esteban, M. E. Orazem, *On the application of the Kramers-Kronig relations to evaluate the consistency of electrochemical impedance data*, J. Electrochem. Soc., 1991, 138, pp. 67–76.

34. A. Sadkowski, M. Dolata, J.-P. Diard, *Kramers-Kronig transforms as validation of electrochemical immitance data near discontinuity*, J. Electrochem. Soc., 2004, 151, 1, pp. E20–E31

35. B. Hirschorn, M. E. Orazem, *On the sensitivity of the Kramers-Kronig relationships to nonlinear effects in impedance measurements*, J. Electrochem. Soc., 2009, 156, 10, pp. C345–351.

36. P. K. Shukla, M. E. Orazem, O. D. Crisalle, *Validation of the measurement model concept for error structure identification*, Electrochim. Acta, 2004, 49, pp. 2881–2889.

37. M. Orazem, *A systematic approach toward error structure identification for impedance spectroscopy*, J. Electroanal. Chem., 2004, 572, 2, pp. 317–327.

38. S. L. Carlson, M. E. Orazem, O. D. Crisalle, L. Garcia-Rubio, *On the error structure of impedance measurements: simulation of FRA instrumentation*, J. Electrochem. Soc., 2003, 150, 10, pp. E477–E490.

39. S. L. Carlson, M. E. Orazem, O. D. Crisalle, L. Garcia-Rubio, *On the error structure of impedance measurements: simulation of PSD instrumentation*, J. Electrochem. Soc., 2003, 150, 10, pp. E491-E500.

40. S. L. Carlson, M. E. Orazem, O. D. Crisalle, L. Garcia-Rubio, *On the error structure of impedance measurements: series expansions*, J. Electrochem. Soc., 2003, 150, 10, pp. E501–E511.

41. J. E. Garland, C. M. Pettit, D. Roy, *Analysis of experimental constraints and variables for time resolved detection of Fourier transform electrochemical impedance spectra*, Electrochim. Acta, 2004, 49, pp. 2623–2635.

42. J. -S. Yoo, S.-M. Park, *An electrochemical impedance measurement technique employing Fourier transform*, Anal. Chem., 2000, 72, pp. 2035–2041.

43. S.-M. Park, J.-S. Yoo, *Electrochemical impedance spectroscopy for better electrochemical measurements*, Anal. Chem., 2003, 21, pp. 455A–461A.

CHAPTER 9

Selected Examples of EIS Applications: Impedance of Electroactive Polymer Films

9.1. The field of electroactive polymers

Extensive studies of electrochemically active polymers during the past two decades have been driven by their special properties, such as flexible solution and melt processability manufacturing, blendability with commodity polymers, good ambient stability, and unconventional electrical and optical properties. The most valuable property of these materials is their unique ability to dramatically and reversibly change electrical conductivity over the full range from insulators to metallic conductors upon partial electrochemical oxidation or reduction, a process commonly referred to as "doping." The "doping" process modifies the electrical properties of the polymer such that either electrochemical oxidation or reduction results in a dramatic change in conductivity. This change is sometimes by as much as 10 to 12 orders of magnitude, from the low value for the noncharged insulating state of the polymer to values with conductivities of $\sigma = 1 - 10^5$ S cm^{-1}, which are comparable to those of metals.

The fundamental observation is that even a rather thick polymer film, where most of the redox sites are 100–10 000 nm away from the metal surface, may be electrochemically oxidized or reduced. According to the classical theory of simple electron-transfer reaction, when the reactants get to the Helmholtz plane close to the electrode surface, the electrons can tunnel over the short distance of a few nanometers between the metal and the electroactive species. In the case of polymer-modified electrodes the active parts of the polymer cannot approach the metal surface, because polymer chains are trapped inside a tangled network, and chain diffusion is usually much slower than the time scale of the transient electrochemical experiment.

The electrochemical oxidation or reduction of the electrochemically active polymers and related charge transport can occur either via a concentration gradient-driven electron exchange reaction between neighboring redox sites (long-distance "electron hopping") if the segmental motions of the polymer chains make it possible or through delocalized electrons that can move through the conjugated systems ("electronic conduction"). The former mechanism is characteristic of redox polymers, and the latter is representative of conducting polymers. The class of redox polymers comprises structures containing immobilized redox active centers, such as polynuclear complexes based on metal hexacyanoferrates, polyoxotungstates and polyoxomolybdates, and ion-exchange polymers with incorporated (electrostatically bound) redox-active cations or anions. Typical conducting organic polymers are based on aniline and its derivatives, simple and substituted five-membered heterocycles (polythiophenes, polypyrroles), and many other polyconjugated, polyaromatic or polyheterocyclic macromolecules. Many conducting polymers can be viewed as electrochemically active organic macromolecules with a conjugated system of π-bonds that undergo oxidations and/or reductions in the doping/undoping processes: charges, "holes," and unpaired electrons are delocalized over a large number of monomer units. A feature of this class of organic polymers is the ability to undergo reversible redox transitions involving the so-called counterion dopants, such as anions and protons. The "doping" process in conjugated polymers is essentially a charge-transfer reaction, resulting in the partial oxidation or reduction of the polymer, which depends primarily on the electrochemical potential and effects of solution ions—both protons and counterions. Electrochemical transformation of the nonconducting form of these polymers leads to reorganization of the bonds of the macromolecule and development of an extensively conjugated system, resulting in very fast and relatively temperature-insensitive electron-transfer processes with unusually high capacitive current and dramatic changes in the film conductivity. As will be shown later, many extensively studied polymers, such as polyaniline (PANI), polypyrrole (PPY), and Nafion®, demonstrate electrochemical features characteristic of both types of polymers and essentially possess a mixed electronic-ionic mechanism of conductivity [1, 2, 3, 4].

Among the family of conducting polymers, PANI and PPY are unique due to their ease of synthesis, environmental stability, and simple doping/dedoping chemistry. Because of their rich chemistry and high electrical conductivity, PANI and PPY have been among the most studied conducting polymers by various optical and electrochemical techniques. These materials can be synthesized and characterized by electrochemical means as standalone polymer films between two solutions or two metal conductors or as surface films deposited on a conductive substrate. The electrochemical synthesis of these polymers allows counteranions from the growth electrolyte to become incorporated into the film in order to balance the net charge buildup during polymerization. In the case of PANI there is an additional protonation doping mechanism, a result of the movement of protons both into and out of the polymer. Thus, PANI conductivity is sensitive to a change in both counteranions and pH, and a sharp transformation of conductivity is exhibited at the onset of oxidation or reduction. PANI has more than one oxidation state, each with an associated level of conductivity. PANI changes at low electrooxidation potentials from nonconductive leucoemeraldine (LE) to highly conducting emeraldine (EM), and at very high oxidation potentials it further changes to insulating perni-graniline (PE) [1].

9.2.   Impedance analysis of electrochemically active polymer films

Electrochemical impedance spectroscopy allows the investigation of charge- and mass-transport kinetics and charging processes taking place within the analyzed material and at the active interfaces of the system. In recent years AC impedance technique has become a primary method of investigation of modified electrodes, and it has proven to be a powerful tool for the characterization of electrochemically active polymer films.

Modeling an electrochemical interface by the equivalent circuit (EC) representation approach has been exceptionally popular in studies of electrodes modified with polymer membranes, although an analytical approach based on transport equations derived from irreversible thermodynamics was also attempted [6, 7]. ECs are typically composed of numerous ideal electrical components, which attempt to represent the redox electrochemistry of the polymer itself, its highly developed morphology, the interpenetration of the electrolyte solution and the polymer matrix, and the extended electrochemical double layer established between the solution and the polymer with variable localized properties (degree of oxidation, porosity, conductivity, etc.).

The elucidation of the nature of charge-transfer processes in electrochemically active polymer films may be the most interesting theoretical problem of the field and a question of great practical importance. A polymer film electrode can be defined as an electrochemical system in which at least three phases are contacted successively in such a way that between an electronic conductor (usually a metal) and an ionic conductor (usually an electrolyte solution) is an electrochemically active polymer layer. The fundamental processes of insertion and transport of charged and noncharged species through this type of electrochemical system are described below and illustrated in Figure 9-1 [8]:

- charge (electron) transfer at the electrode/film interface with associated potential difference ΔV_1. Ions and solvent cannot move through this interface
- transport of species through the active polymer film with associated potential difference ΔV_2
  1. transport of electrons between the electrode/film and film/solution interfaces
  2. transport of ions and water (solvent) through the film

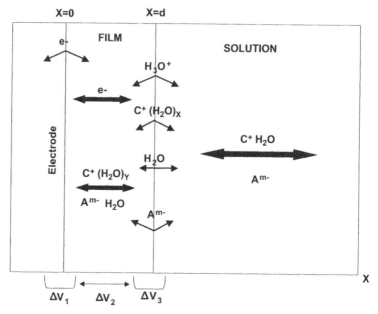

FIGURE 9-1   Scheme of the electrode/electroactive film/solution system

- insertion or expulsion of ions and water at the film/solution interface with associated potential difference ΔV_3. Electrons do not move through this interface
- transport of ions and water in the solution

One of the most challenging aspects of electrochemically active polymers is the occurrence of electron-transfer reactions due to the existence of different oxidation states and simultaneous exchange of solution counterions to compensate for the excess (or lack) of charge, thus maintaining film electroneutrality [9]. The electrical neutrality is maintained by the exchange of protons at low pH and the insertion of anions into the film at higher pH. These systems must therefore be considered as possessing a mixed ionic-electronic conduction where the migration of both electronic and ionic charge carriers contribute to the total charge transport, and any of these steps may become rate-limiting [7, 8, 9, 10].

9.3.   EIS models of conducting polymer films

There has been great interest in in situ characterization of conduction mechanisms and charging/discharging processes in conducting polymers using impedance spectroscopy. When the electrochemical properties of the polymeric films and corresponding EC parameters are investigated, a large number of frequently mutually dependent system parameters should be considered [11, 12], such as:

- applied electrochemical potential and temperature effects on the rate of electron transfer, ionic mass transport, and diffusion coefficient of the chain and segmental motions responsible for the electron hopping process

- chemical composition and concentration of the electrolyte—anionic and cationic content, pH, type of solvent, size and charge of the ions, specific interactions with the polymer

- chemical composition of the film, often influenced by the composition of the solution and film synthesis method used during the electropolymerization as well as the aging effects

- physical properties of the film such as thickness, morphology, permittivity, porosity, density, and swelling

- ionic trapping in the film

In a broad sense a parallel combination of charge transfer resistance and CPE elements, in series with finite diffusion element typically represent the circuit. When potential modulation is introduced, charge-transfer-related impedances decrease with increases in electrochemical potential and capacitance for the metal-polymer interface. The capacitance is usually nonideal due to film or electrode porosity [13] and typically is represented by the CPE element. If the film is formed as a reflective boundary, the angle is sometimes different from $-90°$ because of inhomogeneity of the film and distributed values for diffusion coefficients. If two films are formed on the electrode, two R | CPE semicircles are often observed.

With polymeric film aging several processes can occur. For example, an increase in charge-transfer resistance and a decrease in double-layer capacitance usually indicate that charge transfer becomes more complicated due to a decrease in surface area or the formation of a passive film as a result of electrolyte decomposition and surface deactivation [14]. With time thickness of the film often increases—leading to an increase in film heterogeneity due to oxidation of the polymer. Thickness of the film may also decrease due to time-related degradation, with a corresponding increase in measured interfacial capacitance and a decrease in charge-transfer resistance.

In the case of many electrochemically active films such as PANI, an essential variation of the non-stationary redox response depending on the oxidation state of the polymer was reported: an almost perfect quasi-equilibrium variation of the charge of the film following the instantaneous potential values was observed at high oxidation states, while a pronounced effect of slow charge transport across the film manifests itself at lower bias potentials [12]. Two time constants were observed in the potential region where switching between nonconductive and conductive forms of polyaniline occurs, which was simulated by a parallel combination of a capacitance and a transmission line. The exact nature of these potential-dependent processes remains unclear even today; several thought-provoking possible interpretations of the electrochemically active polymer films physical-chemical identity were suggested:

1. A two-phase bulk polymer structure—this interpretation is typical of dielectric spectroscopy analysis of polymers where "more conducting" and "more insulating" segments are expected to display very different electrical properties. As will be shown later, in the literature this interpretation has often evolved into a number of "spatial distributed" or "temporal distributed" film theories based on differences in the localized electrical properties of various film segments.

2. A single homogeneous phase representation, where a combination of double-layer capacitance and diffusion-controlled Faradaic process is responsible for oxidation reduction of the polymer, resulting in appearance of "transmission line" in the equivalent circuit model. The large capacitances exhibited by conducting polymer electrodes are usually attributed to the double-layer capacitance and pseudocapacitance originating from the redox process of the polymer.

3. A porous phase, exhibiting a pseudocapacitance charging due to a thin layer behavior of the film, a fast electron-transfer process confined to the pores in the film, and double-layer capacitance of a porous material.

As a result of these studies, the general impedance theory of electrochemically active polymer films has been developed [6, 7], and a typical impedance response of electron-ion conducting films in solution-polymer-metal systems can be modeled as (Figure 9-2):

- high frequency impedance response to charge carriers in bulk polymer represented by a parallel combination of film resistance R_{BULK} and

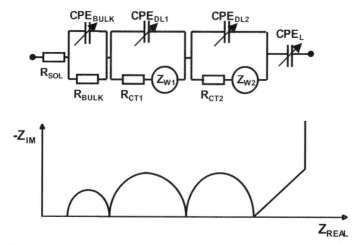

FIGURE 9-2 General equivalent circuit and a typical complex impedance plot representing the electrode/electroactive film/solution system

capacitance C_{BULK} (or more often CPE_{BULK}) in series with uncompensated solution resistance R_{SOL}

- two parallel combinations of charge-transfer resistance R_{CT} and double-layer capacitive CPE_{DL} features at medium frequency representing the electrode-polymer interface (electron transfer) and polymer-solution interface (counterion transfer)

- ionic and electronic diffusion-migration limited segments resulting in $-45°$ Warburg responses (Z_{W1} and Z_{W2}) followed at low frequencies by a $-90°$ capacitive line or a transmission line, representing either reflecting boundary diffusion or pseudocapacitive charging of the film (CPE_L)

When an electrochemically active polymer is bounded by two metal conductors in a metal-polymer-metal arrangement, the system can be represented [15] by (Figure 9-3):

- high-frequency semicircle for R_{BULK} and CPE_{BULK} of the polymer
- one charge-transfer semicircle ($R_{CT} | CPE_{DL}$ parallel segment) for double-layer capacitance and charge-transfer resistance at the electrode-polymer interface (electron transfer)
- due to electron transfer at both metal-polymer interfaces, DC current can flow, and a diffusion-limited low frequency $-45°$ line develops into a finite transmission boundary diffusion Z_O, resulting in the impedance response bending over to the real axis

Depending on exact experimental conditions, some of the segments in the impedance data become predominant, and the others insignificant, and the resulting Nyquist plots and equivalent circuits are simplified. In many cases

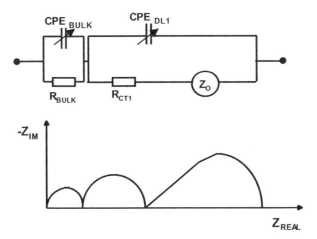

FIGURE 9-3 General equivalent circuit and a typical complex impedance plot representing the electrode/electroactive film/electrode system

the most hindered process is the counterion diffusion, coupled to electron transport. The exact details of proposed kinetic models of conductive polymers differ from one another due to the diversity of opinions about the origin of the experimentally determined parameters. Thus, double-layer capacity has been ascribed either to the metal-polymer [16, 17, 18] or the polymer-solution [19] interfaces. Warburg-like domains have been attributed to diffusion or to conduction control of charge transport [6, 7], whereas CPE_L has been either considered as a bulk [16, 17] or as an interfacial capacitance [20]. Concerning the conductivity of the polymer film, different conduction mechanisms have been considered (metallic, ionic, variable-range hopping, fluctuation-induced tunneling, or due to the electrolyte in the pores of the film) [21, 22, 23]. However, despite seemingly opposite ways of describing the polymer phase in these approaches, the experimental and computational EIS results concerning the responses to DC and AC perturbations often turned out to be similar or even identical [24]. This ambiguity is typical of EIS analysis in general, where one cannot assume that an equivalent circuit that produces a good fit to a data set represents an accurate physical model of the system.

Impedance modeling has seen several major theoretical developments. For example, a "brush-like" porous model was developed based on ordering of the polymeric chains with macropores filled with electrolyte, which represents the capacitive charging and redox reactions at walls of the pores with a transmission line [25]. Very high capacitance dominating in the conductive region of the potential window was explained as originating from diffusion-controlled movements of counterions (protons and anions) leaving or entering the film at the film/solution interface and being blocked at the film/metal interface. Earlier it was proposed that we distinguish two processes during the redox transformation of the polymeric films—a Faradaic type in which the charge carriers are generated in a reversible electron-transfer step and a capacitive type connected to the movement of the carriers to the interface, a charge accumulation without creating new carriers [26]. However, the overlapping oxidation processes and the inseparability of the current components in pure electrical measurements make a clear distinction between these two processes difficult.

Very often modeling of polymeric electrochemically active films has been based on attempts to separate different charging processes in the film, representing an evolution from the first proposed mechanism based on two different segments coexisting inside of the polymer structures. A number of studies were based on representation of these separate charging processes as related to the "fast" vs. "slow" charging properties of the polymer segments determined by their respective morphology, location within the film, chemical composition, and response to the applied AC field. For example, two different diffusion rates are assumed for the doping ion, depending on whether the capacitive doping is related to the ions moving into more closed aggregates of compact polymeric chains or to a bulk phenomenon kinetically controlled by the diffusion of counterions between the large polymeric chains [27, 28, 29]. Another theory [30, 31] considered a similar spatial two-phase distribution with shallowly trapped ions in the bulk of the films and deeply trapped ions in the double layer at the PPY film/electrode interface (Figure 9-4). A more detailed mechanism of spatial and temporal charging was proposed as a result of external oxidation-potential perturbation. It was suggested that polymer oxidation commences at the electrode/film interface and that the reaction front moves along the individual polymer chains toward the film/solution interface, representing more protonic charge compensation as sites deep inside nonconducting film react, while at more positive potentials the mechanism reflects oxidation of outside film layers with more anion injection. As proton expulsion occurs during the oxidation, the inner segments of the polymer film experience an increase in local pH, and the outer regions of the film will continue to be influenced by acidity of the bulk solution. Proton injection into a fully conducting film during reduction is likely to occur first at the outer film/solution interface [32]. Another type of gradient distribution takes into account mixed electron-ion conduction in the polymer, with ions conducting through pores and electrons through the chains in the polymer [33]. If electronic conductivity is much higher than ionic conductivity, the charge "leaks" from the polymer with the higher conductivity toward the electrolyte surface. The exchange reaction begins at the film/solution interface and propagates through the porous coating into the interior to the electrode. Otherwise, if ionic conductivity is much higher than electronic conductivity, the charge penetrates the film along more conductive pores, and oxidation begins at the electrode/polymer interface and gradually propagates into the interior of the polymer coating.

Additionally, it was proposed to assign the current to redox processes connected to a series of polymeric segments of different lengths [34], to the existence of parallel domains of conductive and resistive zones arranged perpendicularly to the electrode [35], and to a hopping motion charge-conduction

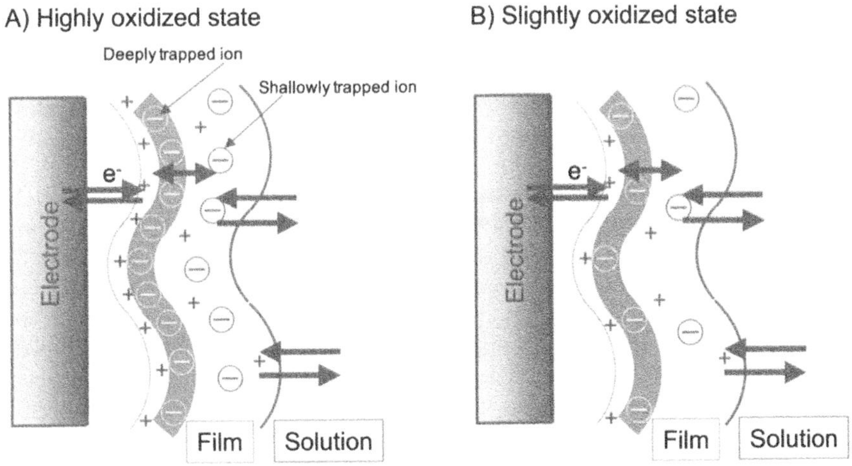

FIGURE 9-4 Ion-transport model in conductive films as a function of DC potential with "deeply trapped ions" responsible for "slow" Faradaic processes and "shallowly trapped ions" responsible for "fast" kinetic processes

mechanism represented by a transmission line [36]. The variation in the impedance spectra was analyzed as a function of characteristic polymer parameters through a computational simulation stressing the effects of both thickness and thickness distribution of the polymer films, suggesting that the ohmic resistance determined from EIS data is connected to ionic conductivity, including the cation incorporation, and not to electronic conductivity [30]. Film thickness, swelling, and expanded porosity are viewed as the predominant factors for higher capacitances, with counterion charging playing a major role [32].

9.4.   The future of electroactive polymers

The use of electrochemically active polymers as new materials in value-added industrial and consumer products is opening up entirely new opportunities for polymeric materials. These include electroactive inks, paints, coatings, and adhesives; electrochromic "smart" windows; electrically conductive transparent and corrosion-protective films; intercalating electrodes in advanced batteries; supercapacitive materials; conductive high-performance fibers, drain-source junctions in MOSFET-like devices; antistatic products; electrocatalytic materials; and ion-specific electrochemical biosensors. Conducting polymers offer a number of potential advantages in electrochemical sensors—for example, as electrode materials not subject to fouling or as substrates for enzyme immobilization. Enzyme-modified conductive polymer (most commonly PPY) electrodes have been reported able to detect glucose using a "ping-pong" mechanism where the polyaniline responds with quantitative conductivity changes to hydrogen peroxide released as a result of interaction between the enzyme and the substrate. PANI-based sensors have been used to detect acidity changes in aqueous and nonaqueous environments based on protonation/deprotonation of PANI. Future possibilities may include conductive and semiconductive shielded cable jackets, light-emitting diodes, microelectronic devices, and conducting fibers.

Clearly, electrochemical impedance analysis of conductive polymer films has provided deeper insight into the mechanisms of conduction, doping, and charge storage and has greatly furthered the theory of the impedance method as a major material-science characterization technique. Although the models that have been used to derive the expression for the impedance differ from one another, reflecting the diversity of opinions about the mechanism of charge-transport processes occurring in these films, it has become clear that a somewhat reliable physical description of experimental impedance data can be obtained. However, no unified approach to data analysis has been developed, most probably because a search for a uniform kinetic model properly designed to fit conducting polymers continues. In almost every case, the behavior of real systems shows deviations from the theoretically expected parameters. Therefore, additional refinements of the models are needed, taking into account interactions between the redox sites, intermolecular and intramolecular charge transport, possible chemical steps associated with charge transfer, nonuniform film thickness and porosity, ionic relaxation processes, and diffusion-coefficient distribution. Recent attempts to find the analytical expression for the transport equations inside the film and the solution, accounting for both ionic and electronic mobile charge carriers inside the film, electron and ion exchanges with the solution, redox couples, and charging of the interfaces [37], can lead to a major new qualitative conclusion about the nature and representation of charging-process dependence on the oxidation level of the film.

Finally, application of other methods of analysis can be recommended. Many of the previous electrochemical studies devoted to conducting polymers were carried out in combination with radiotracer technique, AC electrogravimetry, quartz-crystal microbalance, surface plasmon resonance, and even ellipsometry, and atomic-force microscopy. In this context application of emerging experimental techniques such as local EIS and nonlinear impedance analysis may also be recommended.

References

1. G. Inzelt, *Charge transport in conducting polymer film electrodes,* Chem. Biochem. Eng. Q., 2007, 21, 1, pp. 1–14.

2. G. Inzelt, G. Lang, *Model dependence and reliability of the electrochemical quantities derived from the measured impedance spectra of polymer-modified electrodes,* J. Electroanal. Chem., 1994, 378, 1–2, pp. 39–49.

3. G. Lang, G. Inzelt, *Some problems connected with impedance analysis of polymer film electrodes—effect of the film thickness and the thickness distribution,* Electrochim. Acta, 1991, 36, 5-6, pp. 847–854.

4. G. Inzelt, G. Horanyi, *Some problems connected with the study and evaluation of the effect of pH and electrolyte concentration on the behavior of polyaniline film electrodes,* Electrochim. Acta, 1990, 35, 1, pp. 27–34.

5. E. A. H. Hall, N. G. Skinner, C. Jung, S. Szunerits, *Investigating polymers and conducting metals as transduction mediators or immobilization matrices,* Electroanalysis, 1995, 7, 9, pp. 830–837.

6. M. A. Vorotyntsev, J.-P. Badiali, G. Inzelt, *Electrochemical impedance spectroscopy of thin films with two mobile charge carriers: effects of the interfacial charging,* J. Electroanal. Chem., 1999, 472, pp. 7–19.

7. M. A. Vorotyntsev, L. I. Daikhin, M. D. Levi, *Modelling the impedance properties of electrodes coated with electroactive polymer films,* J. Electroanal. Chem., 1994, 364, 1–2, pp. 37–49.

8. J.J. Garcia-Jareno, D. Gimenez-Romero, F. Vicente, C. Gabrielli, M. Keddam, H. Perrot, *EIS and ACeElectrogravimetry study of PB films in KCl, NaCl, and CsCl aqueous solutions,* J. Phys. Chem. B, 2003, 107, 41, pp. 11321–11330.

9. C. Gabrielli, H. Takenouti, O. Haas, A. Tsukada, *Impedance investigation of the charge transport in film-modified electrodes ,* J. Electroanal. Chem., 1991, 302, 1–2, pp. 59–65.

10. J. H. Jiang, A. Kucernak, *An electrochemical impedance study of the electrochemical doping process of platinum phthalocyanine microcrystals in non-aqueous electrolytes,* J. Electroanal. Chem., 2001, 514, 1–2, pp. 1–15.

11. V. F. Lvovich, *A perspective on electrochemical impedance analysis of polyaniline films on electrodes,* Interface, 2009, 18, 1, pp. 62–66.

12. I. Rubinstein, E. Sabatani, J. Rishpon, *Electrochemical impedance analysis of polyaniline films on electrodes*, J. Electrochem. Soc., 1987, 134, 12, pp. 3078–3083.

13. A. C. Ion, J. -C. Moutet, A. Pailleret, A. Popescu, E. Saint-Aman, E. Siebert, E. M. Ungureanu, *Electrochemical recognition of metal cations by poly(crown ether ferrocene) films investigated by cyclic voltammetry and electrochemical impedance spectroscopy*, J. Electroanal. Chem., 1999, 464, pp. 24–30.

14. R. Tossici, F. Croce, B. Scrosati, R. Marassi, *An electrochemical impedance study on the interfacial behavior of KC$_8$ electrodes in LiClO$_4$ containing electrolytes*, J. Electroanal. Chem., 1999, 474, pp. 107–112.

15. S. Krause, *Impedance methods*, Encyclopedia of Electrochemistry, A. J. Bard (Ed.), Wiley-VCH, Vol. 3, 2001.

16. C. Deslouis, M. M. Musiani, C. Pagura, B. Tribollet, *Determination of kinetic-parameters of Fe3+ reduction mediated by a polyaniline film using steady-state and impedance methods*, J. Electrochem. Soc., 1991, 138, 9, pp. 2606–2612.

17. M. M. Musiani, *Characterization of electroactive polymer layers by electrochemical impedance spectroscopy (EIS)*, Electrochim. Acta, 1990, 35, 10, pp. 1665–1670.

18. S. H. Glarum, J. H. Marshall, *The impedance of poly(aniline) films*, J. Electrochem. Soc.,1987, 134, pp. 142–147.

19. J. Tanguy, N. Mermilliod, M. Hoclet, *Capacitive charge and noncapacitive charge in conducting polymer electrodes*, J. Electrochem. Soc., 1987, 134, pp. 795–802.

20. R. A. Bull, F. R. F. Fan, A. J. Bard, *Polymer films on electrodes*, J. Electrochem. Soc., 1982, 129, pp. 1009–1015.

21. C. Deslouis, M. M. Musiani, B. Tribollet, *Mediated oxidation of hydroquinone on poly(n-ethylcarbazole)—analysis of transport and kinetic phenomena by impedance techniques*, Synth. Met., 1990, 38, 2, pp. 195–203.

22. C. Deslouis, C. Gabrielli, P. Sainte-Rose Fanchine, B. Tribollet, *Electrohydrodynamical impedance on a rotating disk electrode*, J. Electrochem. Soc., 1982, 129, pp.107–118.

23. T. Osaka, K. Naoi, S. Ogano, S. Nakamura, *Dependence of film thickness on electrochemical kinetics of polypyrrole and on properties of lithium/polypyrrole battery*, J. Electrochem. Soc., 1987, 134, pp. 2096–2102.

24. J. -M. Zen, G. Ilangovan, J. -J. Jou, *Square-wave voltammetric determination and AC impedance study of dopamine on preanodized perfluorosulfonated ionomer-caoted glassy carbon electrode*, Anal. Chem., 1999, 71, pp. 2797–2805.

25. G. G. Láng, M. Ujvári, T. A. Rokob, G. Inzelt, *The brush model of polymer films—analysis of the impedance spectra of Au, Pt poly(o-phenylenediamine) electrodes*, Electrochim. Acta, 2006, 51, pp. 1680–1694.

26. S. W. Feldberg, *Reinterpretation of polypyrrole electrochemistry: consideration of capacitive currents in redox switching of conductive polymers*, J. Am. Chem. Soc., 1984, 106, pp. 4671–4674.

27. J. Tanguy, M. Slama, M. Hoclet, J. L. Baudouin, *Impedance measurements on different conducting polymers*, Synth. Met., 1989, 28, 1–2, pp. C145–C150.

28. J. Tanguy, *Modelization of the electrochemical-behavior of conducting polymers*, Synth. Met., 1991, 43, 1–2, pp. 2991–2994.

29. J. Tanguy, J. L. Baudouin, F. Chao, M. Costa, *Study of the redox mechanism of poly-3-methyl-thiophene by impedance spectroscopy*, Electrochim. Acta, 1992, 37, 8, pp. 1417–1428.

30. G. Lang, J. Bacskai, G. Inzelt, *Impedance analysis of polymer film electrodes*, Electrochim. Acta, 1993, 38, pp. 773–780.

31. T. Amemiya, K. Hashimoto, A. Fujishima, *Frequency-resolved faradaic processes in polypyrrole films observed by electromodulation techniques—electrochemical impedance and color impedance spectroscopies*, J. Phys. Chem., 1993, 97, 16, pp. 4187–4191.

32. H. N. Dinh, P. Vanysek, V. I. Birss, *The effect of film thickness and growth method on polyaniline film properties*, J. Electrochem. Soc., 1999, 146, 9, pp. 3324–3334.

33. V. V. Malev, V. V. Kondratiev, *Charge transfer processes in conductive polymer films*, Russ. Chem. Rev., 2006, 75, 2, pp. 166–182.

34. J. Heinze, J. Mortensen, K. Mullen, R. Schenk, *The charge storage mechanism of conducting polymers: a voltammetric study on defined soluble oligomers of the phenylene–vinylene type*, J. Chem. Soc. Chem. Commun., 1987, 1, pp. 701–703.

35. O. Genz, M. M. Lohrengel, J. W. Schultze, *Potentiostatic pulse and impedance investigations of the redox process in polyaniline films*, Electrochim. Acta, 1994, 39, 2, pp. 179–185.

36. W.J. Albery, A. R. Mount, *2nd transmission-line model for conducting polymers*, J. Electroanal. Chem., 1991, 305, pp. 3–18.

37. M. A. Vorotyntsev, *Impedance of thin films with two mobile charge carriers: interfacial exchange of both species with adjacent media: effect of double layer charging*, Electrochim. Acta, 2002, 47, pp. 2071–2079.

CHAPTER 10

Selected Examples of EIS Analysis Applications: Industrial Colloids and Lubricants

10.1. The field of industrial colloids and lubricants

The health of industrial and transportation equipment relies on many things, but unquestionably the lubricants that grease the wheels of industry are at the top of the list. Industrial and automotive lubricants do more than provide an oil film to separate the moving parts of an engine. The lubricants reduce friction, suspend contaminants, neutralize corrosive acids from the combustion process, protect wear surfaces, dissipate heat, and provide other performance-improving features [1, 2]. One of the single largest applications for lubricants is protection of the internal-combustion engines in motor vehicles and powered equipment. On-the-road vehicles, ships, and airplanes also require specialized lubricants for their transmissions and gears. Industrial equipment is another major lubricant user. Industrial-lubricant applications are very diverse, including windmills, stationary hydraulic equipment, transformers, air compressors, gas turbines, and marine cylinder and crankcase systems.

Typical fully formulated lubricating oil is composed of a combination of mineral or synthetic base oils and specialized dispersed additives designed to improve long-term stability and enhance performance in aggressive environments. Any commercial lubricant can be viewed as a nonaqueous highly resistive polymeric colloid with low, primarily ionic electrical conductivity. This system is composed of nonpolar base oil ("continuous phase") and suspended

219

polar molecules and structures ("discontinuous phase"). The suspended phase includes both specialized oil additives blended into base oils to enhance the lubricant performance and polar contaminants such as soot, water, and oxidation products that ingress into the lubricant with age and use. The dipolar nature of nonaqueous lubricants and other industrial colloids permits investigation of their properties using high frequency impedance/dielectric analysis. There is some historic [3, 4, 5] and more recent [6, 7] literature on dielectric analysis of oils and lubricants.

However, significant and interesting adsorption, charge-transfer, and mass-transport activity has been reported at the electrochemical interface between a lubricant film and the metal surfaces of industrial and automotive equipment. A combination of high-frequency dielectric bulk response and interfacial electrochemical potential-modulated kinetics at the low frequency, which can be sampled by both linear and nonlinear EIS, makes lubricant systems an interesting practical and theoretical case study of broadband impedance-analysis application. Lubricants represent exactly the type of systems that are of primary interest in this book—moderately highly resistive samples with bulk-solution capacitive dispersions that can be studied by dielectric spectroscopy while displaying ionic conduction, interfacial adsorption, and charge and mass-transport phenomena that are primary application targets for DC- and AC-modulated low-frequency impedance analysis. The ability to combine electrochemical studies of the bulk-solution colloidal system with interfacial kinetics is a very interesting possibility well worth investigating.

In addition to purely scientific interest, a diagnostics of lubricant degradation is an issue of practical importance to automotive and industrial-equipment manufacturers and end users [8, 9, 10, 11, 12, 13, 14, 15, 16, 17, 18]. There are many chemical components and contaminants present in industrial lubricants that will contribute to the electrochemical properties of the media and may alter the impedance spectra. With this in mind the method of analysis of colloidal dispersions offers the opportunity to characterize engine lubricants and industrial fluids, determine relative effectiveness of oil additives based on their ability to protect the equipment, and provide insights into mechanisms of fluids performance and degradation. EIS has been used for general characterization of engine oils [8, 18], modeling nonaqueous colloidal dispersions [19], studies of lubricants' oxidation [6], monitoring oil degradation due to its contamination by glycol [20], water [21, 22], soot [23], and fuel [24], and development of various electrochemical monitoring devices for lubricants' performance diagnostics [20, 21, 22, 23, 24, 25, 26, 27]. EIS allows the monitoring of lubricant degradation, and by virtue of providing real time online information, keeps automobiles, trucks, and industrial equipment in top operational condition with minimal downtime for maintenance.

The AC impedance technique is compatible with both conventional and microfabrication technologies, allowing production of rugged devices for applications in the harsh, highly resistive, chemically complex, and often aggressive media of industrial lubricants and colloids [25]. EIS presents an opportunity to resolve complicated nonaqueous colloidal system both spatially and chemically, and can be applied to monitoring lubricant's degradation process.

The process of lubricant degradation is relatively slow, as it typically requires many weeks to significantly degrade a fully formulated industrial-grade lubricant. Therefore it can be assumed that the system is stable over a time frame of a single EIS measurement, and the signal changes between fresh and used lubricant are representative of chemical changes [6, 7]. This assumption holds for all types of oil degradation except for the case of a significant and instantaneous water leak.

High resistance of lubricant media mandates application of high-input AC voltage amplitudes (on the order of 1 volt) to produce measurable output currents and acceptable signal-to-noise ratios. Previous linear [6] and nonlinear [7] EIS analysis demonstrated the anticipated system nonlinearity due to the high amplitude of the input AC voltage signal, which was mostly present in the low frequencies. The nonlinearities in the current response can be analyzed by the higher-harmonics nonlinear EIS (NLEIS) method, leading to a better interpretation and quantification of the low-frequency impedance data and resulting in a development of more accurate equivalent circuit models. The reliability of experimental linear impedance data for a wide range of industrial lubricants and oils was validated by confirming its repeatability and by using Kramers-Kronig calculations to compare experimental and theoretical results [6].

As was indicated before, EIS results interpretation often presents a challenge as pattern-recognition problems emerge. One needs to use knowledge of physical processes occurring in the system before an adequate impedance-based model can be proposed. It is often useful to analyze the system using alternative analytical and physical methods to obtain an independent confirmation of the validity of the impedance-analysis results (Chapter 8). Experimental techniques available for chemical and physical characterization of lubricants and oils are limited and often somewhat unreliable, especially for in situ analysis. In spite of that, a number of previous publications established detailed and systematic correlations between EIS lubricant data and their physical and chemical properties [6, 7, 8, 9, 10, 11, 12, 13, 14, 15, 16, 17, 18, 19, 20, 21, 22, 23, 24]. These studies were supported by a knowledge of chemical composition and physical properties of industrial lubricants and by the responses of various segments of the EIS spectrum to variations in the lubricants' aging (degradation), chemical composition, electrochemical potential, temperature, and electrode geometry. The applicability of NLEIS analysis allowed further improvement in the modeling of the lubricant structure, in particular in the low-frequency range where mass- and charge-transfer interfacial processes occur [7].

This chapter presents a summary of the systematic EIS modeling of industrial and automotive lubricants. Initially EIS data interpretation for fresh lubricant, influenced by the effects of the chemical composition, temperature, electrochemical potential, AC frequency, and electrode geometry, is presented. Another important practical aspect is determination of changes in the lubricant's bulk and interfacial impedance model parameters as a result of its degradation by time-dependent oxidation and contamination with soot, fuel, and water at different stages in the exploitation cycle.

10.2. Physical and chemical properties of lubricants

Industrial and automotive lubricants are custom-formulated for various types of vehicle engines (depending on specified requirements from engine manufacturers and end users), drivelines (such as transmissions and gear) and industrial equipment (hydraulic, transformers, compressors) [1, 2]. It is not the intention here to cover the large variety of lubricant formulations in specifically chemical terms, as this area of science and technology is appropriately covered in lubrication-specific reviews and monographs. However, a snapshot of the most important lubricant formulation information in terms of its electrochemical properties is presented below.

Fully formulated lubricants are organic polymeric liquids representing a complex combination of base oils and specialized additives designed to improve long-term lubricant stability and enhanced equipment performance in aggressive exploitation conditions. There is some variability in base oils—which can be both mineral and synthetic but can also be differentiated on the basis of relative content of polar (often aromatic) vs. saturated hydrocarbons. The base oils are essentially insulators with resistivity $\sim 10^{11}$–10^{14} ohm cm. The characteristic dielectric constants (measured at ~ 1 MHz at 25 °C) of base oils are about ~ 2.2–2.3 for mineral and 2.2–2.5 for synthetic base oils, although some synthetic ester-based oils used in industrial lubricants can have dielectric constants as high as 3.5. Permittivity values for fuels are about ~ 2.1, although several percent variability may result from different crude exploration sources, refinery procedures, and levels of impurities.

Industrial and automotive lubricants can contain any number of additives depending upon the application [2]. Chemically and surface-active oil additives include many organic molecules with molecular weights between 100 and 50,000, such as dispersants, detergents (surfactants), viscosity modifiers, oxidation inhibitors, and antiwear agents. Individual oil additives are usually present in amounts between one and five percent of the complete lubricant formulation weight. Oil additives are typically composed of a polar functional group and a long nonpolar hydrocarbon "tail" that allows the additives to become compatible with nonpolar base oil. Polar functional groups of single nanometer sizes are usually composed of oxygen, nitrogen, sulfur, phosphorus, and earth metals. Hydrocarbon "tails" often reach a length of several nanometers to several dozen nanometers. Polar additives often exist in these solutions in the form of inverse micelles [28], with polar head groups forming the internal hydrophilic core and external hydrophobic hydrocarbon tails keeping micelles suspended in base oil. The surface charge on the micelle interface results in an electrical potential on the order of ~ 100mV. Inverse micelles of oil additives (such as detergents and dispersants) can further aggregate, forming a loose additive "network" of larger sizes [29]. Such additives networks typically provide better and longer-lasting lubrication protection properties.

The mechanism by which these additives perform in an engine is quite complex. Simply stated, detergents protect metal surfaces by forming protective films and by neutralizing the acidic products of combustion and thermooxidation. Detergents also actively interact with water, glycol, and other polar contaminants. "Overbased" detergents are typically ~ 500-5000 molecular-weight structures often composed of organic salts of earth and earth-alkaline metals forming the "core" of inverse micelles, surrounded by hydrocarbon nonpolar "tails." The key function of a high molecular-weight ($\sim$10000–50000) dispersant in automotive engine-oil formulation is to reduce oil thickening caused by accumulation and agglomeration of particulate byproducts of combustion and thermo-oxidation, such as soot. Dispersants stabilize colloidal dispersions and prevent contaminant particles from agglomerating. Dispersants and detergents suspend and neutralize contaminants in the bulk lubricant through association via both steric and electrostatic mechanisms. Polar groups of oil additives associate with polar contaminant particles, while the hydrocarbon tails act as physical, or steric, barriers against agglomeration and ensure suspension of the contaminants. The electrostatic mechanism of suspension stabilization is based on charges being created on the surface of the contaminant particles as a result of surface deposition of polar head groups of oil additives. These surface charges lead to mutual electrostatic repulsion of modified contaminant particles and further stabilization of the dispersion [16]. Finally, smaller ($\sim$100–1000 MW) polar molecules such as oxidation inhibitors reduce a lubricant's rate of oxidation, and antiwear agents provide protection against wear by forming protective adsorption films on metal surfaces [14].

Automotive lubricants are typically the most complex, as they are exposed to the most diverse sets of degradation pathways. They are designed to provide long-lasting protection against thermal oxidation as well as water, ethylene glycol coolant, fuel, and soot (for diesel engines) contaminants. Therefore, automotive oil formulations contain the full spectrum of lubricant components—base oil and high (up to 20 percent) combined amounts of all major additives—dispersants to isolate soot particles and prevent them from agglomerating, antioxidants to control oxidation, antiwear agents to provide anticorrosive adsorption films, and detergents to keep engines clean and control water and ethylene glycol contamination entering an engine through deteriorated seals and moisture condensation. As will be shown later, the abundant presence of polar and mobile detergents is the most important electrochemical aspect of lubricant formulation, as their approximate five to seven percent presence results in a decrease of base-oil bulk resistance by a factor of $\sim 10^3$–10^4. A combination of antiwear agents and detergents results in the development of adsorption film on metal interfaces with distinct charge-exchange properties. Typical automotive lubricants possess easily measured bulk impedance on the order of 1 Mohm and low-frequency interfacial impedance of a similar order of magnitude (at 100 °C in a cell with a ratio of surface area A to the electrode gap d of $\sim$100 cm^2/cm). The presence of large nonpolar polymeric molecules such as viscosity modifiers and pour point depressants introduces only superficial effects to the oils' electrochemical properties.

Industrial lubricant formulations are very diverse, reflecting a large variety of target applications. Some of them, such as transformer oils, have very little polar additives, resulting in Tohm levels of material resistance. Hydraulic fluids and air-compressor lubricants typically do not contain detergents and dispersants but have several percent antiwear agents. That results in bulk impedance parameter on the order of $\sim$1 Gohm (at 100 °C) but highly developed potential-dependent adsorption-driven low-frequency interfacial impedance.

Gear oils have a lower presence of polar detergents than automotive oils, resulting in bulk impedance values on the order of ~10–100 Mohm (at 100 °C). Fuels have a very minimal amount of specialized additives, resulting is no surface activity and high bulk impedance at Gohm levels. The danger of sparks and arcing leads to additions of subpercentage amounts of "anti-static" conductivity enhancers. Different levels of sulfur in the base fuel may cause slight (within a factor of two to five) variability in the bulk resistance.

10.3. Degradation modes of lubricants

During its life a lubricant undergoes substantial chemical changes due to oxidative high-temperature degradation and contamination by water, ethylene glycol, fuel, soot, and wear metals. In general the lubricant lifespan depends on initial lubricant quality, quality of the fuel used in the engine, operating environment, and engine design and exploitation conditions. Lubricants perform in equipment over long periods of time (usually several months to several years) without being changed while being subjected to extreme fluctuating temperatures that may go as low as –50 °C to as high as +250 °C, pressures up to 150 psi, constant shear, and equipment vibration. As the requirements for internal-combustion engines and industrial equipment become more stringent, lubricant formulations need to be modified to handle the increased requirements of the engine while still providing equivalent or better performance. There is significant pressure to understand and maintain lubricant life as a function of peak engine performance [15].

The details of degradation mechanisms are as complex and varied as the global engine platforms and the applications that the lubricants service. Industrial and automotive lubricants lose their functionality through high temperature-induced oxidative degradation and/or increasing presence of major contaminants, such as ethylene glycol [20], water [21, 22], soot [23], and fuel [24]. Lubricant degradation often also results from a combination of several interdependent pathways, which result in increased oxidation and nitration, depletion of total base number (TBN) and acid buildup, increased contamination, and viscosity reduction or increase [13]. Degradation of a lubricant by oxidation and contamination mechanisms causes loss of lubricity, loss in pumping ability due to oil thickening, worsening fuel economy, formation of deposits, and increased equipment wear and malfunction [17]. Ultimately, lubricants reach the end of their useful life, leading to deterioration of equipment performance and eventual equipment failure [8, 9].

The complex nature of lubricants and the variability in types of industrial equipment, automotive internal-combustion engines, transmissions, and gear make it difficult to foresee every possible pathway leading to failure. In everyday practice lubricants are changed after a given service time or engine mileage without prior testing; however, assumptions based on miles driven or hours of operation do not provide complete protection. Current methods used to determine lubricant quality, routinely performed by major engine and lubricant manufacturers, are frequent, repetitive, and time-consuming physical and chemical tests. These laboratory tests include determination of oil viscosity, acid and base numbers, insolubles (such as soot) and metals content, fuel and water percent dilution, and glycol contamination [14]. Validity and interpretation of the results of these tests are often ambiguous. Electrochemical methods are generally free from the difficulties associated with current industry testing methods and present an opportunity for a relatively quick, simple, and inexpensive approach, free of temperature limitations and sample preparation issues. Electrochemistry has been previously employed for ex situ and in situ lubricant conditions over the life of engine oils [8, 9, 10, 11, 23, 24].

Degradation by oxidation is a major process of oil deterioration, affecting all types of lubricants and applications. The normal life span of industrial lubricants is limited by the breakdown of the molecular chains at high temperatures as well as by additive depletion, resulting in increased oxidation/nitration of hydrocarbons measured by infrared spectroscopy and increased acidity determined by a TAN titration. It has been demonstrated that lubricants rapidly degrade as the concentration of antioxidant additives decreases. It is a widely accepted view that thermal processing of hydrocarbons in the presence of oxygen, as found in internal-combustion engines or other industrial equipment, leads to the formation of hydroperoxides as a primary oxidation product via a free-radical propagation mechanism catalyzed by metals and accelerated by heat [2]. The primary oxidation products undergo substantial degradation to form various oxidized polar low-molecular-weight compounds such as ketones, alcohols, carboxylic acids, and aldehydes. Lubricant acidity increases due to increasing amounts of carboxylic acids as well as accumulation of combustion products such as NO_x and SO_x. The oxidation products may eventually form higher molecular-weight structures, and further polymerization of oxidation byproducts may form sludge, varnish, and other oil-insoluble compounds, which significantly increase the viscosity of the medium, negatively affecting fuel economy. A viscosity increase above ~15–30 percent when compared with fresh lubricant is considered to be the point when the lubricant has to be replaced. Detergents play an important role in neutralization of carboxylic acids through solubilization, while antioxidants quench hydroperoxides and hence terminate the oxidation chain. Electrostatic potential of detergent micelles plays an important role in solubilization and neutralization of carboxylic acids.

Degradation by contamination includes the effects of soot, fuel, water, ethylene glycol, and wear metals. Soot is a major product of incomplete combustion and thermo-oxidation in diesel internal-combustion engines. Soot mostly consists of aggregated spheroid particles of carbon combined into an extended aromatic network [30, 31, 32, 33]. Characteristics of soot can vary, including the degree of crystallinity, the size of the primary particles, the number of aggregated particles, and the amount and type of functional groups on the surface of the particle [16]. Fuel composition, temperature, and pressure are primary factors in determining these soot characteristics.

Soot formation eventually leads to an increase in oil viscosity and accelerated engine wear. As soot concentration increases beyond a certain level (typically four to eight percent), the dispersants are unable to suspend it, resulting in formation of large soot agglomerates (~ 1 μm), which cause abrasive surface wear. Requirements for reduced diesel-vehicle emissions are leading to changes in engines, such as retarding the engine timing and exhaust-gas recirculation, which further increase soot generation and agglomeration [30].

Unless there are electrical and steric barriers to keep them apart, soot particles will agglomerate as they approach one another. The agglomeration is caused by both Van der Waals forces and electrostatic attraction of the charges present on the surface of the particle. The presence of varying functionalities such as carboxylic acid and phenolic groups on the surface of the soot particles provides locations for the polar head groups of dispersants to interact with soot [16]. Dispersants prevent agglomeration by adsorbing onto the suspended particle surface, providing steric hindrance and an electrostatic barrier to agglomeration [16]. This process results in equilibrium between dispersant micelles, suspended soot with attached dispersants, and a loose "network" of dispersant micelles [29], as illustrated in Figure 10-1 [19].

Fuel dilution on the order of four to five percent is another automotive-lubricant degradation mode [24], causing a decrease in viscosity due to oil thinning and an increase in acidity due to penetration of easily oxidizable low-molecular-weight fuel fragments into oil. The latter is particularly common in the case of biofuels, which often contain relatively easily oxidizable low-molecular-weight components. Biofuel dilution may result in both oil thinning and an accelerated viscosity increase due to fast oxidation of biofuel components and resulting formation of high-molecular-weight polymers.

Contamination by water and/or ethylene glycol caused by coolant leaks from deteriorated seals and moisture from condensation affects all industrial and automotive lubricants, in particular hydraulic equipment. If equipment is

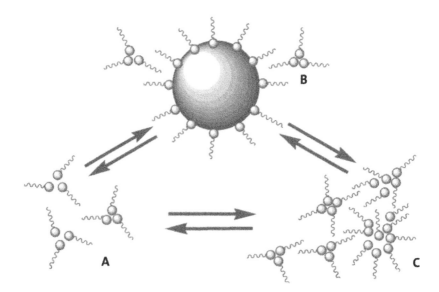

FIGURE 10-1 Dispersant equilibrium states: A. in inverse micelles with other dispersant molecules; B. coating a soot particle; C. as an associated network of dispersant micelles

operated for short periods of time, especially in cold weather, water and other combustion products can be condensed from the blow-by gases. Water interferes with the lubricating properties of the oil through affecting bulk viscosity and through its ability to permanently attach itself to working metal surfaces. Previous studies [21] demonstrated that water has strong affinity for metal interfaces and can be deposited there for long periods of time even at elevated temperatures. Water presence even in ppm amounts promotes wear by increasing the rate of surface corrosion. Major engine manufacturers require close monitoring of water presence, with amounts in excess of 2000 ppm (0.2 percent weight) commonly viewed as a symptom of danger leading to eventual engine operational problems.

Water interacts with oil through various mechanisms, leading to formation of inverse micelles (microemulsions) and free nonbound water [21, 22]. Detergents, considering their strong inorganic surfactant nature, have the highest potential of interacting with water. Detergent-water interactions lead to a formation of inverse micelles in a continuous hydrocarbon oil phase with detergent surrounding the water in microdroplets. A continuous exchange between free nonbound water and water contained inside the inverse micelles occurs over a significant time period [21]. At the same time water-oil balance can be affected by external factors, such as water evaporation at elevated temperatures, oxidative decomposition of oil additives, etc. These processes add an additional degree of complexity to the problem of water quantification in industrial oils and lubricants.

Wear metals are not really "oil contaminants" as much as they are indicators of excessive corrosive or abrasive wear taking place at the operating metal surfaces. Normally functioning undegraded lubricant prevents metal wear by providing the lubricating film between the moving parts. Corrosion of metal surfaces occurs due to their exposure to acidic byproducts of oil oxidation and due to the presence of water. Strongly chemisorbing antiwear agents provide protection against corrosive wear by forming protective adsorption films on metal surfaces.

10.4. Impedance analysis of lubricants

EIS presents an opportunity to resolve a complicated lubricant system both spatially and chemically and to analyze specific parts of that system based on relaxation frequencies. As was indicated above, there are formulation-driven differences that affect electrical properties of lubricants. In this chapter the impedance response of automotive engine oils that typically contain five to seven percent of polar and mobile detergents will be discussed. Automotive lubricants also have a more complex general composition, as they are formulated to respond to the widest range of possible degradation pathways. Hence, the general trends described below for the automotive lubricants impedance data are directly applicable to the other types of lubricants with less complicated formulation specifics and fewer possible degradation modes. For instance, industrial applications typically do not generate soot contamination process.

The impedance diagram of a typical automotive lubricant analyzed at ~100 °C can be separated into at least three regions, representing several independent relaxation phenomena: a very prominent high-frequency region (10MHz-10Hz), a medium-frequency region (10Hz -100 mHz), and a low-frequency region (100 mHz-1mHz) (Figure 10-2) [6]. Identification and characterization of these features, presented below, is based on initial knowledge of lubricants' chemical composition and physical properties. Another important piece of information is the response of the impedance parameters to changes in temperature, surface area of the electrodes, electrode separation distance, electrochemical potential, and degree of oxidative degradation and contamination of a lubricant.

The high-frequency impedance feature is representative of conduction and dielectric relaxation processes occurring in the bulk of nonaqueous colloidal systems [11, 12, 13]. In the bulk several lubricant components can be found, such as individual nanometer-sized free dipoles of oil additives and base-oil fragments, larger agglomerates of inverse micelles of various additives, and contaminants, both as relatively small ~100 nm particles and large soot and water agglomerates with sizes over 1 μm. The impedance plot of a typical lubricant depends on overall additives and contaminants amounts, degree of their agglomeration, and shape of the agglomerates.

Historically, the bulk lubricant has been studied by dielectric spectroscopy and interpreted according to the Debye relaxation theory [3, 4]. In impedance terms the system can also be represented according to a theory of colloidal dispersions or polycrystalline media composed of spheres of vastly different conductivities, where the contaminants become a more conductive phase suspended inside the less conductive additive/base oil matrix [6, 34]. Alternatively, when the contaminants are absent, the polar additives can be considered as a conductive discontinuous phase suspended inside insulating continuous base oil. Initially the description of the impedance representation of the fresh, uncontaminated oil will be provided, and then the effects of oxidative degradation and contaminants will be discussed.

The lubricant base oil (continuous-phase) can be represented by a pure capacitor C_{BASE} placed in parallel with a series of resistance and capacitance of

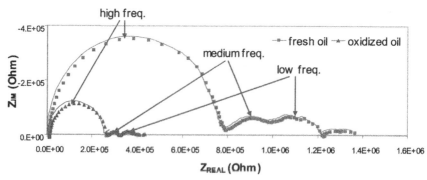

FIGURE 10-2 Impedance spectra for fresh and oxidized oils tested at 120 °C

polar oil additives. The base-oil resistance determined by electrophoretic mobility and abundance of base-oil ionic components should in principle be placed in parallel with the base-oil capacitor. However, this Gohm-Tohm range resistance element R_{BASE} can be largely disregarded, as the impedance of the capacitive segment of the circuit is always lower within the "bulk" oil frequency range.

The polar additives, in particular detergents present at several percentage levels, can be viewed as organic-inorganic macroions with electrophoretic mobilities that exceed those of the base-oil components by several orders of magnitude. As a result, characteristic resistance of polar oil additives $R_{FREE\ ADD}$ even at several percent concentrations is 10^3–10^5 times lower than that of the base oil. Characteristic dielectric constant values of the pure oil additives are in 2.2–4.0 range, similar to that of the base oils. The critical relaxation frequency for polar additives is in the MHz range, often outside of the analysis frequency range. The presence of the polar oil additives results in a significant decrease of the measured bulk-media resistance and has little or no effect on the measured capacitance. The measured dielectric constant of fully formulated fresh oil (containing both base oil and additives) is typically ~ 2.25, with small incremental increases observed when the frequency decreases. The classical Debye circuit that is typically represented by a Cole-Cole plot (permittivity diagram) is not the best description of a typical fully formulated lubricant, as no electrically active components with permittivity values dramatically different from those for the base oil are present in amounts significant enough to become noticeable capacitive current conduits. As will be shown later, at frequencies less than ~100–1000 Hz the interfacial charge-transfer kinetics and double-layer charging processes screen the bulk relaxations in lubricants with ionic conduction.

The high-frequency impedance response of a typical fresh lubricant reveals a single prominent capacitive semicircular feature with a relaxation time τ ~ 1 ms. This high-frequency feature at 100–120 °C can be represented by the bulk resistance R_{BULK} ~1–10 Mohm in parallel with the ideal bulk capacitance C_{BULK} ~20pF/cm^2 (the value of the parameter $\alpha = 1$) for cases where no contaminants are present (Figure 10-3) [3]. For fresh lubricant the bulk capacitance is primarily indicative of base-oil permittivity. The resistance strongly depends on the amount and type of polar additives present in the base oil, in particular ionic detergents. In a typical automotive lubricant, detergents make up five to seven percent of the oil volume content, and the measured bulk resistance decreases from that of the base oil (~10^{10} ohm at 100 °C) by three to four orders of magnitude. If detergents are absent, which is often the case for hydraulic and transformer oils, the bulk-material resistance remains in the Gohm-Tohm range.

The millisecond range-relaxation process is primarily driven by the aggregation of the inverse micelles of the oil additives (Figure 10-1). Networks of micelles exist in equilibrium with free oil additives, ions, and dipoles through diffusion-controlled ejection and recapture of surfactant molecules. The recapture is independent of surfactant type, but the exit rates increase with the shortening of the hydrocarbon tail. The interface between polar and nonpolar components of the surfactant network can become polarized, resulting in the

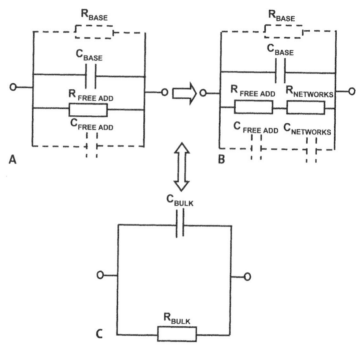

FIGURE 10-3   A. Temperature untreated; B. temperature treated, and C. simplified bulk media model for fresh lubricant. Dashed lines indicate secondary conduction pathways

formation of a weak double-layer structure somewhat similar to that observed in other macro-ions. The equilibrium between the free nanometer-sized additives and the "agglomerated networks" with sizes on the order of 500-1000 nm [29] significantly shifts towards the agglomerated structures after a single elevated temperature (~100–120 °C) treatment, essentially as soon as fresh lubricant becomes heated for the first time in an engine [24]. Fresh automotive lubricant before the first heat treatment typically has $R_{FREE\ ADD}$ ~ 100 kohm with a corresponding high microsecond relaxation time. Following the heat treatment, the formation of larger agglomerated networks composed of structures of generally lower electrophoretic mobility than free oil additives, ions, and dipoles occurs, resulting in higher $R_{NETWORKS}$ ~ 1–10 Mohm. The resulting low kHz range relaxation can be represented by a simple transition from the high-frequency capacitive base oil charging C_{BULK} to the low-frequency ionic conduction through resistor R_{BULK} limited primarily by the oil additives network resistance $R_{NETWORKS}$ (Figure 10-3).

The impedance analysis can be further refined by looking at the modulus representation of the bulk lubricant response. The modulus spectrum sometimes allows determining an additional MHz range relaxation with characteristic capacitance of ~20pF/cm² and resistance in high ohms to low kohms in "model lubricant systems" with abnormally large excesses (15 to 20 percent) of polar additive [19]. This MHz feature was identified as orientational polarization relaxation of free polar dipoles (or ions) of individual additives [35]. The absolute magnitude of the MHz modulus is dependent on the presence and the relative concentrations of polar species such as detergents and dispersants and the type and relative polarity of base oil. This relaxation is being observed due to vastly smaller characteristic resistances of "free" (non-networked) oil additives $R_{FREE\ ADD}$ when present at high concentrations. Small ionic additives usually carry a distributed electrical charge of one to two electrons/micelle and possess some dipole moment, therefore contributing to both capacitive and resistive characteristics, largely due to their relatively high electrophoretic mobility and small size. Individual dipoles have much lower resistance than the large agglomerates of inverse micelles, while the capacitive characteristics of both structures are similar. That explains the ability of the modulus diagram to separate the features on the basis of their very different resistance values in model systems with an abnormally high load of polar additive, while the impedance spectrum usually presents a single capacitive feature (Section 7-1). In realistic situations of fully blended lubricant containing 10 percent additives where the majority are tied up in "networks," the MHz orientation polarization of free dipoles usually cannot be detected.

Fresh lubricant bulk media can be represented by a single capacitive element C_{BULK}~20 pF/cm^2 (as capacitances of base oil, additive "free" ions, and networks of additives are very similar), and by as many as three vastly different resistive characteristics (R_{BASE} ~10^{10} ohm, $R_{NETWORKS}$ ~10^6 ohm, and $R_{FREE\ ADDITIVES}$ ~10^3 ohm, the last two in series with each other). The resistances of base fluid and free additives can be essentially disregarded, and the total ohmic resistance of the system can be represented by a single R_{BULK} element (Figure 10-3).

A migration-driven conduction process in the bulk is dependent on the geometry of the testing cell, concentration and type of charge carrying ionic components, temperature of the system, and the age (oxidation) of the oil, with little dependence on the applied electrochemical potential. With increase in the electrode surface area A and decrease of separation gap d, the characteristic resistance decreases (Eq. 5-3), and conductivity (Eq. 5-4) and capacitance (Eq. 1-3) both increase (Figure 10-4).

Increases in the testing temperature significantly reduce R_{BULK} (Figure 10-5) and increase media conductivity. Conductivity-temperature dependence in industrial lubricants is typical of polymeric systems with ionic conductivity and follows the Arrhenius (Eq. 5-11) or Vogel-Tammann-Fulcher (Eq. 5-12) dependencies [36].

The bulk resistance R_{BULK} and bulk capacitance C_{BULK} show only very minor dependence on the applied electrochemical DC potential with R_{BULK} increasing very slightly at high electrochemical potentials (3-12V) (Figure 10-6). This type of bulk-solution behavior in industrial colloids was reported earlier [37]. In colloidal systems the external electric field between parallel plate electrodes causes the buildup of the surface charge near the electrode, leading to Maxwell-Wagner interfacial polarization. The space charge diminishes the effective electric field in the bulk liquid and slightly increases R_{BULK} (Section 8-4). The diminished electric field causes charged particles far from electrodes to move slowly and appear to have lower electrophoretic mobility than they actually do. Polarization of electrodes is a problem in lubricants where there are

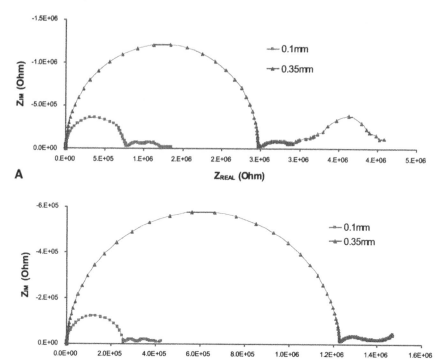

FIGURE 10-4   Impedance spectra of A. fresh and B. oxidized oils tested at 120 °C as a function of electrode separation gap

only relatively few electroactive species that can eliminate polarization through charge transport on electrodes [37].

Oxidation of lubricant leads to an eventual increase in capacitive (C_{BULK}) and a decrease in resistive (R_{BULK}) bulk characteristics as a result of the increase in relative concentration of oxidized ionic degradation products such as car-

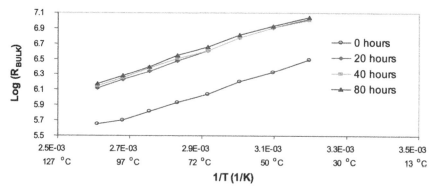

FIGURE 10-5   Bulk resistance vs. inverse of temperature for oil samples taken after 0, 20, 40, and 80 hours of lubricant testing

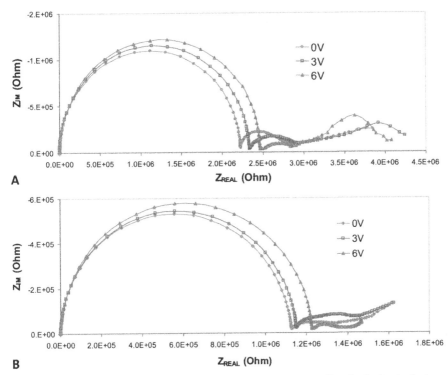

FIGURE 10-6 Impedance spectra of A. fresh and B. oxidized oils tested at 120 °C as a function of applied DC potential

boxylic acids, ketones, and aldehydes. It is interesting to note that these two processes occur at a very different rate and manifest themselves in a different magnitude of changes. As oxidation progresses, the agglomerated network of additives becomes oxidized, with formation of ionic highly mobile degradation products with characteristic capacitances similar to either base oil or polar additives but relatively high conductivities. Production of these polar byproducts of oxidation results in an increase in ionic conduction through the bulk and a decrease in the R_{BULK} but little changes in C_{BULK}. Even at very accelerated oxidation only a minimal change (2 to 5 percent increase) in C_{BULK} is observed, while the R_{BULK} values decrease by 50 to 90 percent.

Very advanced oxidation may eventually lead to the formation of high-molecular-weight polymeric agglomerates of oxidized hydrocarbons, which eventually reverses the R_{BULK} decrease trend. As the oil viscosity increases, ionic mobility decreases, resulting in a slight increase in R_{BULK}. Formation of additional polar structures (polymerization) of strongly oxidized agglomerated additives and oxidation of base oil, which occurs after the additives are "destroyed," results in an accelerated C_{BULK} increase, according to Eq. 7-7 [26].

As discussed above, the presence of soot and water greatly affects the bulk-lubricant region of the impedance diagram. In lubricant colloidal systems containing highly polar or ionic water, soot, and soot agglomerates, additional relaxations in the kHz region can be observed due to interactions of the polar contaminant "particles" with polar oil additives. Characteristic dielectric constants for the oil contaminants are 80 for water and ~3–15 for soot (depending on its compositional specifics, which are dependent on the exact mechanism of soot formation and are very difficult to predict or simulate). The contaminants (soot particles or water droplets) become "inserted" inside the additives network. Additives play the role of counterions to the contaminant particles, surrounding the particles and attempting to prevent the contaminants from coming in contact with one another and agglomerating. Under the influence of an external electric field, the counterions become redistributed along the surface of water agglomerate or soot particles, and the double layer becomes deformed and polarized, leading to interfacial polarization and the resulting relaxation or dispersion (Figure 10-1). This kHz relaxation has been explained according to both Maxwell-Wagner (interfacial polarization between the contaminant particles and the oil phase) and Schwarz (diffusion of dispersant counterions around the contaminant particles) mechanisms [19]. This contaminant-related millisecond range relaxation often overlaps with the kHz relaxation due to the "networks" of oil additives and usually cannot be clearly resolved unless there is a significant amount of contaminants present in the system (Figure 10-7).

For instance, due to soot-additives interactions, an additional polar component represented by $R_{SOOT} | C_{SOOT}$ that can be placed either in series or in parallel with $R_{BULK} | C_{BULK}$ is introduced into the system, and a corresponding kHz range relaxation appears in the impedance data. This additional feature is difficult to separate from the main kHz bulk-lubricant relaxation when soot concentration is under five percent. With an increase in the presence of soot, this kHz bulk-impedance feature turns into a depressed semicircle reflecting two time constants with somewhat similar values. Total high-frequency capacitance and dielectric constant increase, according to the Bruggeman and Schwarz equations (Eqs. 7-1 through 7-7). For low percentage concentrations of contaminants and minimal or no agglomeration, the molar share Φ of polar contaminants (soot and water) can be estimated from a high-frequency permittivity measurement and Eq. 7-4, $\varepsilon_{HF} = \varepsilon_{OIL}(1 + 3\Phi)$. With further additions of contaminants (four to five percent of soot, over one percent of water), the additives lose their ability to control the contaminants. Both water and soot start forming large polar agglomerates. Soot agglomerates behave like elongated clusters of conductive particles oriented in the direction of preferred current flow lines [34]. The α_A and δ_A coefficients in Eq. 7-7 become much larger than unity, resulting in significant deviation from the original linear relationship between ε_{HF} and Φ [22, 23]. With an increase in the presence of contaminants, the kHz bulk-impedance feature eventually separates into several frequency-resolved semicircles representing soot-additive interactions in the upper kHz range, oil-additives network in the low kHz range, and elongated soot aggregates in the low Hz range (Figure 10-7).

The equivalent circuit diagram representing bulk-lubricant transition from fresh, uncontaminated oil to a solution containing both dispersed and aggregated soot can be represented by a Debye-type circuit (Figure 10-8). Contaminated oil contains several conducting segments with large variability in their characteristic capacitive values. According to Eq. 7-8, high-frequency

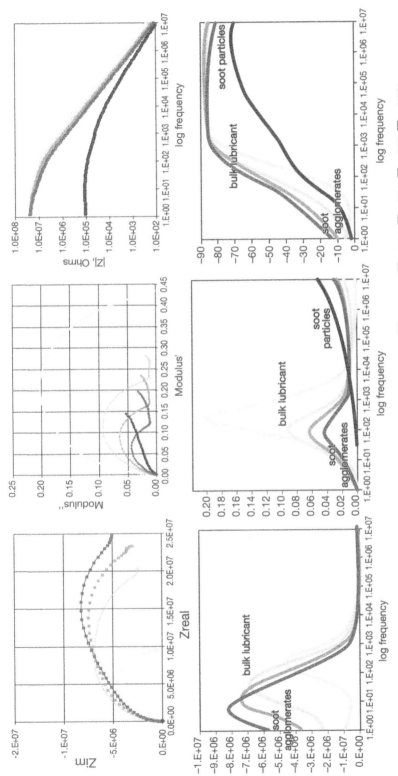

FIGURE 10-7 Evolution of lubricant EIS data with increase in soot content: ☐ 0 = .3%; ▨ = 5%; ▨ = 7%; ▨ = 9%; ▨ = 11%

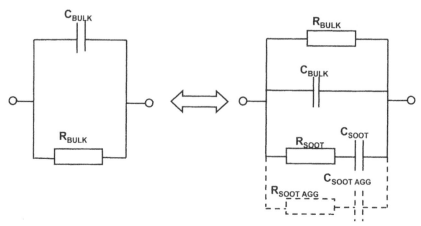

FIGURE 10-8 Evolution of equivalent-circuit diagram for lubricant with increase in soot concentration: A. fresh lubricant; B. lubricant with soot; dashed lines indicate appearance of conduction path through soot agglomerates at higher soot concentrations

capacitance C_{BULK} and low-frequency conductivity increase severalfold. The impedance relaxation for a system with significant agglomeration corresponding to the elongated soot aggregates can be resolved in the Z*-plot and M*-plot, demonstrating that agglomerated contaminants, dispersant clusters, and dispersant dipoles possess not only very different resistive but also very different capacitive properties (Figure 10-7). Further growth in the presence of the contaminants leads to formation of more elongated conductive chains, shortening the circuit with a dramatic decrease in the measured bulk resistance. At that point conduction occurs along the agglomerated clusters, which can be represented by a resistance-capacitance series $R_{SOOT\,AGG} - C_{SOOT\,AGG}$ or parallel combination, depending on the exact mechanism of their formation. In principle, if the chains of conducting particles are complete between the two electrodes, the current flows along the particle surfaces and becomes mostly resistive. If the chains are interrupted, than the base oil separating the neighboring clusters can be represented as a capacitive element in series with the resistance of the clusters.

The dynamics of lubricant interaction with added water is even more complicated. In addition to high-frequency permittivity changes described by Eq. 7-7, the entire impedance spectrum undergoes a complicated pattern of time-dependent changes. The low-frequency impedance changes caused by water will be discussed in a later section of this chapter. In the bulk solution a complex kinetics of water-oil interactions occurs, combining several mutually dependent processes of emulsification of free nonbound water, formation of inverse micelles, and evaporation of free and micellated water [21].

When compared with base oil and oil additives, water is polar and highly mobile, capable of percolating through bulk lubricant. After its introduction into oil, water forms free agglomerates that can support relatively facile ionic conduction, causing an immediate decrease in bulk resistance. As a result of higher operational temperature of the equipment, mixing, and interaction with the oil detergents, these water agglomerates break up and decrease in size over time as water is evaporated and emulsified into inverse micelles. Mutual immobilization of water and ionic detergents as a result of water entrapment inside inverse micelles occurs. Free nonemulsified water remains in equilibrium with the water entrapped inside the inverse micelles. The micelles mobility and electrochemical activity are about 100 times less than those for either free water or detergents [21]. Hence, when even a minor (less than one percent) amount of water is present in oil, detergents are being consumed to isolate water clusters, and the bulk resistance R_{BULK} increases as a function of the effective decrease in both free water and ionic detergents in the system. If additional water is introduced into the system, the changes in R_{BULK} will depend on the relative presence of nonbound free water vs. emulsified water consumed inside inverse micelles. If water is introduced continuously at a rate that surpasses the rate of emulsification, R_{BULK} will rapidly decrease and the circuit will eventually be shortened through water agglomerates just as in the case of agglomerated soot (Figure 10-8). Otherwise, when water is present at subpercent concentrations, R_{BULK} will increase, and a minimal increase in ε_{HF} and C_{BULK} (Eq. 7-4) may be detected only by using the most accurate equipment in tightly controlled experimental laboratory conditions [21].

Fuel presence in automotive lubricants very often leads to a significant decrease in R_{BULK} and a small decrease in C_{BULK}. The addition of fuel to oil increases separation among the oil additive components of the "networks," effectively breaking up the "aggregated networks" of oil additives responsible for the kHz relaxation and leading to a slight decrease in the corresponding capacitance. This process leads to a formation of smaller mobile macro-ions, resulting in higher conductivity and lower R_{BULK}. In addition to that, fuel presence in the oil leads to a decrease in R_{BULK} because of resulting lower oil viscosity ("oil thinning"), allowing for higher mobility of conducting macro-ions and individual dipoles, according to Eq. 5-14. Accelerated oxidation of low-molecular-weight fuel components present in oil results in acidification of the oil media and accelerated breakup of detergents, leading to further increase of ionic content in the bulk oil, higher solution conductivity and lower bulk resistance R_{BULK}.

The medium-frequency-region (10Hz–100mHz) impedance feature is related to charge accumulation due to adsorption of surface-active lubricant additives at the electrode surface. Many oil additives, in particular antiwear agents and some detergents, are extremely surface-active, providing enhanced lubrication by rapidly forming stable adsorption films on metal surfaces [2]. Modeled as a series or a parallel combination of current-limiting resistor R_{ADS} with a CPE_{ADS} representing adsorption pseudocapacitance, adsorption impedance Z_{ADS} dominates the circuit in the 10Hz-100mHz impedance region [6].

This feature is represented by a depressed capacitive semicircle with a relaxation time on the order of 1s. Typical characteristic parameters determined at 100–120 °C for a variety of lubricants are R_{ADS} ~100kohm and capacitance values of ~1µF/cm^2 with the α parameter being between 0.5 and 0.9. Aging of the lubricant causes a slow decrease in the measured resistance and an increase in the measured capacitance (Figure 10-2). R_{ADS} decreases and CPE_{ADS} increases with an increase in electrode surface area (while the effective adsorption capacitance was found to be independent of the electrode separation gap), which is in agreement with Eqs. 5-37 and 5-38 (Figure 10-4). Both CPE_{ADS} and R_{ADS} showed multidirectional temperature variability. With temperature increase, the adsorption capacitance often increases and resistance decreases due to the acceleration of adsorption kinetics [38] and increase in the amount of adsorbed species Γ (Section 5-5). Surface coverages in the range of 10^{-9} to $2*10^{-10}$ mol/cm^2 and the adsorption rate kinetic constants on the order of 10^{-8} to 10^{-11} cm/s were reported for different lubricants and colloids [21]. Values of this magnitude are expected for chemicals in highly resistive lubricant as calculated from the experimental charge-transfer resistance data [6].

The adsorption phenomenon in lubricants is rapid compared to the processes of mass and charge transport. Based on the relative rates of adsorption and desorption of lubricant additives and the presence and rates of the subsequent mass- and charge-transport processes, it is theoretically possible for R_{ADS} and C_{ADS} to become negative and show inductive behavior [40, 41, 42]. However, this phenomenon is very rarely observed in lubricants, as the reaction rates of adsorption significantly exceed the rates of desorption and charge-transfer processes (Eq. 7-64).

Adsorption processes in lubricants typically show non-linear dependence on DC voltage, as adsorption/desorption may occur abruptly at given potentials. For lubricants an increase in electrochemical potential often does not affect the medium-frequency resistance but may sometimes gradually significantly increase its capacitance (Figure 10-6). The resistance R_{ADS} corresponds to charging pseudocapacitance through an electron-transfer process that is typically sluggish. In polymers electrochemical oxidation and aging give rise to a substantial charge accumulation on the interface without constant current. Capacitive semicircles corresponding to adsorption processes are often depressed [39]. This dispersion results from electrode inhomogeneity and surface roughness, variability in thickness of surface films and coatings, slow and uneven adsorption processes, and nonuniform potential and current distribution. In lubricants some of these processes become more prominent at higher DC potentials. Consequentially, the α parameter decreases from 0.9 to as low as 0.5 with an increase in the DC potential and oil aging.

Experimental and modeled data allow simplifying lubricant adsorption kinetics to the effect of two types of surface adsorption-driven processes [7]. Although lubricants may have a large spectrum of surface-active fresh and degraded components with a wide range of rate constants, it was shown that the total number of different adsorbing species can be limited to only two groups. Different groups of surfactants and their breakdown products often have similar electrochemical properties, such as mobility, molecular size, diffusion coefficients, etc. [7]. Therefore, one process is related to the adsorption of at least two different types of surface-active oil additives at the interface, which were assumed to be low-molecular-weight ionic antiwear agents and higher-molecular-weight detergents. The second type of adsorption process is related to temperature-driven interfacial reactions of these surface-active oil additives, resulting in the formation of adsorbed high-molecular-weight complexes. These conclusions are in agreement with generally accepted views of the interactions involving the complexation of surface-active lubricant additives (detergents and antiwear agents) at the metal inter-faces responsible for surface protective properties of the lubricants [2]. The individual adsorbed species and their complexes ultimately oxidize and de-grade, with a corresponding decrease in total and imaginary adsorption im-pedances as smaller, more polar-adsorbed species are formed. It is also likely that more complex adsorption processes occur, such as multiple-component adsorption, adsorption with incomplete charge transfer, and interactions of various adsorbing species forming multiple adsorption layers.

The low-frequency-region (100mHz–1mHz) impedance response accounts for diffusion and charge-transfer processes at the electrode-solution interface [25]. This impedance feature is represented by one or two depressed capacitive semicircles or sometimes a combination of a depressed capacitive semicircle and a Warburg-type "transmission" line (Figure 10-2). The Warburg-type line often reveals $-45°$ phase angle, but sometimes the angle is less than that, termi-nating in a depressed semicircle. In the case of two separate semicircles, the first semicircle typically has a relaxation time of ~50s, and the second one a relaxation time of ~200s. Characteristic parameters of these features, deter-mined at 100–120 °C for a variety of lubricants, are resistance values of 100 and 10 Mohm, capacitance values of ~5µF/cm^2 and ~20µF/cm^2, and α parameters between 0.4 and 0.9, respectively. A constant phase element (CPE_{DL}) represent-ing the double-layer capacitance is included in models of low-frequency im-pedance region [6]. In lubricants the electrochemical double-layer has capaci-tance values on the order of ~1 µF/cm^2, which is lower than those in aqueous systems due to lower permittivity and thicker compact layer [39].

Increases in electrode surface area and decreases in the spacing between the electrodes (Figure 10-4), increases in temperature, and oxidative degrada-tion (Figure 10-2) all cause decrease in the resistive and increase in the capaci-tive characteristics of the low-frequency impedance region. Changes in the separation gap affect the shape of the low-frequency features, with very small gaps frequently causing $-45°$ Warburg transmission lines to change into semi-circular features. Increase in the electrochemical potential (Figure 10-6) causes a decrease in impedance by several orders of magnitude, especially at the low-est frequencies.

Diffusion resistance (Z_{DIFF}) to mass transport in the system is represented as semi-infinite or finite Warburg impedance, depending on the assumptions about the thickness of the diffusion layer and its boundary conditions, with reactions or trapping giving rise to various diffusion-limited structures, as was presented in Section 5-6 [43]. In lubricants both absorbing and reflecting boundary cases of finite diffusion may exist. The nature of the encountered boundaries is dependent on the chemical composition of the lubricant system and experimental conditions. Deposition of nonconductive films of lubricants and their degradation byproducts may result in the appearance of a "reflecting boundary" condition and the appearance of a –90° capacitive line following the initial $-45°$ semi-infinite diffusion feature [39]. If the number of diffusing particles is not conserved [43], diffusion with an "absorbing boundary" takes place, and the system is represented by a parallel combination of resistance and a CPE giving a distorted semicircle at low frequency, following the initial rising 45° response [44, 45, 46].

The critical frequency for finite diffusion ω_c is determined as $\omega_c = D / L_D^2$ where D is the diffusion coefficient, and L_D is the thickness of the diffusion layer [44]. Typical diffusion coefficients for oil additives are $\sim 10^{-8} - 10^{-9}$ cm^2/s, and the diffusion-layer thicknesses are on the order of $10^{-2} / 10^{-4}$ cm, leading to critical frequencies of ~ 0.1Hz to 1mHz [6, 7]. The concentrations (C_O, C_R) and the diffusion coefficients (D_O, D_R) of oxidized and reduced species can be determined from the experimental data (Section 5-6).

The last portion of the low-frequency impedance model includes a charge-transfer resistance (R_{CT}), a nonlinear element controlled by a Volmer-Butler relationship related to the distance between the species and the electrode, and the activation energy of electrochemical reactions (Section 5-4). This parameter describes all electron exchanges not covered by the adsorption-driven electron-exchange mechanism. Experimentally, R_{CT} is very dependant on the electrode potential (Figure 10-6) and the oil's state of degradation (Figure 10-2). Molecules composing a typical lubricant have sluggish charge- and mass-transfer rates due to their large size, low mobility, diffusion coefficients of 10^{-8} to 10^{-12} cm^2/s, small ionic charges of large organic molecules, and delocalization of electrical charges [6]. Electron transfer in and out of micelles at the interface separating adsorbed micelles and solid electrodes is very slow due to steric hindrance caused by a surfactant layer several nanometers thick [21, 28] and occurs at the very lowest frequencies even at elevated temperatures. Typically the value for this resistance, corresponding to a relaxation near 1 mHz, can be experimentally determined only at high temperatures [6]. If there is no electron transfer, R_{CT} is very large, and the electrode is polarizable with a poorly defined potential. If there is a reaction, R_{CT} becomes smaller and connects in parallel with the capacitance of the double layer C_{DL}.

Major lubricant contaminants have different effects on medium- and low-frequency impedance characteristics. For example, no evidence of impedance changes caused by soot or fuel has been reported. On the other hand, the presence of even ~ 0.1-0.2 percent water strongly affects the interfacial (medium- and low-frequency) ranges of the impedance spectrum [21]. Immediately following the subpercent water injection the interfacial real impedance virtually disappears, and a large increase in low-frequency capacitance is observed (Figure 10-9).

A short-term electrophoretic-type separation of water and hydrocarbon oil occurs in the external electric field in the vicinity of the electrode surface. Under even a moderate electric field the mass transport of small water dipoles with dielectric constant of ~ 80 is much more facile than that of polymers with dielectric constant of ~ 2. High local concentration of water dipoles may be created in the vicinity of the electrode interface as a result of water redistribution and percolation in an external electric field. Relatively high water conductivity and the facile nature of charge-transfer reactions carried out through water dipoles located in the interfacial region explain the dramatic decrease in the low-frequency impedance immediately following the water injection. Deposition of conductive layers of water on the electrode/lubricant interface and significant presence of water dipoles in the diffusion layer replace both specifically adsorbed and electroactive lubricant additive species at the lubricant-electrode

INTERFACE

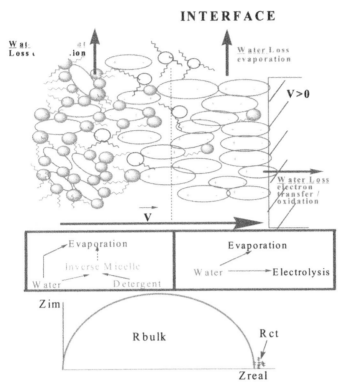

FIGURE 10-9   Water-oil separation several hours after the water leak with representative Nyquist diagram

interface. Rapid decrease of the low-frequency impedance is therefore a valuable diagnostic tool for very sensitive detection of ppm amounts of water.

If the water is introduced continuously at a rate that surpasses the rate of emulsification, the interfacial impedance will continue to remain extremely low. Contribution of adsorbing film lubricant additives to charge-transfer processes becomes virtually nonexistent until the water is depleted from the surface due to electrolysis, emulsification, and evaporation [21]. Negative effects of a water leak into automotive and hydraulic lubricants, such as impairment of the lubricant film, oil-additive precipitation, formation of water pockets on the surfaces, and accelerated corrosion have all been previously reported [2].

Long-term water-detergent (surfactant) interactions in systems where the water leak into the oil was stopped (or limited water presence occurred due to humidity condensation, which subsides when the oil becomes hot during the equipment exploitation) result in the formation of inverse micelles. It was also shown that the electrochemical electron-transfer reactions (either as direct redox charge transfer or through adsorption-mediated processes) for separated detergent and water are several orders of magnitude faster than the electrochemical activity of water/detergent inverse micelles produced through emulsification. Emulsification results in relatively electrochemically "passive" inverse micelles forming layers on the surface of the electrode, suppression of more facile electron-exchange reactions involving pure water or detergents, and gradual increase in low-frequency impedance in systems after the water ingress into the oil is discontinued. This explanation supports the properties of detergents as corrosion-inhibiting water-adsorbing agents [21].

10.5. Equivalent-circuit model of lubricants

An equivalent-circuit model describing the impedance response of a typical lubricant has to take into account processes divided spatially into bulk lubricant and interfacial electrochemical (adsorption, mass transport, and charge transfer) zones [6]. Separation between these two zones can be achieved following the distributed impedance analysis in Section 6-3. The bulk resistance R_{BULK} represents a lossy part of the overall bulk-media relaxation mechanism [3, 4], always existing in parallel with a bulk-media capacitance C_{BULK} as:

$$Z_{HF} = \left(\frac{1}{R_{BULK}} + j\omega C_{BULK} \right)^{-1}$$

(10-1)

A typical lubricant's impedance diagram reveals a single well-defined semicircle with a millisecond time constant [6]. More involved analysis of the impedance data and correlation with known chemical composition of lubricants demonstrated that the bulk region in fact is composed of a serial combination of two parallel $R \,|\, C$ circuits, although the MHz relaxation due to orientation polarization of small dipoles is invisible under most practical circumstances and can be essentially disregarded. The impedance profile in this region becomes more complicated when polar contaminants (soot and water) are introduced, presenting an opportunity to study this region as a complex composite medium (Figure 10-8).

The interfacial zone consists of adsorption, diffusion, and charge-transfer processes of oil additives and contaminants. A linear EIS [6] and NLEIS analysis [7] have led to a refinement of this model, taking into consideration the best fit of kinetic parameters to both nonlinear and fundamental-frequency experimental impedance data. The low-frequency impedance data are modeled as a combination of a CPE_{DL} representing double-layer charging processes, adsorption (Z_{ADS}), diffusion (Z_{DIFF}), and charge-transfer resistance (R_{CT}) impedance segments.

Adsorption impedance Z_{ADS} may be represented by a parallel combination of resistance R_{ADS} and CPE_{ADS}, which represents pseudocapacitance created by charge accumulation at the electrode surface. The adsorbed species become involved in complex surface-mediated kinetics, such as temperature-driven complexation of the surface-active oil additives [7]. The experimental and modeled data confirmed the presence of two facile adsorption-driven processes—adsorption of at least two different groups of surface-active oil additives (likely to be detergents and antwear agents), and their mutual complexation.

Mass-transport impedance Z_{DIFF} also includes both capacitive and resistive parameters. Specific arrangement of these parameters in finite or semi-infinite diffusion impedance depends on the nature of diffusion boundaries. Because the diffusion hinders the discharge of adsorbed species, the adsorption and diffusion components could be placed in series. Consumption of diffusing species at the electrode surface through adsorption-driven phenomena is facile compared to the rate of their replenishment through diffusion. The adsorbing species change their chemical substance fast while reacting with the metal surface and producing full or partial electron exchange. This type of kinetics usually leads to a condition where excess concentration of diffusing species is instantly drained out, resulting in the finite absorbing boundary condition for a small electrode gap. At the lowest frequencies a sluggish charge-transfer resistance R_{CT}, related to the electrochemical reactions involving all electron exchanges not covered by the adsorption-driven electron-exchange mechanism, is placed in parallel with adsorption-diffusion and C_{DL} portions of the circuit. In principle there should be another Warburg diffusion element in series with R_{CT} to account for impedance to mass transport of charged nonadsorbed species (which likely are dispersants and friction and viscosity modifiers) to the electrode. However, this diffusion element should appear at an even lower frequency range than the charge-transfer impedance (μHz), clearly outside of typical experimental measurement range. As a fair degree of 2D and 3D dispersions in low-frequency capacitance are anticipated, the capacitive part of the double layer can be represented by a CPE_{DL}.

In the "best-fit" model, the low-frequency portion of the impedance was represented by a series of complex adsorption and finite transmission boundary diffusion, placed in parallel with the charge transfer and the double layer CPE_{DL}:

$$Z_{LF} = \left(Q(j\omega)^{\alpha} + \frac{1}{R_{CT}} + \frac{1}{Z_{DIFF} + Z_{ADS}} \right)^{-1} \qquad (10\text{-}2)$$

The Z_{LF} circuit is then placed in series with the bulk circuit (Figure 10-10). The developed model is in good agreement with the general representation of

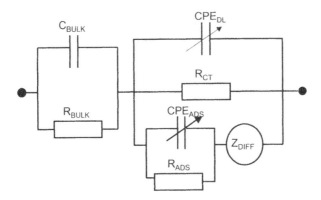

FIGURE 10-10 Equivalent-circuit model of lubricant

polymeric and organic colloidal bulk-solution structures reported in publications based on impedance and dielectric studies [3, 4, 5].

References

1. R. M. Mortier, S. T. Orszulik, *Chemistry and technology of lubricants*, Blackie Academic & Professional, London, 1992.
2. S. Q. A. Rizvi, *Lubricants and aubricant additives*, Lubrizol Corp., 1995.
3. K. Asami, *Evaluation of colloids by dielectric spectroscopy*, HP Application Note 380-3, 1995, pp. 1–20.
4. T. Hanai, *Electrical properties of emulsions* Emulsion Science, P. Sherman (Ed.), Academic Press, London, 1968, pp. 467–470.
5. G. Schwarz, *A theory of the low-frequency dielectic dispersion of colloidal particles in electrolyte solution*, J. Phys. Chem.,1962, 66, pp. 2636–2642.
6. M. Smiechowski, V. Lvovich, *Impedance characterization of Industrial lubricants*, Electrochim. Acta, 2006, 51, pp. 1487–1496.
7. V. F. Lvovich, M. F. Smiechowski, *Non-linear impedance analysis of industrial lubricants*, Electrochim. Acta, 2008, 53, pp. 7375–7385.
8. S.S. Wang, H.S. Lee, D.J. Smolenski, *The development of in situ electrochemical oil-condition sensor*, Sens. Act. B, 1994, 17, pp. 179–185.
9. S. Wang, *Road tests of oil condition sensor and sensing technique*, Sens. Act. B, 2001, 73, pp. 106–111.
10. Lee, H-S., Wang, S. S., Smolenski, D. J., Viola, M. B., and Klusendorf, E. E., *In situ monitoring of high-temperature degraded engine oil condition with microsensors*, Sens. Act. B, 1994, 20, pp. 49–54.
11. R. Kauffman, *Development of a remaining useful life of a lubricant evaluation technique. Part III: cyclic voltammetric methods*, J. Soc. Tribol. Lubr. Eng., 1988, 44, 6, pp. 709–716.
12. R. E. Kauffman, *On-line and off-line techniques for monitoring the thermal and oxidative degradations of aircraft turbine engine oils. Part 1:lLaboratory evaluations*, Lubr. Eng., 1995, 11, pp. 914–921.
13. A. Agoston, C. Otsch, B. Jacoby, *Viscosity sensors for engine oil condition monitoring—application and interpretation of results*, Sens. Act. A, Physical, 2005, 121, pp. 327–332.
14. M. Johnson, S. Korcek, R. Djensen, A. Gangopadhyay, E. Soltis, *Laboratory assessment of the oxidation and wear performance capabilities of low phosphorus engine oils*, SAE Technical Paper 013541, 2001.
15. T. Smith, V. Kersey, T. Bidwell, *The effect of engine age, engine oil age and drain interval on vehicle tailpipe emissions and fuel efficiency*, SAE Technical Paper 013545, 2001.
16. R. E. Kornbrekke, P. Semanic, T. Kirchner-Jean, M. Galic, E. A. Bardasz, *Understanding soot mediated thickening part 6: base oil*, SAE Paper 982665, 1998.
17. D. Cipris, A. Walsh, and T. Palanisamy, *Sensor for motor oil quality*, The Electrochemical Society Proceedings Series, D.R. Turner (Ed.), PV 87-9, Pennington, NJ, 1990, p. 401.
18. V. Lvovich, F. Boyle, *Method and apparatus for on-line monitoring of quality and/or condition of highly-resistive fluids*, US Patent 6577112 B2, 2003.
19. M. F. Smiechowski, V.F. Lvovich, *Characterization of non-aqueous dispersions of carbon black nanoparticles by electrochemical impedance spectroscopy*, J. Electroanal. Chem., 2005, 577, pp. 67–78.
20. S.S. Wang, H.-S. Lee, *The application of A C impedance technique for detecting glycol contamination in engine oil*, Sens. Act. B, 1997, 40, pp. 193–197.

21. M. Smiechowski, V. Lvovich, *Electrochemical monitoring of water-surfactant interactions in industrial lubricants*, J. Electroanal. Chem., 2002, 534, pp. 171–180.

22. B. Jakoby, M. J. Vellekoop, *Physical sensors for water-in-oil emulsions*, Sens. Act. A, 2004, 110, pp. 28–32.

23. F. P. Boyle, V. F. Lvovich, B. K. Humphrey, S. A. Goodlive, *On-board sensor systems to diagnose condition of diesel engine lubricants—focus on soot:*, SAE paper 2004-01-3010, 2004.

24. F. P. Boyle, V. F. Lvovich, H. Leye, *On-board diagnosis of lubricant performance condition—a better way to make maintenance decisions*, 2005 Annual Meeting of Society of Tribologists and Lubrication Engineers, Las Vegas, May 16, 2005.

25. V. Lvovich, M. Smiechowski, C. C. Liu, *Optimization and fabrication of planar interdigitated impedance sensors for highly resistive non-aqueous industrial fluids*, Sens. Act. B, 2006, 119, pp. 490–496.

26. V. F. Lvovich, D. B. Skursha, F. P. Boyle, *Method for on-line monitoring of quality and condition of non-aqueous fluids*, US Patent 6861851, 2005.

27. M. Smiechowski, V. Lvovich, *Iridium oxide sensors for acidity and basicity detection in industrial lubricants*, Sens. Act. B, 2003, 96, pp. 261–267.

28. J. F. Rusling, *Electrochemistry in micelles, microemulsions, and related microheterogeneous fluids*, Electroanalytical Chemistry, A Series of Advances, A. J. Bard (Ed.),Vol. 18, Marcel Dekker, New York, 1994, pp. 1–88.

29. C. A. Bearchell, J. A. Edgar, D. M. Heyes, S.E. Taylor, *Dielectric spectroscopic and molecular simulation evidence of aggregation of surfactant-stabilized calcium carbonate nanocolloids in organic media*, J. Colloid Interface Sci., 1999, 210, pp. 231–240.

30. J. A. McGeehan, *New diesel engine oil category for 1998: API CH-4*, SAE Paper 981371, 1998.

31. M. A. A. Nazha, and R. J. Crookes, *Design and of a high-pressure combustion system for study of soot formation*, SAE Paper 922206, 1992.

32. M. Kawamura, T. Ishiguro, H. Morimoto, *Electron microscopic observation of soots in used engine oils*, Lubrication Engineering, 1987, 7, pp. 572–575.

33. J. B. Donnet, R. C. Bansa, M. J. Wang, Carbon Black—Science and Technology, Second edition, Marcel Dekker, 1993.

34. D. G. Han, G. M. Choi, *Simulation of impedance for 2-D composite*, Electrochim. Acta, 1999, 44, pp. 4155–4161.

35. G. Schwarz, *Dielectric relaxation of biopolymers in solution*, Adv. Mol. Relaxation Processes,1972, 3, pp. 281–295.

36. H. Nakagawa, S. Izuchi, K. Kuwana, Y. Nikuda, Y. Aihara, *Liquid and polymer gel electrolytes for lithium batteries composed of room-temperature molten salt doped by lithium salt*, J. Electrochem. Soc., 2003, 150, 6, pp. A695–A700.

37. R. E. Kornbrekke, I. D. Morrison, T. Oja, *Electrophoretic mobility measurements in low conductivity media*, Langmuir ,1992, 8, pp. 1211–1217.

38. V. Lvovich, M. Smiechowski, *AC impedance investigation of conductivity of industrial lubricants using 2- and 4-electrode electrochemical cells*, J. Appl. Electrochem., 2009, 39, 12, pp. 2439–2452.

39. A. A. O. Magalhaes, B. Tribollet, O. R. Mattos, I. C. P. Margarit, O. E. Barcia, *Chromate conversion coatings formation on zinc studies by electrochemical and electrohydrodynamic impedances*, J. Electrochem. Soc., 2003, 150, 1, pp. B16–B25.

40. M. V. ten Kortenaar, C. Tessont, Z. I. Kolar, H. van der Weijde, *Anodic oxidation of formaldehyde on gold studied by electrochemical impedance spectroscopy: an equivalent circuit approach*, J. Electrochem. Soc., 1999, 146, 6, pp. 2146–2155.

41. L. Bai, B. Conway, *Three-dimensional impedance spectroscopy diagrams for processes involving electrosorbed intermediates, introducing the third electrode-potential variable—examination of conditions leading to pseudo-inductive behavior*, Electrochim. Acta, 1993, 38, 14, pp. 1803–1815

42. C. Gabrielli, P.P. Grand, A. Lasia, H. Perrot, *Investigation of hydrogen adsorption-absorption into thin palladium films*, J. Electrochem. Soc., 2004, 151, pp. A1925–1949.

43. J. Bisquert, A. Compte, *Theory of the electrochemical impedance of anomalous diffusion*, J. Electroanal. Chem., 2001, 499, pp. 112–120.

44. M. A. Vorotyntsev, J.-P. Badiali, G. Inzelt, *Electrochemical impedance spectroscopy of thin films with two mobile charge carriers: effects of the interfacial charging*, J. Electroanal. Chem., 1999, 472, pp. 7–19.

45. M.J. Yang, Y. Li, N. Camaioni, G. Casalbore-Miceli, A. Martinelli, G. Ridolfi, *Polymer electrolytes as humidity sensors: progress in improving an impedance device*, Sensors and Actuators B, 2002, 86, pp. 229–234.

46. J. Bisquert, G. Garcia-Belmonte, F. Fabregat-Santiago, P. R. Bueno, *Theoretical model of AC impedance of finite diffusion layers exhibiting low frequency dispersion*, J. Electroanal. Chem., 1999, 475, pp. 152–163.

CHAPTER 11

EIS Analysis Applications: Cell Suspensions, Protein Adsorption, and Implantable Biomedical Devices

11.1. The field of biomedical impedance applications

The number of electrochemistry-based biomedical devices for clinical applications, including various diagnostic tools, implants, and biosensors, has been steadily increasing since the 1960s. Many of these devices are fundamentally based on the application of AC frequency dielectric and impedance analysis to biological and clinical systems.

Electrochemical-based methods are often higher-quality substitutes for fluorescent staining, magnetic counting, microdialysis, plate-culture techniques, and other clinical laboratory methods. For example, electrochemical biosensors offer a number of advantages over optical, ultrasonic, magnetic and other diagnostic principles employed in clinical and biomedical settings. The increased interest in applications of electrochemical technology to point-of-care biomedical diagnostic devices and sensors arises from its high sensitivity, selectivity, picomolar detection limit, temporal and spatial resolution, rapid response, simplicity of rapid screening procedures, label-free noninvasive sensing, cost effectiveness, versatility, flexibility of design, ease of integration, compatibility with microfabrication technology, high throughput screening, and ultimately an ability (either real of potential, depending on biocompatibility) to perform in vivo and respond adequately to the dynamic nature of living systems [1]. The small size of electrochemical devices allows them to be used in microfluidic products and sensor arrays when simultaneous detection of several analytes present in low-volume samples is required. One of the examples of in vivo application of electrochemical technology is the use of fast scan voltammetry with microelectrodes, which produces a unique fingerprint for dopamine with excellent selectivity and sensitivity [2]. Microelectrode arrays can often reliably record neural activity for several months after implantation.

Since the pioneering work of Schwan in the 1950s [3], the foundation was laid for dielectric analysis and interpretation of biological cell dispersions. Since then many electrochemical researchers have characterized biological colloidal suspensions [4, 5, 6, 7], developed "Coulter" counters and capacitive cytometers for bioparticle detection [8, 9, 10], designed enzyme-based biosensors for glucose monitoring [11, 12], and practiced electrophoretic and dielectrophoretic separations of drugs, proteins, cells, DNA, and pathogenic bacteria [13, 14, 15, 16, 17]. Dielectric spectroscopy has been used to study biomedical and pathogenic cell cultures, which is extremely useful for both medical diagnosis of many major clinical complications and early detection and prevention of infectious diseases.

The field of electrochemical biosensing has also become one of the most important methods in the detection of bacterial and blood cells. For instance, the need for pathogen detection arises in areas as diverse as the food industry, water or sludge treatment, and even national security. Prevention and early detection leading to prompt treatment are essential, since minor infections can rapidly turn life-threatening. Diagnostic systems therefore play an integral role in facilitating an effective response against these infections. In most of these cases electrochemical techniques were used to record changes in impedance or capacitance near the electrode surface. Overall the contribution of electrochemical techniques to the detection of clinical and pathogenic cells is mainly present in the detection step. However, that contribution not only involves the generation of an electrochemical signal but is also utilized in steps such as immobilization of bacteria on the electrode surface and cell lysis [14, 15, 16, 17, 18]. In an attempt to deliver higher selectivity and specificity and speed up the detection process, either the sensor surface has been immunomodified to help the capturing of cells or dielectrophoretic trapping or manipulation steps have been used to preconcentrate the target particles (Section 7-3).

In an aqueous solution, charge is carried between any two electrodes. The presence of particulate matter (such as blood or bacterial cells) physically obstructs the movement of these charge-carrying ions and thereby leads to higher impedance between the electrodes. Typical bacterial cells (~1000 cells/ml) have a volume fraction of ~10^{-12}, and consequently the corresponding impedance change is not significant. It may, however, be discernable if the low-volume suspension is made to pass between two electrodes separated by a narrow slit only slightly larger than the size of a cell or if the target cells can be made to adhere to or congregate at a surface of the electrodes. Concentrations of 10^5 cells/ml can be detected using the impedance technique with a detection time of two to three hours [17, 18]. However, the above-mentioned electrical detection techniques, such as the "Coulter" counter, suffer from low throughput and slit clogging.

Long-term stability, biocompatibility, and selectivity in complex biochemical environments are often an issue with electrochemical devices. The biomedical industry is continuously searching for materials and devices that can maintain their functionality for prolonged periods of time (days to years) after being implanted. Optimization and performance improvement of electrochemical sensors are usually based on development of modified surfaces, often utilizing enzyme membranes combined with electropolymerization [19]. The biosensor typically contains a biologically sensitive element, such as a protein or peptide and is placed in contact with a tissue. Biofouling, adsorption of biomolecules, and temperature instability of enzyme-based membranes contribute to a decrease in biomedical-device performance [20]. In order for biosensors and other implantable medical devices to function properly, the mutual interactions of the device and the surrounding tissue must not influence the performance of the device.

For all in vivo measurements, the implanted device perturbs the environment and initiates inflammatory response in tissue, resulting in encapsulation of the implant [21]. The acute inflammatory response starts immediately after the implantation, when fluid carrying plasma proteins and inflammatory cells migrate to the site of the implant and adsorb at the implant interface. These cellular events result in the development of a compact sheath of cells and accumulations of extracellular protein matrix material surrounding the implant. The tissue damage following the implantation may result in alteration of functionalities of the damaged cells in plasma and changes in local concentration of the analytes (such as glucose or oxygen) in the vicinity of the implant, resulting from a wound-healing process. The extracellular environment around implanted electrodes changes due to insertion-related damage and sustained response promoted by the presence of the device. This tissue encapsulation can cause changes in the electrical properties of the tissue adjacent to an implant. This encapsulation tissue was found to typically have a higher resistance than normal tissue. In case of biosensors this tissue response often leads to a significant modification of sensor functionality, making its response difficult or impossible to interpret, and in the case of medical implants leading to implant performance degradation due to corrosion, surface modification, and tissue modification/damage around the implant. These issues are particularly felt in the biosensor industry, where a reliable implantable glucose-monitoring sensor with a lifetime even on the order of days is yet to be developed [12]. Attempts to reduce biofouling have been primarily concentrated on development of specialized coatings (such as hydrogels, Nafion™, polyethylene oxide/polyethylene glycol, etc.) on the outermost membrane surface, which inhibits protein adhesion to the surface.

Considering these mechanisms of interaction between an electrochemical device and a tissue, it is necessary to investigate overall tissue and tissue-implant interfacial impedance responses for several reasons. Firstly, studying interfacial impedance responses allows determination of the dynamics of bio-fouling, which is usually related to the protein and cell adsorption kinetics at foreign-body surfaces of sensors and implants. This is important in studies of implantable in vivo sensors and device performance, reliability, and interactions with the tissue. Considering that, for instance, glucose and other enzyme-based sensor performance is based on multilayered membranes including an additional biocompatible protection layer, the resulting structure presents itself to an impedance analysis as a complicated electrochemical surface-kinetics system. This system may combine electron transfer through several mediating enzyme layers and an external protection layer, porosity-modified electron and mass transport, and the effects of protein adsorption on bare metallic surfaces of the medical implants or over the sensing surfaces of the sensors. Secondly, investigations of bulk-solution and interfacial impedance responses open a window into an interesting study of cellular colloidal suspensions containing components that are quite useful as biomarkers for clinical diagnostics of such diseases as cancer, diabetes, and thrombosis. These studies are also essential for establishing technologies for preconcentrating pathogenic bacteria and developing portable field devices capable of rapid detection at low concentrations.

Electrochemical impedance analysis, due to its AC-frequency-dependent nature, allows combining in a single measurement the interfacial studies of membrane-mediated electron and mass-transport kinetics, protein and cell surface adsorption, processes at the implant surface (such as corrosion and surface deactivation), and bulk-solution "dielectric" studies of biological and bacterial cell colloidal suspensions. EIS ability to study temporary and spatial details of bulk media and interfacial kinetics—either as a method of solution-species characterization, biofouling monitoring, or membrane degradation, is an important advantage in in vivo clinical applications. Application of EIS allows characterization of different types of biological cells, microparticles, proteins, and bacteria by estimating their concentrations, sizes, diffusion coefficients, and chemical changes. The two-electrode, three-electrode and four-electrode [13, 22, 23, 24] methods with reference electrodes to measure and exclude nonspecific changes in the test module (Section 8-4) have been routinely used for impedance characterization of biological membranes, adsorption, and colloidal phenomena. Impedance sensors have an advantage of being able to be effectively packaged with the DEP and EP electrodes in a single device (Section 7-3). The reference module can serve as a control for temperature changes, evaporation, changes in amounts of dissolved gases, and degradation of culture medium during incubation. The construction of a realistic model of the analyzed system is important for extraction of relevant cellular characteristics such as concentration, diffusion coefficients, sizes, and membrane properties. Impedance monitoring of protein adsorption is essential for qualification of embedded sensors and other implantable devices. This chapter is primarily dedicated to these selected aspects of impedance applications to the biomedical universe.

11.2. Analysis of biological cell suspensions by dielectric, impedance, and AC electrokinetic methods

There is a 50-year history of active broadband impedance studies of such biological materials as cell suspensions, amino acids, polypeptides, proteins, and tissues. Dielectric analysis has been applied in to the investigation of biological cells in colloidal suspensions, monitoring enzyme and immunological reactions, estimation of microbial biomass, detection of microbial metabolism, and protein adsorption to supported lipid bilayers [20, 21, 22, 23, 24, 25, 26, 27]. Accurate quantification and detection of major clinical and pathological cells provide important diagnostic information about the state of a patient and allow rapid screening for a variety of diseases, such as cancer, diabetes, thrombosis, and other cardiovascular complications. With a long-standing tradition in characterizing dielectric properties of biological cells, these systems are often better understood than many artificial colloidal suspensions [5]. Studies of cell suspensions also present an additional interesting opportunity for combining dielectric studies of bulk-solution relaxations at high AC frequencies (which are often conducted by using AC electrokinetics) and interfacial impedance studies at low AC frequencies [26, 27, 28, 29, 30, 31, 32].

Adult humans have roughly 5-6 million red blood cells (RBC), 4000–11,000 white blood cells, and about 150,000–400,000 platelets in each microliter of human blood. RBCs are the most common type of blood cell and the organism's principal means of delivering oxygen to the body tissues via blood flow through the circulatory system. A typical human RBC is a disk that lacks a cell nucleus and has a diameter of 6–8 μm and a thickness of 2 μm. These cells' cytoplasm is rich in negatively charged iron-containing hemoglobin that can bind oxygen and is responsible for the blood's red color. Along with hemoglobin anion RBCs contain disequilibria of Na^+ and K^+—creating a sodium-potassium pump. The RBC membrane comprises a typical lipid bilayer (phospholipids and cholesterol in equal proportions by weight) and proteins. Lipid composition is important, as it defines many physical properties such as membrane permeability and fluidity. Additionally, the activity of many membrane proteins is regulated by interactions with lipids in the bilayer. The proteins of the membrane skeleton are responsible for the deformability, flexibility, and durability of the red blood cell, enabling it to squeeze through capillaries and recover the discoid shape as soon as these cells stop receiving compressive forces.

Platelets are 2-3 μm-diameter, irregularly shaped cells that do not have a nucleus containing DNA. Platelets' membrane receptors are responsible for procoagulant activity. If the number of platelets is too low, excessive bleeding can occur. However, if the number of platelets is too high, blood clots can form (thrombosis), which may obstruct blood vessels and result in such events as a stroke, heart attack, pulmonary embolism, or blockage of blood vessels to other parts of the body, such as the extremities of the arms or legs. Endothelial cells are very flat cells that are about 1–2 μm thick and some 10–20 μm in diameter. The endothelium is the thin layer of cells that lines the interior surface of blood vessels. These nucleus–containing surface-active cells form flat, pavement-like patterns on the inside of the vessels lining the entire circulatory system from the heart to the smallest capillary. These cells reduce turbulence in blood flow, allowing the fluid to be pumped further and providing anticoagulant, antithrombotic and anti-inflammatory states. White blood cells, or monocytes, are immune-system cells defending the body against both infectious disease and foreign materials. Monocytes contain a kidney-shaped nucleus and abundant cytoplasmc and have a diameter of 14–17 µm. Monocytes eventually leave the bloodstream to become tissue macrophages which remove dead cell debris as well as attacking microorganisms.

Bacterial cells typically range between 1-5 µm in length. As in other cells, the bacterial membrane provides structural integrity to the cell. The bacterial membrane differs from that of all other organisms by the presence of peptidoglycan (poly-N-acetylglucosamine and N-acetylmuramic acid), which is located immediately outside the cytoplasmic membrane. Porous peptidoglycan is responsible for the rigidity of the bacterial membrane and its cell shape.

Idealized cell structure (as spherical particles with membranes) is represented in Figure 11-1. In principle the "live" cells can be represented by multiplayer shells, with the internal cellular fluid (cytoplasm) having permittivity (ε_{CP}) and conductivity (σ_{CP}) characteristics often similar to those of suspending media such as extracellular fluid (ε_M and σ_M), while the nanometer-thick membrane has much lower permittivity (ε_{MBR}) and very low conductivity ($\sigma_{MBR} \sim 10^{-8}$ S/cm). For "live" cells $\varepsilon_{CP} \sim \varepsilon_M > \varepsilon_{MBR}$, $\sigma_{CP} > \sigma_M >> \sigma_{MBR}$, while for "dead" cells the membranes become "lossy" and do not present any noticeable resistance to current flow ($\sigma_{MBR} \to \infty$).

A cell with a typical radius of $R_C = 2$ µm is contained in a ~ 5–10 nm thick phospholipid bilayer membrane. Each cell membrane has hydrophobic external and hydrophilic internal layers. The cell can control its cytoplasm composition by controlling its membrane permeability. In a cell membrane there are

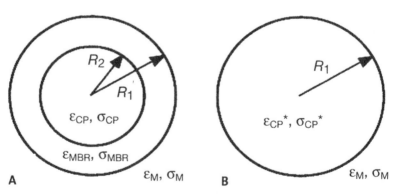

FIGURE 11-1 Multilayered shells: A. spherical particle with one concentric shell; B. equivalent homogeneous particle; as long as the physical scale of the nonuniformity of the imposed field is much larger than the particle radius R_1, the external field of the equivalent particle is indistinguishable from that of the heterogeneous particle

many potential-dependent ion channels that are able to reduce membrane resistance instantaneously per application of external potential, causing the components of cytoplasm to leak out and reduce or oxidize at the electrode surface. Adsorbed electroactive glycoproteins and globular proteins spanning the entire membrane greatly enhance the functionality of the membrane and provide for species transport across the membrane. Some proteins ("channel proteins") facilitate development of channel structures for certain types of ions (Na^+, K^+, Cl^-, Ca^{2+}). Because there are many voltage-sensitive channels, they are mostly responsible for nonlinear properties of cell membranes at low voltages. Cell polarization is generated by these channels, which function as "ion pumps" when polarization of a cell by transmembrane movement of ions occurs; the exterior of the cell (cytoplasm) has a potential of about -70mV with respect to extracellular electrolyte. If -70 mV potential difference exists between cytoplasm and extracellular fluid at membrane thickness of ~ 7 nm, the potential difference is ~ 10 kV/mm. These potential changes across the membrane control cellular activities such as admission or release of certain types of ions. The membrane is very thin, and the resulting membrane capacitance is high as $C_{MBR} \sim 1$ $\mu F/cm^2$. To create V $= -70$ mV potential, the needed charge is $q = CV$. For a cell with a radius $R_C \sim 2$ μm and surface area $4\pi R_C^2 = 5*10^{-7}$ cm^2, that translates into $q = 0.07V*10^{-6}F/cm^2*5*10^{-7}cm^2 \sim 3.5*10^{-14}C$ or $\sim 10^6$ unit charges. This value is supported by an independent estimation [32] that every cell cytoplasm contains ~ 0.15 mM ions or free charges.

The membrane breaks down when the potential difference across the membrane increases to $\sim -200-300$ mV. External electric fields generating transmembrane potentials of this order cause rupture of the lipid bilayer-forming pores, leading to irreversible damage or "lyses" after sufficient time of exposure. For exposure times longer than a few microseconds the cells exposed to an external field of several kV/cm are killed, as the membranes become leaky and large transport of matter in and out of the cells occurs. An external electric field of 2 kV/cm represents a 2V drop over the length of a 10 µm cell, indicating that at the margins of the cell closest to the electrodes, there will be a 1V drop across the cell membrane (assuming far greater electrical resistance across the membrane than through the cell cytoplasm, thus creating a constant potential in the cell interior).

When placed in an external electric field, blood cells suspended in dispersing media can show several frequency-dependent relaxations, which were formulated by Schwan [33, 34]. The resulting dielectric response, which yields valuable information about the structural and functional properties of the system, is influenced by many factors. Some of them are associated with the bulk electrical properties of the media, some with the cellular particle surface properties, and some with the geometry of the dispersed cells. Depending on the applied frequency, live biological cells may exhibit two extreme cases of polarizability. Being insulated by the membrane at frequencies less than 10 kHz, cell polarizability is usually low. At higher frequencies cell polarizability becomes dominated by the cytoplasm conductivity, which is often higher than that of the dispersing media. Impedance spectra of biological cell suspensions typically show α-, β-, and γ-dispersions. The γ-dispersion due to dielectric relaxation of bulk-dispersing media exists in GHz range, such as Debye dispersion in water at about 17 GHz (Figure 11-2).

The β-dispersions found between 10 kHz and 100 MHz are interpreted in terms of cell membrane interfacial polarization or the Maxwell-Wagner effect. The β-dispersion is due to structural relaxations related to membrane capacitive bridging (β1) at low kHz, above which current starts flowing directly through the cell interior. Debye dispersion in cell cytoplasm (β3) may occur between kHz and MHz relaxations if conduction through cellular cytoplasm switches from conductive to capacitive but the current still flows through the cells. The conduction through cells switches to capacitive conduction through the suspending media in the low MHz range (β2) (Figure 11-3).

The β1 relaxation process is due to capacitive membrane bridging in low kHz range, which is changing the conductive (resistive) current from predominantly flowing around the cell to predominantly flowing through its highly conductive cytoplasm and capacitive membrane. The effective impedance of the membrane becomes progressively smaller with the frequency increase, as it is inversely proportional to frequency-capacitance product. At sufficiently high frequency the membrane impedance is bridged through its capacitive component. For a membrane with capacitance $C_{MBR} \sim 10^{-6} F/cm^2$, at ~10 kHz the membrane will be shorted through its capacitive impedance with $Z_{MBR} \sim (\omega C_{MBR})^{-1} \sim 10$ ohm. Cell cytoplasm is typically more conductive that the extracellular media ($\sigma_{CP} > \sigma_M$). The resulting current flows through the cells, producing the β1-relaxation at ~ 10kHz. The relaxation time $\tau_{\beta1}$ of a membrane conduction process for a cell with radius R_C becomes (R_{SOL} is solution or media resistance in ohms):

$$\tau_{\beta1} = C_{MBR} R_C (1 / \sigma_{CP} + 1 / 2\sigma_M) \sim C_{MBR} R_{SOL} \qquad (11-1)$$

With further frequency increase, the cytoplasm can produce dispersion due to a switch from resistive to capacitive conduction in kHz-low MHz range (β3), which can be detected if the characteristic frequency of this relaxation is between β1 and β2 [4]. However, the presence of a significant cytoplasm-capacitive component is often disregarded, and the total cytoplasm imped-

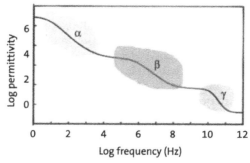

FIGURE 11-2 Dielectric relaxations in bulk solution suspensions of blood cells

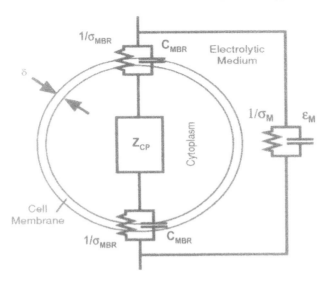

FIGURE 11-3 Circuit model showing transconductance through a cell membrane ($1/\sigma_{MBR}$ | C_{MBR}) and internal cytoplasm (Z_{CP}) vs. the parallel conducting path through the media ($1/\sigma_M$ | ε_M)

ance is simplified as the ohmic resistance $R_{CP} = Z_{CP}$. With further frequency increase, the suspending media-capacitive current determined by the bulk-media permittivity starts superseding the conductivity currents through the cells. Media permittivity ε_M is typically slightly higher than permittivity of the "effective cell," which includes both cytoplasm and membrane contributions, leading again to current flowing primarily through the suspending media and not through the cells. This conduction-path switch results in the β2 relaxation at ~ 100MHz. The relaxation time $\tau_{\beta 2}$ of this high-frequency relaxation process becomes:

$$\tau_{\beta 2} \sim R_{CP} C_M \qquad (11\text{-}2)$$

In the simplified $R_{CP} - C_{MBR}$ representation of the cell in parallel with the solution, the relative values of R_{CP} and C_{MBR} affect the relaxation times $\tau_{\beta 1}$ and $\tau_{\beta 2}$ and their corresponding critical frequencies. Increase in C_{MBR} and R_{SOL} values allows current to pass through the cells at lower frequencies for the β1-relaxation as $\tau_{\beta 1}$ increases per Eq. 11-1. Lower values of R_{CP} allow for lower values of $\tau_{\beta 2}$ and higher frequency for the β2-relaxation.

Figure 11-3 shows both the transconductive path through the cell and the parallel path through the media around the cell. In general both current paths exist, but usually one or the other dominates in its influence on the effective dielectric response of a particle. Figures 11-4A and 11-4B demonstrate complete and simplified EC models for the β–dispersion. The β–dispersion mechanisms are dependent on concentration and conductivity of the suspending media. Analysis of the β-dispersion processes (such as done by the "Coulter" counter technique) allows determining sizes of contributing species but often

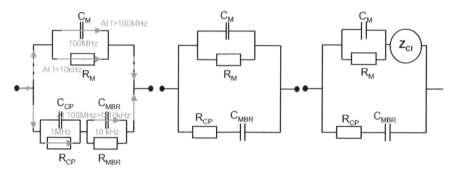

FIGURE 11-4 Diagram showing conductive paths in the cellular colloidal suspension: A. complete model; B. simplified model; C. model B with Schwarz counterion diffusion path Z_{Cl}

is lacking selectivity and accuracy, especially for low concentrations of analyzed particles. The mechanism becomes activated only when the cell concentration exceeds the percolation threshold, allowing enough current to be channeled through the cells. It is also easier to characterize the cell's β–dispersion mechanism in media with lower conductivity where $\sigma_{CP} \gg \sigma_M$ ($R_M \gg R_{CP}$) [35], as in suspending media with high conductivity σ_M the current will likely not flow through the cells at all, and therefore their interactions with the field will not be determined.

Schwan's [34] expression Eq. 7-10 describing increments in permittivity due to the β1-relaxation process (Figure 7-7) was modified by Asami [6] for low concentrations of cells N_C (cells/mL) as their volume fraction $\Phi = 4/3\pi R_C^3 N_C$ as:

$$\Delta\varepsilon_{\beta1} = \frac{9}{4\varepsilon_0} \frac{\Phi R_C C_{MBR}}{\left[1 + R_C G_{MBR}(1/\sigma_C + 1/2\sigma_M)\right]^2} = \frac{3\pi R_C^4}{\varepsilon_0} C_{MBR} N_C \qquad (11\text{-}3)$$

For one percent of cells in the suspension, the permittivity increment becomes only $\Delta\varepsilon_{\beta1} \sim 25$. The sensitivity of this method for cell detection depends on the 4th power of the cell diameter. In the case of cells with radius $R_C \sim 1$ μm, permittivity constant $\varepsilon_0 = 8.85*10^{-14}$ F/cm, membrane thickness $\delta = 5$ nm, membrane permittivity $\varepsilon_{MBR} = 6$, membrane conductivity for "live" cell $\sigma_{MBR} = 10^{-9}$ S/cm, effective capacitance of cell membrane $C_{MBR} = \varepsilon_0\varepsilon_{MBR}/\delta \sim$ 1μF/cm², effective conductance of membrane for "live cell" $G_{MBR} = \sigma_{MBR}/\delta \sim$ 2*10⁻³ S/cm², media permittivity at MHz frequency $\varepsilon_M = 78$, media conductivity $\sigma_M = 10^{-5}$ S/cm, and cytoplasm conductivity $\sigma_C = 5*10^{-3}$ S/cm, the detection limit is ~ 10⁴ cells/μL [6].

Schwarz [36, 37] postulated that kHz permittivity increments calculated from the β-relaxation theory are significantly lower than experimentally observed values and proposed the counterion α-relaxation theory with its corresponding mechanism (Eq. 7-15). The permittivity increment from this theory, following the same concentration-related modification, is:

$$\Delta\varepsilon_\alpha = \frac{9}{4\varepsilon_0} \frac{\Phi}{(1+\Phi/2)^2} \frac{e_0^2 R_C q_0}{K_B T} \sim \frac{3\pi R_C^4}{\varepsilon_0} \frac{e_0^2 q_0}{K_B T} N_C \qquad (11\text{-}4)$$

For fewer than one percent suspended cells the permittivity increment becomes $\Delta\varepsilon_\alpha \sim 3000$, closer to typical experimental values. That also suggests the potentially much higher sensitivity of this method for quantification of the cells, compared to Eq. 11-3. For elementary charge $e_0 = 1.6 \ast 10^{-23}$ Coulombs, Avogadro constant $N_A = 6.02 \ast 10^{23}$, Boltzmann constant $K_B = 1.38 \ast 10^{-23}$ J/K, absolute temperature $T = 298$ K, average charge density on a typical cell surface $q_0 \sim 10^{13}$ cm^{-2}, and other parameters as in Eq. 11-3, a detection limit of ~ 10 cells/ μL can be feasibly achieved in controlled laboratory conditions for an instrument with the permittivity resolution of 0.01 unit.

The hydrodynamic counterion-diffusion mechanism is based on diffusion-controlled redistribution of counterions in the electrochemical double layer of micron-sized cells and is expected to occur at $\sim$ 1kHz [37]. The counterion double-layer polarization mechanism through counterion diffusion tangential to the cell surface is possible in media where ionic counterions are present in high concentrations. In biological cells, where the electrical properties of the interfaces induce different polarization mechanisms, causing dielectric dispersions, the contribution of the bulk ion diffusion in the inner and outer medium must be appropriately taken into account [31]. Owing to the diffusion of ions, a finite thickness charge distribution on both sides of each cell shell arises, influencing the overall interfacial polarization. The Schwarz diffusion polarization-relaxation mechanism can be used to evaluate this additional polarization and the related parallel current path around the cell [37] (Figure 11-4C).

As was discussed in Section 7-3, in cases of plentiful counterions the Schwarz mechanism is expected to produce a low kHz range α–relaxation within the same frequency range as the membrane bridging β1-relaxation. The Maxwell-Warner mechanism is expected to produce at least two β-relaxations, while the Schwarz mechanism also results in two relaxations in similar frequency ranges—a high MHz relaxation in the bulk-media solution due to orientation polarization of counterions and a kHz relaxation due to diffusion of counterions in the double layer of micron-sized cells. Both sets of relaxations occur in the same general frequency regions, and it is left to an investigator to validate how realistic both models might be (Figure 11-4). In principle two mechanisms may or may not coexist. The relative magnitude of the models' contribution to the experimental data can be verified by a consideration of the relative amount of cells vs. counterions and conductivity of the media σ_M. If the counterions are plentiful, resulting in higher media conduction, the current bypasses the cells and no significant β-relaxation process will occur. However, the cells still present diffusion polarization impedance to the counterions' current due to migration, and the Schwarz diffusion model remains valid [38].

In addition to a purely counterion-related relaxation process, the α-dispersion that appears below 10 kHz was attributed to various cell irregularities such as agglomeration, nonspherical elongated shapes, cell positioning vs. the field, osmotic holes in membranes, and interfacial impedance of cell lysed components [3, 4, 5, 6]. For various cell types, such as bacteria (E. coli), platelets, skeletal muscle, and erythrocytes, the intensity ratios $\Delta\varepsilon_\alpha/\Delta\varepsilon_\beta$ of the α-dispersion to the β-dispersion have been reported in the 1-30 range. The characteristic frequencies for these types of α-dispersions fall between 100Hz and 10 kHz.

Figure 11-5 demonstrates the effects of membrane holes, irregular cellular shapes, cell agglomeration, and alignment with the external electric field on dielectric spectra. When cellular irregularities such as elongation, agglomerates, or membrane holes are aligned with the field, measured low-frequency permittivity of the system increases. Dielectric spectra for the orientation of the cell irregularities perpendicular to the field demonstrated no change in permittivity, irrespective of the degree of the irregularities [6, 31, 39]. General conclusions from these studies with respect to permittivity increases due to particle agglomeration, nonspherical shapes, and alignment with the field are similar to those presented in Section 7-2.

Analysis of specific electrochemical cellular properties provides a unique and recognizable signature with high selectivity and sensitivity and detection limits as low as 10–100 cells/μl and detection ranges between 10^2 and 10^{10} cells/μl. For many types of cell suspensions significant dielectric increments $\Delta\varepsilon_\alpha$ ~1000-100,000 are frequently experimentally observed in the low kHz range. In contrast, the dielectric constant of aqueous solutions ε_M~ 80, and if so the presence of even a few cells should greatly alter the measured low-frequency dielectric values. Unfortunately, the double-layer capacitance and "interfacial polarization" at the electrode-solution boundary effectively screen this cellular capacitance, especially when operating at frequencies below 1 kHz [40]. However, by modulating the geometry and positioning of the electrodes in a manner that increases the bulk-media resistance (R_{SOL}) and hence increases the time constant of the medium $\tau_M = R_{SOL}C_M$, the measured permittivity signal (or "imaginary" impedance) can be made sensitive to the cell-induced changes in the measured system capacitance. This approach has been used in "Coulter" counters. Signature presence of cells of clinical and pathological importance, such as bacteria and their proliferation at concentrations low enough (10–1000 CFU/ml) to be applicable to many "real world" problems have been detected [20, 21, 22].

In cell biology AC electrokinetics is used for analysis of cellular dispersions and determination, screening, or pursuit of changes in membrane capacitance, conductance, and cytoplasmic properties of cells and cells byproducts [41, 42, 43, 44]. AC electrokinetic representation and Clausius-Mossotti [β]-factor frequency dependence analysis are typically employed to predict the mechanism of conduction in the media, determine relaxation frequencies, identify the changes in the cell biology ("dead" cells vs "alive" cells), and evaluate cell separation possibilities for dielectrophoresis. Studying AC frequency variability of the [β]-factor is important for understanding which components of a studied system are involved in the conduction process, in particular in the bulk-solution relaxations. When [β] >0 the current is transported predominantly through cells, as the particles' characteristic impedance is lower than that of the supporting media. When [β] <0 the current is transported through a solution medium (extracellular fluid), which offers the path of least resistance to the current (Section 7-3). The above statements, however, are not always correct, as the relative concentration of the particles in the media is also important. If the particles are highly conductive but their concentration is very low, then at least part of the current will have to be supported by the

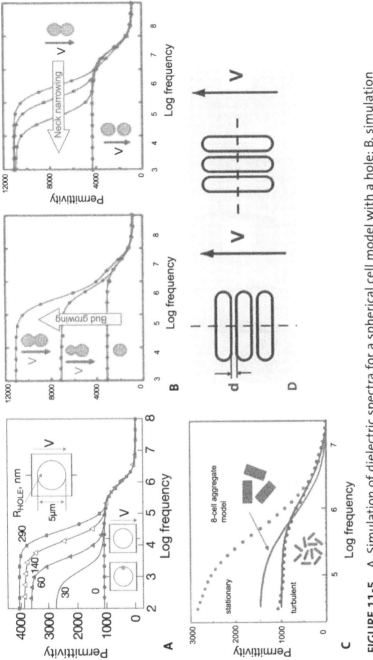

FIGURE 11-5   A. Simulation of dielectric spectra for a spherical cell model with a hole; B. simulation of dielectric spectra for a cell of various shapes; C. simulation of dielectric spectra for turbulent (no agglomeration) vs. stationary (agglomerated) human blood cells; D. simulation of the effect of elongated cells perpendicular and parallel to the electric field

migration of other ionic species in the supporting electrolyte. Typically, high-frequency AC electrokinetics or dielectric characterization principles operate reliably for reasonable concentrations of the analytes in the higher-resistance supporting medium to maximize the positive impedance mismatch [41, 42, 43, 44, 45].

For the case of "dead" cells, the membrane becomes irreversibly permeable, and the Clausius-Mossotti factor $[\beta]_{DC}$ becomes:

$$[\beta]_{DC} = RE\left[\frac{Z_M{}^* - Z_{DC}{}^*}{Z_M{}^* + 2Z_{DC}{}^*}\right] = \frac{\varepsilon_0^2(\varepsilon_{DC} - \varepsilon_M)(\varepsilon_{DC} + 2\varepsilon_M) + (\sigma_{DC} - \sigma_M)(\sigma_{DC} + 2\sigma_M)/\omega}{\varepsilon_0^2(\varepsilon_{DC} + 2\varepsilon_M)^2 + (\sigma_{DC} + 2\sigma_M)^2/\omega^2}$$

(11-5)

In case of "live" cells with loss-less membranes the situation becomes more complex and is treated as a case of "particles with thin shells" [4]. The formula for the $[\beta]$-factor becomes more elaborate. The first approach was developed for a simplified case assuming a completely loss-less membrane with infinitely low $\sigma_{MBR} \to 0$ as [41]:

$$[\beta]_{LC} = -\frac{[\omega^2\varepsilon_0(\dfrac{\varepsilon_M C_{MBR}R_C - \varepsilon_C C_{MBR}R_C}{\sigma_M\sigma_C}) - 1][\omega^2\varepsilon_0(\dfrac{2\varepsilon_M C_{MBR}R_C + \varepsilon_C C_{MBR}R_C}{\sigma_M\sigma_C}) - 2]}{[\omega^2\varepsilon_0(\dfrac{2\varepsilon_M C_{MBR}R_C + \varepsilon_C C_{MBR}R_C}{\sigma_M\sigma_C}) - 2]^2 + \omega^2[\dfrac{C_{MBR}R_C}{\sigma_M} + \dfrac{2\varepsilon_0\varepsilon_M}{\sigma_M} + \dfrac{C_{MBR}R_C}{\sigma_C}]^2} -$$

$$\frac{\omega^2[\dfrac{C_{MBR}R_C}{\sigma_M} - \dfrac{\varepsilon_0\varepsilon_M}{\sigma_M} - \dfrac{C_{MBR}R_C}{\sigma_C}][\dfrac{C_{MBR}R_C}{\sigma_M} + \dfrac{2\varepsilon_0\varepsilon_M}{\sigma_M} + \dfrac{C_{MBR}R_C}{\sigma_C}]}{[\omega^2\varepsilon_0(\dfrac{2\varepsilon_M C_{MBR}R_C + \varepsilon_C C_{MBR}R_C}{\sigma_M\sigma_C}) - 2]^2 + \omega^2[\dfrac{C_{MBR}R_C}{\sigma_M} + \dfrac{2\varepsilon_0\varepsilon_M}{\sigma_M} + \dfrac{C_{MBR}R_C}{\sigma_C}]^2}$$

(11-6)

For this and the previous example the following typical values are assumed: permittivity of air $\varepsilon_0 = 8.85*10^{-14}$ F/cm, cell radius $R_C \sim 2$ μm, membrane thickness $\delta = 5$ nm, membrane permittivity $\varepsilon_{MBR} = 6$, membrane conductivity for "live" cell (initially assumed to be $\to 0$) $\sigma_{MBR} = 10^{-9}$ S/cm, membrane resistivity $\rho_{MBR} \sim 10^9$ ohm cm^2, effective capacitance of cell membrane $C_{MBR} = \varepsilon_0\varepsilon_{MBR}/\delta \sim 1$μF/cm^2, effective conductance of membrane for "live cell"; $G_{MBR} = \sigma_{MBR}/\delta \sim 2*10^{-3}$ S/cm^2, media permittivity $\varepsilon_M = 78$, media conductivity $\sigma_M = 10^{-5}$ S/cm, cytoplasm permittivity for "live cell" and effective permittivity of "dead cell" $\varepsilon_C = \varepsilon_{LC} = \varepsilon_{DC} = 60$, and cytoplasm conductivity $\sigma_C = 5*10^{-3}$ S/cm.

An alternative approach [4] calls for a so-called "perpendicular" model tangential to the cell surface. It starts with an expression for the effective permittivity of a layered particle (Figure 11-1B), accounting for the conductive loss:

$$\varepsilon_{EFF}^* = \varepsilon_{MBR}^* \frac{(\dfrac{R_C}{R_C - \delta})^3 + 2\dfrac{\varepsilon_C^* - \varepsilon_{MBR}^*}{\varepsilon_C^* + 2\varepsilon_{MBR}^*}}{(\dfrac{R_C}{R_C - \delta})^3 - \dfrac{\varepsilon_C^* - \varepsilon_{MBR}^*}{\varepsilon_C^* + 2\varepsilon_{MBR}^*}}$$

(11-7)

In a realistic case when the cell radius is much larger than the cell membrane $R_C \gg \delta$ the equation simplifies to:

$$\varepsilon^*_{EFF} = \frac{R_C(C_{MBR} + \frac{G_{MBR}}{j\omega})(\varepsilon_C + \frac{\sigma_C}{j\omega})}{R_C(C_{MBR} + \frac{G_{MBR}}{j\omega}) + (\varepsilon_C + \frac{\sigma_C}{j\omega})} \tag{11-8}$$

In this derivation a condition $G_{MBR} > 0$ is taken into consideration, unlike Eq. 11-6, where $G_{MBR} = 0$ was assumed. A combination of this "effective" cell-complex permittivity, which depends both on the electrical properties of the cytoplasm and the membrane, and the media-complex permittivity results in a modified Clausius-Mossotti expression. An independent derivation followed through several intermediate steps involving parameters yields:

$$
\begin{aligned}
A^* &= \omega^2 R_C(C_{MBR}\omega^2\varepsilon_C\varepsilon_0 - G_{MBR}\sigma_C)(R_C C_{MBR} + \varepsilon_C\varepsilon_0) \\
B^* &= \omega^2 R_C(C_{MBR}\sigma_C + G_{MBR}\varepsilon_C\varepsilon_0)(R_C G_{MBR} + \sigma_C) \\
C^* &= \omega R_C(C_{MBR}\omega^2\varepsilon_C\varepsilon_0 - G_{MBR}\sigma_C)(R_C G_{MBR} + \sigma_C) \\
D^* &= -\omega^3 R_C(C_{MBR}\sigma_C + G_{MBR}\varepsilon_C\varepsilon_0)(R_C C_{MBR} + \varepsilon_C\varepsilon_0) \\
E^* &= \omega^4(R_C C_{MBR} + \varepsilon_C\varepsilon_0)^2 + \omega^2(R_C G_{MBR} + \sigma_C)^2
\end{aligned}
\tag{11-9}
$$

And the final expression for the Clausius-Mossotti factor for a live cell suspended in a medium with permittivity ε_M and conductivity σ_M becomes:

$$[\beta]_{LC} =$$

$$\frac{\omega^2(A^* + B^* - E^*\sigma_M)(A^* + B^* + 2E^*\varepsilon_M) + (\omega(C^* + D^*) + E^*\sigma_M)(\omega(C^* + D^*) - 2E^*\sigma_M)}{\omega^2(A^* + B^* + 2E^*\varepsilon_M\varepsilon_0)^2 + (\omega(C^* + D^*) - 2E^*\sigma_M)^2}$$

$$\tag{11-10}$$

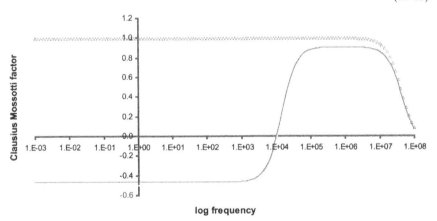

FIGURE 11-6 Examples of frequency dependence of Clausius-Mossotti factor for dead cells using Eq. 11-5 and live cells using Eqs. 11-6 and 11-10

Figure 11-6 offers a simulation of the variability of Clausius-Mossotti factor results for "live" and "dead" cells (using the same conductivities and permittivities of all components of the system) using Eqs. 11-5 through 11-10. For the "dead cells" the [β]-factor is always positive below β2 critical relaxation frequency, as membrane resistance to the current is not a factor. The "dead" cells become predominant current conductors in a uniform AC or DC electric field and can be separated by electrophoresis. Alternatively, in a nonuniform field the "dead" cells will move to points of higher field intensity. For the "live cells" the [β]-factor becomes positive above the β1– relaxation critical frequency and below the critical frequency of β2– relaxation. It can be noticed that there is a small difference between Eq. 11-6 and Eq. 11-10 calculations for live cells at low frequency due to differences in assumptions about membrane conductance (assumed to be $G_{MBR} = 0$ in Eq. 11-6 and $G_{MBR} = 2*10^{-3}$ S/cm^2 in Eq. 11-10). According to the dielectric suspension model, biological cells are suspended in solution and there is no migration movement of the cells, as they are nonionic due to a membrane of very low conductivity. Biological cells indeed are very poor conductors at low frequencies below 10 kHz and therefore force electrical currents to bypass them. When the [β]-factor becomes negative in the nonuniform field, the cells are transported to lower field-intensity areas. In an external field "live" cells become current conductors in the AC frequency window between the β1– and β2– relaxation critical frequencies. When cells are polarized at frequencies between 10 kHz and 100 MHz, they develop electrophoretic velocities in a uniform field or move in a nonuniform electric field to points of higher field intensity.

This is a simpler model for spherical cells containing uniform cytoplasm and covered with membrane, which is applicable, for example, to nuclei-free cells such as red blood cells and platelets. More complex "double-shell" models requiring a computer-assisted approach should be considered for cells with nuclei, such as monocytes and endothelial cells. For nonspherical cellular shapes geometrical corrections can be introduced into the Clausius-Mossotti formula [9].

The impedance ("Bode plot") and permittivity data are sometimes as revealing as the AC electrokinetic analysis in understanding the bulk-solution relaxations. The experimental data for monocyte blood cell solutions demonstrated the α– and β– bulk-solution relaxations in MHz and kHz regions and low-frequency interfacial processes (Figure 11-7) [38].

At low frequencies a near DC electrophoresis and interfacial electrochemistry processes may occur in cellular solutions, especially for adherently grown "cultured" cells on interdigitated electrode structures or attracted to the electrode by DEP, magnetic, or other means. The simplest way to represent ionic charge transport is to employ an ohmic model. Largely because of their structure and high internal electrical conductivity, this approximation usually suffices within the cells. On the other hand, an ohmic model is sometimes less successful in representing electrolytic media in which cells are commonly suspended. The problem is that double-layer phenomena introduce the complication of mobile space charge outside but directly adjacent to the cell wall [4].

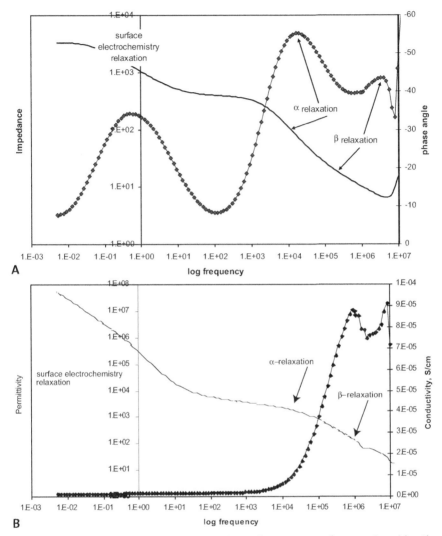

FIGURE 11-7 Analysis of monocyte colloidal suspension (6.5 ·10⁻⁸ mol/cm³) by impedance methods: A. phase angle and impedance; B. permittivity and conductivity data

As the double layer is formed outside the membrane in the external electric field due to interfacial effects, including interactions of the media with polar lipids and proteins at the outside membrane, the cell becomes polarized and deformed. The basic building block of all biological membranes is that of the lipid bilayer, where the lipid molecules are oriented with their polar groups facing outwards into the aqueous environment and their hydrophobic hydrocarbon chains pointing inward to form the membrane interior. The cell may become transformed into an externally polarizable macro-ion that can approach the electrode and become deposited at the surface. Cells can adhere to other cells or to solid electrode surfaces despite being apparently separated by a gap of 10-20 nm. A direct molecular connection can also be established between the cell surface, presumably through the glycocalyx (extracellular carbohydrate-linked segments of integral membrane glycoproteins) and the substrate. The contact zone can be viewed as an aqueous compartment spanned by the polycationic segments of membrane macromolecules and can act as a molecular sieve to the diffusion of aqueous molecules [23]. This results in the interfacial impedance change. Cell density, growth, and long-term behavior of cells on the electrodes also change the measured impedance. Depending on the type of cells, the response can be purely capacitive, Faradaic, or a mixture of the two. Leakage of ions through the membrane (lyses) resulting from this polarization can introduce additional factors, such as a measurable increase in media conductivity σ_M determined at high Hz range. This effect potentially allows detection of the cells in concentrations less than 10 units/μL [32, 35]. Other possible factors (Figure 11-8) include cell adsorption at the interfaces mediated by proteins present in their membranes (Z_{ADS}) and redox reactions of the membrane proteins, lipids, and cytoplasm components at the electrode (Z_{REDOX}) [38].

The impedance measurements on the biosensor can provide information about the type, spreading, attachment, and morphology of the cultured cells. Typically, as membranes are insulative, the "adsorption-type" capacitive response develops at the interfaces of the electrodes in the absence of significant externally applied electrochemical DC potential V. The main effect of cells on the sensor signal is due to the insulating property of the cell membrane. The presence of intact cell membranes on the electrodes and their distance to the electrodes determine the current flow and thus the sensor signal. If cells grow directly on an electrode, this effectively reduces the electrode area reached by the solution ionic current, and the interfacial impedance increases. The aqueous gap between the cell membrane and the substrate prevents direct influence of the cell-membrane capacity on the interface impedance of the electrodes. Nevertheless, an increase in interfacial impedance should be expected. Effects of unspecific protein adsorption from the medium should be small compared to a closely packed lipid bilayer. The difference between dead and live cells was observed using interfacial impedance changes [35].

As was already mentioned, the voltage-sensitive "channel proteins" facilitate ionic transport in and out of the extracellular electrolyte, resulting in reduction or oxidation activity of the released ionic components of the cell cytoplasm. Exposure to high electrochemical potentials or special "lysing" media results in lyses where even large macromolecules (such as proteins) pass in and out of the cells. These species come in contact with the electrode, resulting in additional reduction or oxidation Faradaic activity of ionic and organic components of the cell cytoplasm that are specific to a particular cell type. This interfacial process often combines charge-transfer resistance and finite or semi-infinite diffusion. Rich voltage-dependent surface electrochemistry (Figure 11-7A), primarily driven by charge-transfer resistance $R_{CT} \sim 100$ ohm and high finite diffusion resistance at ~ 3000 ohm (with at average cell size $R_c \sim 2$ μm, $D \sim 10^{-6}$ cm^2/sec), with realistic double-layer capacitance 50 μF was reported [38]. This model takes into consideration realistic conductivity and permittivity values of the system components for the relaxation mechanisms in extracellular media ($\beta 2$) at ~ 10 MHz (for R_{CP}~6 ohm, ε_M ~80), cytoplasm relaxation ($\beta 3$) at ~ 15 MHz (R_{CP}~6 ohm, ε_{CP} ~60), membrane capacitive bridging ($\beta 1$) at ~2kHz (ε_{MBR}~8; R_M ~10 ohm; R_{MBR}~ 400 ohm), and parallel Schwartz counterion α–relaxation process at 1kHz. The bulk solution can be analyzed using AC electrokinetic principles. For this modeled system for live cells with active membranes, positive values of the [β]-factor were predicted only for frequencies between 50 kHz and 5 MHz, indicating predominant current conduction through the cells. At high MHz frequencies orientation polarization of counterions becomes predominant. The model is still not ideal due to variability in surface and volume factors for the cells and counterions, as their concentration is probably less that that of total saturation when the cells are touching each other, therefore the exact surface area/distance factors are often unpredictable.

Although in many cases the measured impedance data are well represented by equivalent-circuit models incorporating empirical "constant phase angle" elements, it must always be borne in mind that biological tissues and interfaces are distributed in space and inevitably possess microscopic properties that vary throughout the sample under investigation [38]. It is unlikely that the distributed properties of a system can be isolated and represented by one, albeit nonstandard, circuit element. As biological tissues and interfaces are distributed in space with microscopic properties that vary throughout the sample, circuits incorporating lumped circuit elements such as CPE (ideal or otherwise) will inevitably prove less than optimal. In the absence of detailed,

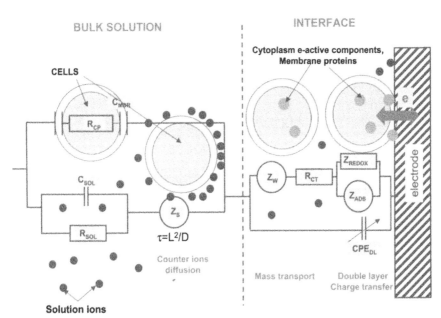

FIGURE 11-8 Impedance model showing conductive paths in the solution, transcellular path, counterion diffusion path, and typical interfacial impedance

many-bodied, physicochemical models, we must be content with simpler continuum models, such as equivalent circuits, when seeking to represent and understand the observed phenomena. However, such models should be used circumspectly, preferably only after extensive study of the system under different conditions [46, 47, 48, 49].

11.3. Impedance analysis of protein adsorption kinetics

Proteins are the most abundant macromolecules in living cells. Sixty-five percent of the protein mass of the human body is intracellular. The structure of low-nanometer-sized proteins often are organized around metal ions (Zn, Co, Fe), and possess a molecular weight of hundreds of kD. In the dry state some proteins lose their ionic nature and may become electronic conductors. When in the bodily fluid, proteins form aqueous ionic colloidal suspensions. Proteins typically carry a small positive or negative charge (dipole moment is 15-50 Debye units), and their presence increases measured sample permittivity (ε) compared to that of pure supporting media (ε_M) by 20–200 units of permittivity. The increase is proportional (δ) to the concentration of proteins C^* as $\varepsilon = \varepsilon_M + \delta C^*$. Proteins readily interact to form complexes with other molecules such as fatty acids and drugs, and their interaction with several different electrode materials has been studied and confirmed. Proteins possess electrophoretic mobility, have isoelectric point, and demonstrate electrochemical activity combining redox and adsorption processes [50].

A number of voltammetry and impedance studies were dedicated to protein adsorption processes [51, 52, 53, 54, 55]. Investigation of the effects of protein interactions with metal surfaces and simulation of protein adsorption on solid-state implanted devices constitute one of the most prominent and continuously researched areas of bioelectrochemistry. This information is important for demonstrating the feasibility of constructing implantable electronic biosensors based on metal electrodes and developing microfluidic systems that will ultimately enable portable and rapid analysis of protein markers for point-of-care medical diagnostics. Characterization of many proteins is of clinical relevance in detecting biofouling in hemofiltration equipment, biofilm growth on prosthetic devices, or hemolysis within a prosthetic or therapeutic device. For example, an indwelling catheter that could self-report a changing biofilm layer or a prosthetic joint that could be interrogated regarding metal-bone interface status could simplify surveillance for device failure and achieve significant cost savings by avoiding expensive imaging modalities. Cell lyses could be detected in real time by surveillance for certain intracellular proteins normally absent from plasma, such as lactate dehydrogenase (LDH). Presently, detection of these processes involves laboratory processing of patient samples (blood tests for hemolysis), radiographic techniques (MRI/CT/plain films of prosthetic joints), or surgical removal of the device (removal/replacement of catheter). These approaches are inherently expensive or morbid and are only ordered after a pathologic process has caused enough damage to be clinically significant. "Smart" medical devices that prospectively monitor their status and report deterioration could lead to significant reductions in diagnostic ambiguity, unnecessary procedures, and patient distress [56].

The need to prevent fouling of medical implants, biosensors, microfiltration membranes, and industrial surfaces used in food processing, combined with the desire for methods to identify, separate, and purify proteins, has fostered several studies of electrochemical detection of proteins and determination of their concentration in solution. A variety of techniques from radioactive labeling [57, 58, 59] to infrared reflection-adsorption spectroscopy [59], electron microscopy [60], ellipsometry [61], cyclic voltammetry [62, 63], and electrochemical impedance spectroscopy [64, 65, 66, 67, 68, 69, 70, 71, 72, 73] have been used to characterize proteins in solutions and at the interface. Results from these studies showed that electroactive proteins often exhibit strong adsorption phenomena on metal and nonmetal surfaces (such as titanium oxide, platinum, iridium, iridium oxide, 316 L stainless steel, gold, and glassy carbon), hindering electron-transfer rates at the electrode-solution interface and sometimes disabling implanted medical devices. Adsorption is affected by a number of factors, including temperature, pH, ionic strength, bulk concentration and conformation of proteins in solution. Protein resistivity can be as high as $\sim 10^8$ ohm cm, resulting in a measurable interfacial resistance of ~ 1kohm for a nanometer-thick single monolayer film. EIS has been employed to treat the protein-electrode system as an example of electron-transfer kinetics of adsorbed electroactive layers as a function of applied electrochemical overpotential [74].

Usually the most prominent attribute of the protein-containing systems analyzed by EIS is the low-frequency impedance related to protein adsorption. Often a large depressed capacitive semicircle and/or Warburg semi-infinite linear diffusion characteristics are observed at low frequency [56]. General Faradaic processes involving other electroactive ionic species in the analyzed system can also be affected by the presence of bulk-solution proteins. These kinetic processes are affected by the applied electrochemical potential; types of electrode; solution pH; concentration of the proteins and other electroactive species; degree of protein adsorption at the interface; and formation of proteins films of various thickness, porosity, levels of insulation and catalytic activity. Considering these multiple parameters, development of a universal equivalent-circuit model for the protein system is a somewhat ambiguous task.

In this chapter a few of the most likely outcomes for protein-containing electrochemical redox systems are articulated. Electrode surface coverage for electroactive proteins can theoretically vary between 0 and 100 percent and often depends on electrochemical potential. The model takes into account solution resistance R_{SOL}, the low-frequency feature representing the double-layer capacitance CPE_{DL}, charge-transfer resistance R_{CT}, and mass-transfer (diffusion) impedance Z_{DIFF}. The adsorption-limited interfacial kinetics of electroactive proteins can be described by a formation of a barrier to current with a resistance R_{ADS}, accounting for the partial charge transfer from strongly adsorbed proteins and capacitance C_{ADS} both placed in parallel with the Faradaic impedance of electroactive solution species (Figure 11-9). Both solution- and protein-dependent elements of the circuit are concentration- and applied electrochemical potential-dependent.

There are several options to consider. If the protein coverage is less than 100 percent and the proteins do not participate in redox reaction catalysis, then a simple model of proteins partially blocking the electrode surface should be most realistic. Nonelectroactive protein adsorption increases R_{CT} and decreases CPE_{DL} for discharging solution electroactive species as the electrode surface area available for the reaction decreases. For that case, parallel interfacial electrochemical processes are being developed, including electroactive solution species discharge (where double-layer capacitance represented by CPE_{DL} is in parallel with Faradaic process $R_{CT} - Z_{DIFF}$) and protein adsorption represented by $R_{ADS} \mid CPE_{ADS}$ (Figure 11-9A).

In a simplified circuit (Figure 11-9B) the low-frequency capacitive feature is described by a combined constant phase element CPE_{ADS} that accounts for charging of the double layer by electroactive ionic species (CPE_{DL}) and protein adsorption pseudocapacitance C_{ADS}, in parallel with a series combination of ionic charge transfer resistance R_{CT} and diffusion impedance Z_{DIFF}. This low-frequency feature is also in parallel with the protein adsorption resistance R_{ADS}, accounting for partial electron exchange between the electrode and the adsorbed proteins, which in many cases may become very large and can be eliminated from the circuit. Due to the distribution of relaxation times on the electrode surface, a CPE_{ADS} is a more appropriate representation for the adsorption of electroactive species and charging of the double layer than a pure capacitance. This distribution is generally caused by different active sites on the proteins and irregularities in the electrodes, causing inhomogeneities at the electrode-solution interface. This experimental situation will result in the appearance of two capacitive semicircles in the Nyquist plot—one at high frequency due to transport-impeding effects of the protein film and the low-frequency feature due to the charge transfer of electroactive species at the electrode.

If electrode surface coverage by the adsorbed proteins approaches 100 percent, the entire low-frequency feature may effectively become reduced to a parallel combination of $R_{ADS} \mid CPE_{ADS}$, as no free space will remain for the other

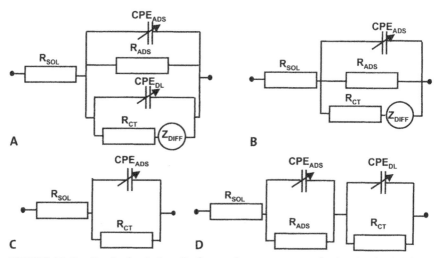

FIGURE 11-9 Equivalent circuits for surface coverage of adsorbed proteins in ionic solution: A. partial coverage; B. simplified model for partial coverage; C. full coverage; D. alternative model for full coverage

electroactive species to discharge [64, 65]. Depending on the electrochemical potential, both resistive components R_{CT} and R_{ADS} can be combined into effective R_{CT}, resulting in an appearance of a single depressed resistive-capacitive semicircle with possible sluggish diffusion at low frequency (Figure 11-9C). If the applied electrochemical potential is not sufficient to promote electron exchange, then charge transfer becomes very sluggish with very large $R_{CT} = R_{ADS}$ ~ 1 kohm per adsorbed protein monolayer.

Typically the process of protein adsorption is facile, resulting in near unity coverage at the interface. In the spontaneous adsorption processes of a globally uncharged protein molecule (at a pH value equal to the isoelectric potential) on a positive-charged TiO_2 surface [75], the driving force of the adsorption process may arise from interactions produced by structural modifications of the protein in contact with the metal electrode as well as specific chemical interactions. Depending on the electrochemical potential, the electroactive solution species may still undergo charge transfer at the electrode interface after they penetrate the adsorbed layer of proteins at the surface. For example, strong impedance dependence on pH evidences the participation of hydrogen ions either from the bulk solution or from exchange between the adsorbed protein layer and the (hydr)oxide surface groups in the electrode. The discharging species will have to be transported through the protein film and discharge, resulting in two semicircles representing the protein film impedance to transport of the electroactive species and their charge transfer at the interface (Figure 11-9D).

Figure 11-10A [75, 76] shows the Nyquist spectra at +1.00 V at TiO_2 electrodes in solution containing electroactive $[Fe(CN)_6]^{3-/4-}$ ions (pH 4.7) with and without the addition of HSA into the bulk solution. In the absence of the protein, the equivalent circuit used to fit the experimental data is reduced to a typical Randles circuit containing Warburg diffusion and $[Fe(CN)_6]^{3-/4-}$ charge-transfer resistance elements in parallel with a double-layer capacitance (Figure 11-10A). In the presence of a full protein layer, additional processes related

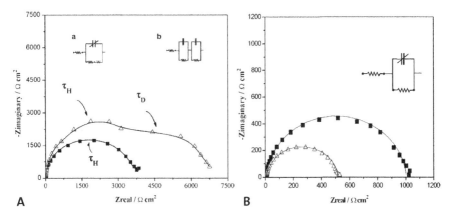

FIGURE 11-10 TiO_2 electrode impedance measured in 0.1 M NaCl + $1 \cdot 10^{-2}$ M $[Fe(CN)_6]^{3-/4-}$ ion, pH 4.7 with (open triangle) and without (solid square) 0.1 g/L HSA addition obtained at : A. +1.00 V; B. -0.185 V [75, 76] (with permission – copyright Elsevier, 2005)

to the effect of the protein adsorption layer appear. The impedance modulus is significantly higher in the presence of protein, indicating that the $[Fe(CN)_6]^{3-/4}$ charge-transfer-related process is inhibited. The proposed equivalent circuit (Figure 11-10A) accounts for the two time constants that are evidenced by the two capacitive loops in the Nyquist plot, which is analogous to Figure 11-9D. The resistive element R_{SOL} is associated with the solution resistance, which is in series with two parallel combinations of a capacitance and a resistance, assigned to the charge-transfer process at the interface and the resistance of the protein film to mass transport of the $[Fe(CN)_6]^{3-/4+}$ ions through the film. The different behavior observed could be explained in terms of changes in the distribution of the potential drop between the electrolyte and the electrode. The electrochemical response of the system is mainly determined by the blocking effect of the protein-adsorbed layer on the electrode surface. This general effect was also observed for the adsorption of 125I-radiolabeled Immunoglobulin G on a TiO_2 electrode surface by EIS at positive potentials [76]. Activation in a narrow potential range and a general blocking effect of the adsorbed protein layer were also observed. Similar models describing adsorption of electroactive species at the electrode interfaces were reported earlier [77].

However, at some electrochemical potentials adsorbed protein-mediated catalysis of solution electroactive species redox reactions may take place. This results in a more facile charge transfer, current increase, and decrease in R_{CT} with the formation of a single Faradaic-process-related impedance feature, such as in the example of $[Fe(CN)_6]^{3-/4+}$ discharge at -0.185 V at TiO_2 electrodes (Figure 11–10B) [76]. In this example, the Nyquist complex plane analysis showed that the initial high impedance values were lowered after the HSA protein was added to the electrochemical cell. For the protein-oxide interface the changes observed in the equivalent circuit elements as a function of electrode potential and solution pH are determined by the physicochemical processes occurring in the Helmholtz compact portion of the double layer. The experimental observable properties of the adsorbed protein layer could be explained using chemical catalytic reactions involving acid-base groups from the protein and the oxide surface. The CPE element is associated with a capacitance of the protein layer CPE_{ADS}, and the R_{CT} element was associated with the charge-transfer-related processes. In the presence of proteins, R_{CT} decreased almost by half, the CPE capacitance increased, and the α–parameter decreased, indicating more facile kinetics and pronounced distributed behavior with an adsorbed layer of protein on the electrode surface (Figure 11-9C).

The protein film resistance R_{ADS} and corresponding capacitance CPE_{ADS} can be related to the amount of protein adsorbed onto the electrode surface [64]. The inverse of adsorption resistance, R_{ADS}^{-1}, is directly proportional to the surface-charge density of the protein-adsorbed layer. The direct relationship between the amount of adsorbed protein Γ (mol cm²) and the surface-charge density q_{ADS} (C cm⁻²) is shown in Eq. 11-11:

$$\Gamma = \frac{q_{ADS}}{zF} \tag{11-11}$$

The direct relationship of C_{ADS} to Γ is displayed as [74]:

$$C_{ADS} = \frac{F^2 A\Gamma}{4R_G T} \qquad (11\text{-}12)$$

where A is the area of the electrodes. The interfacial capacitive charging represented in the equivalent circuit by CPE_{ADS} is in strict sense a frequency-dependent combination of the double-layer capacitance and adsorption pseudocapacitance [78] (Figure 11-9B). As was mentioned above, the relative contributions of adsorption and double-layer capacitive features are often difficult to deconvolute, and the analysis is simplified by assuming that the process of adsorption of electroactive proteins is a contributor to total measured interfacial capacitance C_{ADS}. The CPE values can be transformed into a pure capacitance C_{ADS} by using Eq. 3-6.

As expected, with an increase in relative concentration of proteins, the amount of electroactive protein adsorption at the interface increases, leading to increases in C_{ADS} and q_{ADS}. Langmuir adsorption (Eq. 11-13) provides a relationship between the concentration of the protein in solution C^* and the amount of material adsorbed on the surface [60]:

$$\frac{C^*}{\Gamma} = \frac{1}{\gamma_{ADS}\Gamma_{MAX}} + \frac{C^*}{\Gamma_{MAX}} \qquad (11\text{-}13)$$

where γ_{ADS} = the adsorption coefficient and Γ_{MAX} = the maximum amount of material that can adsorb on the surface. Plotting C^*/Γ vs. C^* yields a straight line, and Γ_{MAX} and γ_{ADS} can be derived from the slope and intercept, respectively. B_{ADS} is related to the affinity of the protein to adsorption sites on the electrode surface at the measured temperature. This value can be related to the Gibbs energy of adsorption, ΔG_{ADS} [79] through:

$$\gamma_{ADS} = \frac{1}{55.5}\exp\left(\frac{-\Delta G_{ADS}}{RT}\right) \qquad (11\text{-}14)$$

where 55.5 represents the molar concentration of water (mol L^{-1}) when it is used as the solvent. Large negative values of ΔG_{ADS} indicate spontaneous protein adsorption onto the electrode surface. Determination of the amount of adsorbed protein, surface-charge density, and other adsorption characteristics from the EIS data at different concentrations of protein permits calculating the adsorption coefficient and the maximum amount of protein material that could be present at the electrode surface.

11.4. Impedance monitoring of implanted devices

Traditionally electrode impedance has been utilized for testing of potency of the electrode insulation after implantation. This was done by measuring the magnitude and phase at a single frequency of ~1 kHz, mainly because this is the fundamental frequency of a neuronal action potential corresponding to a time period of 1 ms. It was believed that this would be the frequency component of the electrode impedance that would primarily affect the recorded signal from surrounding neurons. Due to the fact that most traditional neurophysiology electrode-impedance monitoring equipment is set up to measure the impedance magnitude at 1 kHz only, it would be useful to be able to correlate 1 kHz impedance to the state of tissue around an implanted electrode. Typically these types of implant monitoring demonstrate a general increase in the measured impedance magnitude at 1 kHz, which reaches a peak several days post-implant. Histological analyses suggest that tissue reactions are confined to localized regions of the implants around the monitoring electrodes [20].

The broadband impedance spectroscopy data was used to develop a more comprehensive model for a typical impedance response of an implant (Figure 11-11) [20]. The media around an implant consists of biological cells (Section 11-2) and adsorbing proteins (Section 11-3). A three-component model consisting of an electrode–tissue interface, an encapsulation (protein adsorption) region, and a neural cellular tissue component is sufficient to represent an implant-tissue system. The impedance profiling of the system reveals several semicircular arcs in the Nyquist space, typical of a distributed network of lumped parallel resistive and capacitive pathways. Due to the commutative property of electrical components in series, models based upon impedance measurements cannot be used to explicitly determine spatial distribution, although when paired with histological observations spatial distributions of model components may be inferred.

The electrode–tissue interface of this model, simplified as $Z_{INTERFACE}$, has been well established. It consists of a capacitive element C_{DL} [46, 80] representing double-layer interfacial charging in parallel with a Faradaic component representing a poorly defined charge-transfer resistance R_{CT} and diffusion impedance Z_{DIFF} [81]. From the low-frequency impedance data in Figure 11-11, it appears that a transmission line (represented by a CPE with relevant $Z_{INTERFACE}$ impedance) may be in fact another representation of the interfacial impedance. The CPE component reflects 3D current distribution due to variability in length of conducting passes between the electrode surface and electroactive components in the tissue, electrode surface roughness, and porosity effects. Diffusion impedance to mass transport of discharging species to the interface appears at the lowest frequencies. Additionally, a shunt capacitance or inductance (C_{SH}), apparent only at very high frequencies, may arise from the finite capacitive coupling between the electrode lead wires and the surrounding electrolyte solution. In experiments using microwire arrays, the uncompensated impedance of the cables is only apparent at frequencies exceeding 1 MHz (Section5-1). C_{SH} will have a more substantial effect when using thin film electrodes, which have relatively thin dielectric layers and close lead spacing. These interfacial-impedance kinetics and cell-geometry-related values (CPE, R_{CT}, Z_{DIFF}, and C_{SH}) can be experimentally determined based upon preimplant impedance tests in a physiological buffer solution.

As the current propagates toward ground in vivo, it would next encounter the protein-rich encapsulation (adsorption) layer. This medium-frequency impedance region is composed primarily of extracellular matrix proteins and is considered to be largely resistive [82, 83, 84, 85]. As was shown in Section 11-3, this zone can be modeled as a parallel combination of protein-adsorption resistance (R_{ADS}) and capacitance (C_{ADS}) but also can be simplified as encapsulation resistance $R_{EN} = R_{ADS}$. The protein-adsorption capacitance is often masked by its combination with other electrode-surface charging processes, represented here as a generic transmission line CPE element describing the electrode interfacial impedance (Figure 11-9B).

Subsequently, at higher frequencies the current would propagate outward and come across the bulk medium, which can be represented by a parallel combination of a resistive extracellular fluid and cell membranes. As was discussed above in Section11-2, the "bulk"-solution current at high frequencies can be conducted along two pathways, either around or between cells in extracellular media or through and across cells (intracellular). The extracellular medium is assumed to be principally resistive and is represented by a purely resistive element (R_{EX}). At low cell densities, there are sufficient numbers of low-resistance pathways for current flow to follow, such that the capacitance contribution is negligible. As the cell density increases, the extracellular space decreases, thus diminishing the low-resistance pathway. For a positive value of the frequency-dependent Clausius-Mossotti factor, the current flows through a capacitive (membrane) and resistive (cytoplasm) cellular compartments, resulting in the β-relaxation. An impedance of this pathway is simplified by impedance Z_{CELL}, which is characterized by membrane conductance (G_{MBR}) and capacitance (C_{MBR}) scaled by the total cell membrane area (A_{MBR}). The conductance and capacitance are constant parameters drawn from the literature [81]. Qualitatively, the impedance spectral signatures match those reported for chronic implantation of various types of electrodes in vitro [86, 87, 88, 89].

The impact of the surrounding tissue on the impedance implant measurement is heavily influenced by the volume immediately surrounding the electrode (primarily protein adsorption), as the voltage drops off in all directions as it spreads through the tissue volume to the ground. Therefore, the influence of reactive cells and proteins will have a more significant effect as the cell density immediately adjacent to the electrode surface increases and clear impedance spectral signatures arise in cases where there is significant accumulation of dense reactive cells and proteins in the immediate volume adjacent to the electrode. The model also implies the infiltration of reactive cells and increased cellular density in close proximity to the electrode site. These results suggest that changes in impedance spectra are directly influenced by cellular distributions around implanted electrodes over time and that impedance measurements may provide an online assessment of cellular reactions to implanted devices [20].

In this model, one of the fundamental assumptions is that the impedance of the electrodes does not change significantly over time. Alternatively, it could be proposed that the interaction of cells and tissue with the electrode surface fundamentally alters the properties of the electrode–surface interactions. In the literature, models have been proposed that suggest that the interaction between electrodes and cells could possibly affect the properties of both the electrode and the cell membrane [90]. There are observations that support these

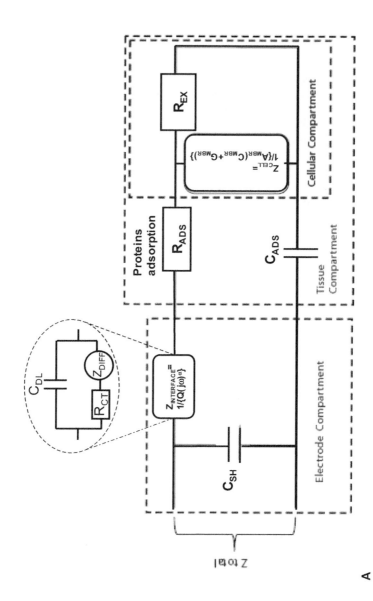

A

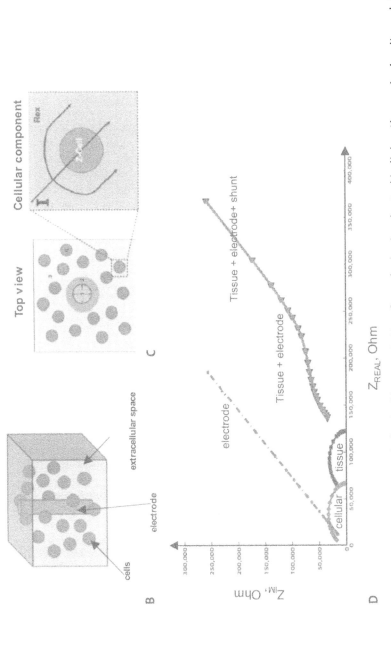

FIGURE 11-11 Model circuit diagram and physical interpretation for a device implanted in living tissue: A. circuit model representing the impedance variation commonly found during in vivo impedance spectroscopy; B. three-dimensional and C. two-dimensional rendering of theorized physical components of the model; D. Nyquist representation of expected model results from a typical in vivo impedance spectrum [20] (with permission from IOPP)

assumptions, namely that the change in the electrode impedance in saline from pre-implantation conditions is not significant compared to the changes in the impedance spectra observed over the implant duration. This is also supported by experimental studies in culture that suggest that intimate cell-electrode interactions did not significantly alter electrode parameters [81]. While this does not necessarily imply that interactions between the electrode tissue and the adjacent tissue do not alter the fundamental properties of the electrode characteristics, it appears that the changes in the electrical properties of the surrounding volume and frequent formation of scar tissue contribute more significantly to the observed changes in impedance over time. What is not understood are the long-term effects of implantation on the electrode state, since the bulk of this study involved experiments with relatively short implant durations. Further experiments and a better understanding of the mechanisms involved in electrode-tissue interactions will be useful for future electrode-tissue model development and refinement.

References

1. G. S. Wilson, R. Gifford, *Biosensors for real-time in vivo measurements*, Biosens. Bioelectr., 2005, 20, pp. 2388–2403.
2. R. M. Wightman, P. Runnels, K. Troyer, *Analysis of chemical dynamics in microenvironments*, Anal. Chim. Acta,1999, 400, 1, pp. 5–12.
3. T. P. Bothwell, H. P. Schwan, *Studies of deionization and impedance spectroscopy for blood analyzer*, Nature, 1956, 178, 4527, pp. 265–266.
4. T. Jones, *Basic theory of dielectrophoresis and electrorotation*, IEEE Eng. Med. And Biol., 2003, Vol. 6, pp. 33–42.
5. J. Gimsa, *Characterization of particles and biological cells by AC electrokinetics*, Interfac. Electrokinetics and Electrophoresis, 2001, 13, pp. 369–400.
6. K. Asami, T. Yonezawa, H. Wakamatsu, N. Koyanagi, *Dielectric spectroscopy of biological cells*, Bioelectrochem. Bioelectr., 1996, 40, pp. 141–145.
7. T. Hanai, *Electrical properties of emulsions*, Emulsion Science, P. Sherman (Ed.), Academic Press, London, 1968, pp. 467–470.
8. L. L. Sohn, O. A. Saleh, G. R. Facer, A. J. Beavis, R. S. Allan, D. A. Notterman, *Capacitance cytometry: Measuring biological cells one by one*, Proceedings of the National Academy of Sciences, 2000, 97, 20, pp. 10687–10690.
9. Z. Gagnon, J. Gordon, S. Sengupta, H.-C. Chang, *Bovine red blood cell starvation age discrimination through a glutaraldehydeamplified dielectrophoretic approach with buffer selection and membrane cross-linking*, Electrophoresis, 2008, 29, pp. 2272–2279.
10. Y. Qiu, R. Liao, X. Zhang, *Impedance-based monitoring of ongoing cardiomyocyte death induced by tumor necrosis factor-A*, Biophys. J., 2009, 96, pp.1985–1991.
11. A. Caduff, F. Dewarrat, M. Talary, G. Stalder, L. Heinemann, Yu. Feldman, *Non-invasive glucose monitoring in patients with diabetes: a novel system based on impedance spectroscopy*, Biosens. Bioelectr., 2006, 22, 5, pp. 598–604.
12. C. W. Chia, C. D. Saudek, *Glucose sensors: toward closed loop insulin delivery*, Endocrinology & Metabolism Clinics of North America, 2004, 33, pp. 175–195.
13. M. A. C. Brett and A. M. O. Brett, *Electrochemistry: Principles, Methods and Applications*, 1993, New York: Oxford Science.

14. M. Varshney, L.J. Yang, X.L. Su, Y.B. Li, *Magnetic nanoparticle-antibody conjugates for the separation of Escherichia Coli O157: H7in ground beef*, J. Food Prot.,2005, 68, pp. 1804–1812.

15. M. Varshney, Y.B. Li, B. Srinivasan, S. Tung, *A label-free, microfluidics and interdigitated array microelectrode-based impedance biosensor in combination with nanoparticles immunoseparation for detection of Escherichia coli O157:H7 in food samples*, Sens. Act. B: Chemical, 2007, 128, pp. 99–107.

16. I. H. Boyaci, Z. P. Aguilar, M. Hossain, H. B. Halsall, C. J. Seliskar, W. R. Heineman, *Amperometric determination of live Escherichia coli using antibody-coated paramagnetic beads*, Anal. Bioanal. Chem., 2005, 382, pp.1234–1241.

17. L. Yang, R. Bashir, *Electrical/electrochemical impedance for rapid detection of foodborne pathogenic bacteria*, Biotech. Adv.,2008, 26, pp. 135–150.

18. D. Ivnitski, I. Abdel-Hamid, P. Atanasov, E. Wilkins, *Biosensors for detection of pathogenic bacteria*, Biosens. Bioelectr., 1999, 14, pp. 599–624.

19. V. F. Lvovich, A. Scheeline, *Amperometric sensors for simultaneous superoxide and hydrogen peroxide detection*, Anal. Chem., 1997, 69, pp. 454-462.

20. J. C. Williams, J. A. Hippensteel J. Dilgen, W. Shain, D. R. Kipke, *Complex impedance spectroscopy for monitoring tissue responses to inserted neural implants*, J. Neural. Eng., 2007, 4, pp. 410–423.

21. R. Pethig, *Dielectric and Electronic Properties of Biological Materials*, J. Wiley & Sons, New York, 1979.

22. M. Yang, X. Zhang, *A novel impedance assay for cardiac myocyte hypertrophy sensing*, Sens. Act A, 2007, 136, pp. 504–509.

23. R. Ehret, W. Baumann, M. Brischwein, A. Schwinde, K. Stegbauer, B. Wolf, *Monitoring of cellular behaviour impedance measurements on by interdigitated electrode structures*, Biosens. Bioelectr., 1997, 12, 1, pp. 29–41.

24. M. DeSilva, Y. Zhang, P. J. Hesketh, G. J. Maclay, S. M. Gendel, J. R. Stetter, *Impedance based sensing of the specific binding reaction between Staphylococcus enterotoxin B and its antibody on an ultra-thin platinum film*, Biosens. Bioelectr., 1995, 10, pp. 675–682.

25. C. M. Harris, D. B. Kell, *The estimation of microbial biomass*, Biosensors,1985, 1, pp. 17–84.

26. E. Palmqvist, C. Berggren Kriz, M. Khayyami, B. Danielsson, P.- O. Larsson, K. Mosbach, D. Kriz, *Development of a simple detector for microbial metabolism, based on a polypyrrole resistometric device*, Biosens. Bioelectr., 1994, 9, pp. 55 1–556.

27. M. Stelze, E. Sackmann, E., *Sensitive detection of protein adsorption to supported bilayers by capacitance measurements*, GBF Monogr., 1989, 13, pp. 339–346.

28. E. Gheorghiu, *Measuring living cells using dielectric spectroscopy*, Bioelectrochem. Bioelectr,. 1996, 40, pp. 133–139.

29. R. Lisin, B. Z. Ginzburg, M. Schlesinger, Y. Feldman, *Time domain dielectric spectroscopy study of human cells I*, Biochim. Biophys. Acta, 1996, 1280, pp. 34–40.

30. Y. Polevaya, I. Ermoline, M. Schlesinger, B. Z. Ginzburg, Y. Feldman, *Time domain dielectric spectroscopy study of human cells II*, Biochim. Biophys. Acta, 1999, 1419, pp. 257–271.

31. A. DiBiasio, C. Cametti, *Effect of the dielectric properties of biological cell suspensions*, Bioelectrochem., 2007, 71, 2, pp. 149–156.

32. X. Cheng, Y. Liu, D. Irimia, U. Demerci, L. Yang, L. Zamir, W. Rodriguez, M. Toner, R. Bashir, *Cell detection and counting through cell lysate impedance spectroscopy in microfluidic devices*, Royal Soc. Chem., 2007, 7, pp. 746–755.

33. H. P. Schwan, H. J. Morowitz, *Electrical properties of the membranes of the pleuropneumonia-like organism A 5969*, Biophys. J., 1962, 2, pp. 395–407

34. H. P. Schwan, S. Takashima, V. K. Miyamoto, W. Stoeckenius, *Electrical properties of phospholipid vesicles*, Biophys. J., 1970, 10, pp. 1102–1119.

35. T. Houssin, J. Follet, A. Follet, E. Dei-Cas, V. Senez, *Label-free analysis of water-polluting parasite by electrochemical impedance spectroscopy*, Biosens. Bioelectr., 2010, 25, 5, pp. 1122–1129.

36. G. Schwarz, *A theory of the low-frequency dielectic dispersion of colloidal particles in electrolyte solution*, J. Phys. Chem., 1962, 66, pp. 2636–2642.

37. G. Schwarz, *Dielectic relaxation of biopolymers in solution*, Adv. Mol. Relaxation Processes, 1972, 3, pp. 281–295.

38. V. Lvovich, S. Srikanthan, R. L. Silverstein, *A novel broadband impedance method for detection of cell-derived microparticles*, Biosens. Bioelectr., 2010, 26, 2, pp. 444–451.

39. S. Arndt, J. Seebach, K. Psathaki, H. -J. Galla, J. Wegener, *Bioelectrical impedance assay to monitor changes in cell shape during apoptosis*, Biosens. Bioelectr., 2004, 19, 583–594.

40. C. J. Felice, M. E. Valentinuzzi, *Medium and interface components in impedance microbiology*. IEEE Transactions on Biomed. Eng., 1999, 46, pp. 1483–1487.

41. R. Diaz, S. Payen, *Biological cell separation using dielectrophoresis in a microfluidic device*, robotics.eecs.berkeley.edu/~pister/245/project/DiazPayen.pdf.

42. M. –W. Wang, *Using dielectrophoresis to trap nanobead/stem cell compounds in continuous flow*, J. Electrochem. Soc., 2009, 156, 8, pp. G97–G102.

43. Y. Kang, B. Cetin, Z. Wu, D. Li, *Continuous particle separation with localized AC-dielectrophoresis using embedded electrodes and an insulating hurdle*, Electrohim. Acta, 2009, 54, pp. 1715–1720.

44. P. K. Wong, T.-H. Wang, J. H. Deval, C. –M. Ho, *Electrokinetics in micro devices for biotechnology applications*, IEEE Trans. Mechatronics, 2004, 9, 2, pp. 1–12.

45. P. J. Burke, *Nanodielectrophoresis: electronic nanotweezers*, Encyclopedia of Nanoscience and Nanotechnology, H. S. Nalva (Ed.), 2003, 6, pp. 623–641.

46. E.T. McAdams, J. Jossinet, *Problems in equivalent circuit modelling of the electrical properties of biological tissues*, Bioelectrochem. Bioengin., 1996, 40, pp. 147–152.

47. E. T. McAdams, J. Josinette, *Tissue impedance: a historical overview*, Physiol. Meas., 1995, 16, p. A1–A13.

48. E. T. McAdams, A. Lackermeier, T. McLaughlin, D. Macken D, J. Jossinet, *The linear and non-linear electrical properties of the electrode-electrolyte interface*, Biosensor. Bioelectron., 1995, 10, pp. 67–74

49. G. Leung, H. Tang, R. McGuinness, E. Verdonk, J. Michelotti, V. Liu V., *Cellular dielectric spectroscopy: a label-free technology for drug discovery*, J. Ass. Lab. Autom., 2005, 10, 4, pp 258–269.

50. R. Cecil, *The Proteins*, H. Neurath, Ed., Vol. 1, Ch.5. Academic Press, New York,1963.

51. M. Stankovich, A. J. Bard, *The electrochemistry of proteins and related substances: part III. Bovine Serum Albumin*, J. Electroanal. Chem., 1978, 86, pp.189–199.

52. Y. Wu, X. Ji., S. Hu, *Studies on electrochemical oxidation of azithromycin and its interaction with bovine serum albumin*, Bioelectrochemistry, 2004, 64, pp. 91–97.

53. J. G. E. M. Fraaije, W. Norde, J. Lyklema, *Interfacial thermodynamics of protein adsorbtion and ion co-adsorption. III. Electrochemistry of bovine serum albumin adsorption on silver iodide*, Biophys. Chem., 1991, 41, pp. 263–276.

54. M. Liu, G. A. Rechnitz, K. Li, Q. X. Li, *Capacitive immunosensing of polycyclic aromatic hydrocarbon and protein conjugates*, Analytical Letters, 1998, 31, 12, pp. 2025–2038.

55. M. Katterle, U. Wollenberger, F. Scheller, *Electrochemistry of hemoglobin at modified silver electrodes is not a redox-process of iron protoporphyrin*, Electroanalysis,1997, 9, pp. 1393–1396.

56. M. F. Smiechowski, V. F. Lvovich, S. Roy, A. Fleischman, W. H. Fissell, A. T. Riga, *Electrochemical detection and characterization of proteins*, Biosens. Bioelectr., 2007, 22, 5, pp. 670–677.

57. T. Arnebrant, B. Ivarsson, K. Larrson, I. Lundstrom, T. Nylander, *Bilayer formation at adsorption of proteins from aqueous solutions on metal surfaces*, Prog. Colloid Polym. Sci., 1985, 70, pp. 62–66.

58. T. Arnebrant, T. Nylander, *Sequential and competitive adsorption of -lactoglobulin and -casein on metal surfaces*, J. Colloid Interface Sci., 1986, 111, pp. 529–533.

59. B. B. Damaskin, O. A. Petrii, V. V. Batrakov, Adsorption of Organic Compounds on Electrodes, Plenum, New York, 1971.

60. R. G. Lee, S. W. Kim, *Adsorption of proteins onto hydrophobic polymer surfaces: adsorption isotherms and kinetics*, J. Biomed. Mater. Res., 1974, 8, 5, pp. 251–259.

61. B. A. Ivarsson, P. Hegg, K. I. Lundstrom, U. Jonsson, *Adsorption of proteins on metal surfaces studied by ellipsometric and capacitance measurements*, Colloids Surf., 1985, 13, pp. 169–192.

62. S. G. Roscoe, K. L. Fuller, K.L., *Interfacial behaviour of globular proteins at a platinum electrode*, J. Colloid Interface Sci., 1992, 152, pp. 429–441.

63. S. M. MacDonald, S. G. Roscoe, *Electrochemical studies of the interfacial behavior of insulin*, J. Colloid Interface Sci., 1996, 184, pp. 449–455.

64. J. E. I. Wright, N. P. Cosman, K. Fatih, S. Omanovic, S. G. Roscoe, *Electrochemical impedance spectroscopy and quartz crystal nanobalance (EQCN) studies of insulin adsorption on Pt*, J. Electroanal. Chem., 2004, 564, pp. 185–197.

65. S. E. Moulton, J. N. Barisci, A. Bath, R. Stella, G. G. Wallace, *Studies of double layer capacitance and electron transfer at a gold electrode exposed to protein solution"*, Electrochim. Acta, 2004, 49, pp. 4223–4230.

66. N. P. Cosman, K. Fatih, S. G. Roscoe, *Electrochemical impedance spectroscopy study of the adsorption behaviour of α-lactalbumin and β-casein at stainless steel*, J. Electroanal. Chem., 2005, 574, pp. 261–271.

67. S. Omanovic, S.G. Roscoe, *Electrochemical studies of the adsorption behavior of bovine serum albumin on stainless steel*, Langmuir,1999, 15, 23, pp. 8315–8321.

68. R. K. R. Phillips, S. Omanovic, S. G. Roscoe, *Electrochemical studies of the effect of temperature on the adsorption of yeast alcohol dehydrogenase at Pt*, Langmuir, 2001, 17, 8, pp. 2471–2477.

69. S. G. Roscoe, *Electrochemical investigations of the interfacial behavior of proteins*, in: Modern Aspects of Electrochemistry, J. O. M. Brockis, B. E. Conway, R. E. White, Eds., Vol. 29, Plenum Press, New York (1996).

70. S. G. Roscoe, K. Fuller, G. Robitalle, *An electrochemical study of the effect of temperature on the adsorption behavior of β-Lactoglobulin*, J. Colloid Interface Sci., 1993, 160, pp. 243–251.

71. R. Rouhana, S. M. Budge, S. M. Macdonald, S. G. Roscoe, *Electrochemical studies of the interfacial behaviour of α–lactalbumin and bovine serum albumin*, Food Res. Int., 1997, 30, 5, pp. 303–310.

72. K. Hanrahan, S. M. MacDonald, S. G., Roscoe, *An electrochemical study of the interfacial and conformal behaviour of cytochrome c and other heme proteins*, Electrochim. Acta, 1996, 41, pp. 2469–2479.

73. F. Y. Olivia, L. B. Avalle, V. A. Macagno, C. P. De Paulli, *Study of human serum albumin-TiO2 nanocrystalline electrodes interaction by impedance electrochemical spectroscopy*, Biophys. Chem., 2001, 91, 2, pp. 141–155.

74. A. J. Bard, L. R. Faulkner, Electrochemical Methods, Fundamentals and Applications, John Wiley & Sons, New York, 2001.

75. F. Y. Oliva, L. B. Avalle, O. R. Camara, *Electrochemical behaviour of human serum albumin–TiO2 nanocrystalline electrodes studied as a function of pH: Part 1. Voltammetric response*, J. Electroanal. Chem., 2002, 534, pp. 19–29.

76. L. B. Avalle, O. R. Camara, F. Y. Oliva, *Electrochemical behavior of human serum albumin–TiO$_2$ nanocrystalline electrodes studied as a function of pH. Part 2: Presence of [Fe(CN)6]$^{3-/4-}$ redox couple in solution*, J. Electroanal. Chem., 2005, 585, pp. 281–289.

77. J. R. MacDonald, *Impedance Spectroscopy*, John Wiley & Sons, New York, 2005.

78. V. D. Jovic, B. M. Jovic, B.M., *EIS and differential capacitance measurements onto single crystal faces in different solutions*, J. Electroanal. Chem.,2003, 541, pp. 1–21.

79. G. K. Gomma, M. H. Wahsan, M.H., *Effect of temperature on the acidic dissolution of copper in the presence of amino acids*, Mater. Chem. Phys., 1994, 39, 2, pp. 142–148.

80. J. R. Macdonald, *Impedance spectroscopy*, Annu. Biomed. Eng., 1992, 20, pp. 289–305.

81. J. R. Buitenweg, W. L. C. Rutten, W. P. A. Willems, van J. W. Nieuwkasteele, *Measurement of sealing resistance of cell-electrode interfaces in neuronal cultures using impedance spectroscopy*, Med. Biol. Eng. Comput., 1998, 36, pp. 630–637.

82. S. S. Stensaas, L. J. Stensaas, *Histopathological evaluation of materials implanted in the cerebral cortex*, Acta Neuropathol., 1978, 41, pp. 145–155.

83. D. J. Edell, V. V. Toi, V. M. McNeil, L. D. Clark, *Factors influencing the biocompatibility of insertable silicon microshafts in cerebral cortex*, IEEE Trans. Biomed. Eng., 1992, 39, pp. 635–643.

84. J. N. Turner, W. Shain, D. H. Szarowski, M. Andersen, S. Martins, M. Isaacson, H. Craighead , *Cerebral astrocyte response to micromachined silicon implants*, Exp. Neurol., 1999, 156, pp. 33–49.

85. W. M. Grill, J. T. Mortimer, *Electrical properties of implant encapsulation tissue*, Ann. Biomed. Eng., 1994, 22, pp. 23–33.

86. M. D. Johnson, K. J. Otto, D. R. Kipke, *Repeated voltage biasing improves unit recordings by reducing resistive tissue impedances*, IEEE Trans. Neural. Syst. Rehabil. Eng., 2005, 13, pp. 160–165.

87. M. D. Johnson, K. J. Otto, J. C. Williams, D. R. Kipke, *Bias voltage at microelectrodes change neural interface properties in vivo*, 26th Annu. Int. Conf. IEEE EMBS, September 2004.

88. A. H. Kyle, C. T. Chan, A. L. Minchinton, *Characterization of three-dimensional tissue cultures using electrical impedance spectroscopy*, Biophys. J., 1999, 76, pp. 2640–2648.

89. X. Liu, D. B. McCreery, R. R. Carter, L. A. Bullara, T. G. Yuen, W. F.Agnew, *Stability of the interface between neural tissue and chronically implanted intracortical microelectrodes*, IEEE Trans. Rehabil. Eng., 1999, 7, pp. 315–326.

90. M. Grattarola, S. Martinoia, *Modeling the neuron-microtransducer junction: from extracellular to patch recording*, IEEE Trans. Biomed. Eng., 1993, 40–41, pp. 35–41.

CHAPTER 12

Selected Examples of Impedance-Analysis Applications

12.1. Impedance analysis of insulating films and coatings

Metal parts exposed to outdoor or chemically aggressive environments are frequently protected from the effects of corrosion and other degradation processes by insulating film or paint. Such parts are represented in experimental laboratory conditions by metalic working electrodes covered with insulating film placed in corrosive solutions. An equivalent circuit for an electrode covered with intact ideal protective coating is represented by a series of solution resistance and very high-impedance insulating film (Figure 12-1) [1, 2].

The circuit for a purely capacitive coating is shown in Nyquist (Figure 12-2A) and Bode plots (Figure 12-2B) for the values of R_{SOL} = 500 ohm (realistic for a moderately conductive solution) and geometrical coating capacitance C_{FILM} = 200 pF/cm^2 (realistic for a 1 cm^2 sample, a 25 µM coating, and ε = 6). The capacitance of the intact coating C_{FILM} typically has a value in the pF-nF/cm^2 range, which is much smaller than a typical ~µF/cm^2 double-layer capacitance. The capacitance value can be determined from the imaginary portion of a Bode plot by using a curve fit or from an examination of the data points. The intercept of the curve with the real axis gives an estimate of the solution resistance R_{SOL}. The film-coating capacitance is in parallel with the coating resistance R_{FILM}, which for many insulating organic films may be as high as 10^7–10^{10} ohm/cm^2, which is close to the limit of measurement of most impedance analyzers and therefore may not even reveal itself in the plotted data. In this case the plot reveals a capacitive −90° line at low frequencies instead of a resistance-capacitance semicircle (Figure 12-2).

281

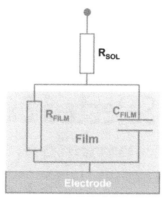

FIGURE 12-1 Representative equivalent circuit for fresh, intact coating

Most coatings eventually degrade with time, resulting in more complex behavior. For example, after a certain amount of exposure to corrosive environments, water penetrates through pores in the coating. As a result of this process, a new liquid/metal interface is formed under the coating. Additional corrosion-related impedance phenomena can also occur at this new interface [2]. Water uptake into the film is usually a fairly slow process that can be monitored by taking EIS spectra at set time intervals. EIS can determine the kinetics of water interaction with polymer film and thus the stages of film degradation [1, 2], based on the following principles:

1. If a simple layer-by layer-dissolution of a polymer takes place, the film thickness will decrease, the measured film capacitance will increase,

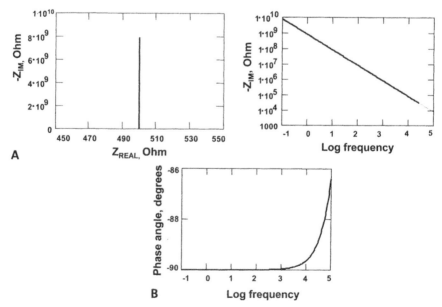

FIGURE 12-2 Typical A.; Nyquist; B. Bode plots for fresh, intact coating

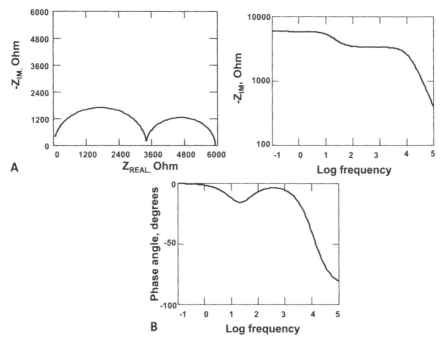

FIGURE 12-3   A. Nyquist; B. Bode plots for degraded coating

and the film resistance R_{FILM} will decrease. This process results in simple changes of the film capacitance C_{FILM} and resistance R_{FILM}, while the shape of the EIS diagram will stay the same (Figure 12-2).

2. Water uptake may also lead to formation of water-filled pores and coating failure, with additional charge- transfer-related features appearing at low frequencies (Figure 12-3). The film resistance R_{FILM} will be effectively replaced by a pore resistance R_{PORE}, which is the resistance of ion-conducting paths that develop in the coating, and the film capacitance may be replaced by CPE_{FILM} as a result of higher film heterogeneity.

The resulting Nyquist and Bode plots (Figure 12-3) were generated for a 1 cm² sample of metal coated with a 12 μM film and five delaminated areas corresponding to one percent of the total metal area. The pores in the film that accesses these delaminated areas are represented as solution cylinders with a 30 μM diameter. $C_{FILM} = 400$ pF/cm², film permittivity $\varepsilon = 6$, $R_{PORE} = 3400$ ohm (calculated assuming $\sigma_{SOL} = 0.01$ S/cm), $R_{SOL} = 20$ ohm; $C_{DL} = 0.4$μF (calculated for one percent of 1 cm² area and assuming standard double layer capacitance of 40 μF/cm²), and $R_{CT} = 2500$ ohm (calculated for one percent of 1 cm² area using polarization resistance of $R_{POL} = 25$ ohm). The two time constants are very well resolved in the Nyquist plot (Figure 12-4) [3, 4, 5].

For degraded porous films, the discussion in Section 7-5 is generally valid. The high-frequency semicircle is largely related to the film porosity and independent of the kinetics of the Faradaic process at the electrode, while the

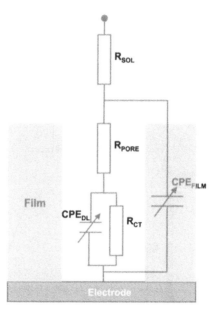

FIGURE 12-4   Equivalent circuit for degraded coating

low-frequency semicircle is determined by the Faradaic kinetics [4]. It is assumed that on the metal side of the pore an area of the coating has delaminated and a porous pocket filled with an electrolyte solution has formed. This electrolyte solution resistance inside the pore R_{PORE} can be very different from the bulk solution R_{SOL} outside the coating. The porosity of the coating can be estimated from the ratio of the resistance values of the "damaged" and original "intact" coating R_{PORE}/R_{FILM}. Most researchers agree that this model can be used to evaluate the quality of a coating.

Several additional physical processes potentially exist that can create somewhat more complicated equivalent circuit diagrams, as is evident for example from discussions in Chapter 10 on conductive polymer films. For instance, additional macrodefect corrosion and diffusion effects may develop between the coating and the surface, with an additional low frequency relaxation becoming visible in the Nyquist plot. The interface between a pocket of solution and the bare metal is modeled as a double-layer capacitance in parallel with a kinetically controlled charge-transfer resistance R_{CT}, which can also often include the diffusion element Z_{DIFF} associated with corrosion products in series with R_{CT}. If the diffusion element represents a finite diffusion, an additional $R_{DIFF}|CPE_{DIFF}$ element appears and a third relaxation at low frequencies, representing diffusion impedance to transport through metal-oxide films formed on the electrode as a result of corrosion [6]. Alternatively, the film delamination can be represented by a transmission line composed of many CPE-R combinations in parallel with $R_{CT}|CPE_{DL}$. The presence of significant and deep scratches and deformities in the protecting layer may introduce additional parallel conducting paths through these irregularities, provided that the scratch geometry is very different from that of the pores. The geometry of deep scratches may have a strong effect on the mass- and charge-transfer processes inside the scratch, which are different from these processes inside of the pores in the barrier coating. It may result in a more complicated model with several unknown parameters, such as several parallel Faradaic processes inside the scratch [7].

12.2. Impedance analysis of metallic paints

Metallic paints, often based on Zn or Cu, represent another important case study in impedance applications to the analysis of coatings and their degradation. As was shown in Section 12-1, many intact organic paints and films can be modeled as porous insulating polymeric coatings with a fairly homogeneous chemical composition. "Metallic" paints represent a different example of such systems. These types of paints are essentially solid composite materials with a very high content of dispersed conducting metal particles suspended in continuous insulating polymeric media or "binders." Hence, impedance monitoring of metallic paints before and during their exposure to water and corrosive electrolytes should take into account charge-transfer (corrosion) processes on the metal particles in addition to the effects of insulating coating delamination described in Section 12-1. For instance, zinc particles provide corrosive sacrificial protection to the metallic substrate as a result of a cathodic oxygen-reduction process on the particles ($O_2 + 2H_2O + 4e \rightarrow 4OH^-$) and anodic particles dissolution ($Zn \rightarrow Zn^{2+} + 2e$) [8, 9, 10].

As the first approximation, the impedance structure of the paints can be simplified by $R_{SOL} - (R_{FILM} \mid C_{FILM})$ for nondegraded "dry" film (Figure 12-1) and $R_{SOL} - R_{FILM} \mid C_{FILM} - R_{CT} \mid C_{DL}$ for a degraded porous film exposed to an electrolyte such as concentrated NaCl (Figure 12-4). However, in degraded films of metallic paints a low-frequency relaxation represented by a complex transmission line or by a series of mass-transport and Faradic interfacial impedances due to charge transfer processes on metal particles can often be observed. A comprehensive model describing the EIS response dynamics in metal-rich paints before and after the paint is exposed to the ionic corrosive solution was developed [10, 11] using Gabrielli's derivations for electroactive beds [12].

Metallic paint containing 61.8 percent (by volume) of ~ 6 μm diameter Zn particles with no agglomeration and no direct electrical particle-particle or particle-electrode contacts was studied before and after the exposure to NaCl corrosive electrolyte [8, 9, 10]. Before exposure to the electrolyte, a "dry" paint is essentially a dielectric film with often very large (~50 to 70 percent by volume, up to 90 percent by weight) concentrations of well-distributed and well-dispersed conducting particles, such as zinc or copper epoxy. This system is similar to the one discussed in Section 7-1 for suspended high concentrations of particles, which are expected to display either one, two, or three semicircles in the Nyquist plot, depending on the exact molar share Φ of the particles.

The impedance analysis of the dry paint revealed two prominent capacitive relaxations—at 10 kHz (with a corresponding capacitance of 50 pF and resistance ~20 kohm) and a ~300 kohm impedance relaxation at 10 Hz [9]. The high-frequency relaxation is explained by the particles' percolation-driven conduction through the "binder" and the low-frequency relaxation by contact impedance between the Zn particles. The polymeric binder is a dielectric with capacitance C_{BINDER} (and permittivity ε ~2.9) and alpha value α_1~1. The high-frequency total sample permittivity was ~ 5-10 times higher that that of a pure binder as a result of the Bruggerman's modification coefficient for high load Φ of conductive particles (Eqs. 7-8 and 7-9). C_{BINDER} is effectively in parallel with resistance $R_{PARTICLE}$ ~20kohm, which is electronic resistance through the external oxide layer of zinc particles, or resistance to the particle percolation through the binder for lower concentrations of Zn particles, where they can move in an electric field. A separate ~Mohm-Gohm range resistance R_{BINDER} is geometric resistance of the insulating binder film, which typically does not affect the measurement, as the portion of the circuit becomes shortened through much lower $R_{PARTICLE}$. The rest of the structure is similar to the case of highly concentrated colloids with particle conduction [8]. The low-frequency contact impedance between the adjacent particles can be described by a parallel combination of capacitance $CPE_{CONTACT}$ ~ 50 nF/cm^2 and resistance $R_{CONTACT}$~300 kohm with the alpha value α_2~0.5 or by a transmission line. The combined model can be represented by two parallel paths slightly more complicated than the Debye circuit (Section 4-4). At high frequency the conduction is through the binder capacitance, C_{BINDER}, below 10 kHz through the resistance of the particles zinc oxide layer $R_{PARTICLE}$, and at lower frequencies through the contact resistance $R_{CONTACT} | CPE_{CONTACT}$ (Figure 12-5).

The model for the $R_{PARTICLE}$, $R_{CONTACT}$ and $C_{CONTACT}$ elements can be validated by assuming that uniform metallic particles in the paint are cylinders with diameter ~6 μm, length of ~10 μm, and typical surface conductance of G_{SC}~10^{-9} Sm, with the oxide layer being approximately one percent of the total cross-section area of the particle. The latter assessment was confirmed [9] in a case where a surface oxide layer with a thickness of 69 nm was estimated for 6 μm diameter zinc particles. For a filling close to 100 percent, the electrode area $A = 1$cm^2, and a separation gap between the electrodes $d = 100$ μm can be filled by up to 10 closely packed electrode-to-electrode particle chains. A number of particles at the electrode surface, and therefore total particle-to-particle chains across the gap, becomes 1cm^2/($\pi \cdot (3 \cdot 10^{-4}cm)^2$)~3$\cdot 10^6$. The resistance of the conducting thin layer of zinc oxide is $R = 1/G_s = 10^9$ ohm, or

$$R = \rho \frac{d}{A_{LAYER}} = \rho \frac{0.01\text{cm}}{3 \cdot 10^6 \cdot \pi \cdot (3 \cdot 10^{-4}\text{cm} \cdot 0.01)^2},$$ with a resistivity value for the oxide surface layer ρ~8*10^6 ohm/cm. Therefore, the experimentally measured high-frequency resistance becomes $R_{PARTICLE} = \rho \frac{d}{A} =$

$$\frac{8 \cdot 10^6 \text{ohm/cm} \cdot 0.01 \text{ cm}}{1 \text{ cm}} = 8 \cdot 10^4 \text{ohm},$$ a value of the same order as ~ 20 kohm reported in [9]. The remaining discrepancy can be accounted by considering a clear overestimation of the particles' filling efficiency and other assumptions.

It is interesting to note that contact capacitance $C_{CONTACT}$, which is partially due to the zinc-oxide layer and partially to the binder between the particles, was ~1000 times higher than that of the pure binder with permittivity ε ~2.9. At the same time the contact resistance $R_{CONTACT}$ is ~ 1000 times lower than that

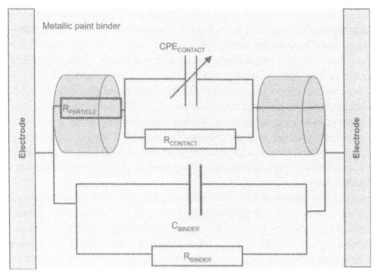

FIGURE 12-5 Impedance diagram representing dielectric behavior of dry film of metallic paint

expected for the binder resistance R_{BINDER}. For the electrode-to-electrode gap of $d = 60$ µm and particle size~ 6µm, the particle-to-particle contact gap was estimated at ~ 70nm, which is roughly 1000 times less than the total electrode-to-electrode separation $d = 100$ µm [9]. For the same binder permittivity and resistivity values, the measured contact capacitance increases and contact resistance decreases as the inverse of the distance between the particles when compared to that of the electrode-to-electrode gap.

Taking into account the particle electronic resistance $R_{PARTICLE}$, contact particle to particle impedance $Z_{CONTACT}$, and dielectric properties of the binder C_{BINDER}, the equations describing the dry-paint impedance response Z and its contact impedance component $Z_{CONTACT}$ were derived as [9]:

$$Z = \cfrac{R_{PARTICLE}}{(j\omega R_{PARTICLE}C_{BINDER})^{\alpha 1} + \cfrac{R_{PARTICLE}}{R_{PARTICLE} + \cfrac{R_{CONTACT}}{1 + (j\omega R_{CONTACT}C_{CONTACT})^{\alpha 2}}}} \qquad (12\text{-}1)$$

$$Z_{CONTACT} = \frac{R_{CONTACT}}{1 + (j\omega R_{CONTACT}C_{CONTACT})^{\alpha 2}} \qquad (12\text{-}2)$$

The same paint subjected to corrosive electrolyte for several days showed significantly different impedance characteristics. Additional effects of cathodic and anodic charge-transfer reactions at the particle interfaces come into prominence, with changes in the particle-particle contact impedance $Z_{CONTACT}$, and large reductions in the binder impedance as many electrolyte-filled pores are created in the binder. An indirect electrical contact between the Zn particles due to percolation and between the particles and the steel substrates (electrodes) were considered to be the main factors affecting the corrosion cathodic protection. A transmission-line model accounting for a distribution of Zn particles within the wetted coating was proposed to represent the low-frequency impedance process in the paint. This model is largely similar to the one describing a porous coating system (Section 7-5).

A "wet paint" impedance model can be developed as a parallel combination of impedance to the ionic conduction through the binder $R_{BINDER} | C_{BINDER}$ and impedances to electrochemical processes occurring on the particles. These impedances include the particle electronic resistance $R_{PARTICLE}$, contact impedance $Z_{CONTACT}$ between the adjacent particles (Eq. 12-2), and impedance of interfacial reactions of the electrolyte and oxygen at the particle interfaces $Z_{INTERFACE}$ (Figure 12-6).

The total binder impedance decreases as a result of electrolyte penetration through the pores and a large decrease in the R_{BINDER} value. The geometrical film capacitive impedance $\sim 1/CPE_{BINDER}$ may be completely bypassed if porosity is high and the resulting R_{BINDER} becomes low. Zinc particles become dissolved in the anodic process, and a relatively high zinc-oxide electronic resistance $R_{PARTICLE}$ is bypassed through this electrochemical interfacial process with impedance $Z_{INTERFACE}$. This interfacial impedance to the zinc anodic dissolution process is composed of a parallel combination of interfacial resistance $R_{INTERFACE}$, representing ion-exchange kinetic resistance at the interface between the particles, and the electrolyte and the double-layer capacitance $CPE_{INTERFACE}$ as:

$$Z_{INTERFACE} = \frac{R_{INTERFACE}}{1 + (j\omega R_{INTERFACE} C_{INTERFACE})^{\alpha 3}} \tag{12-3}$$

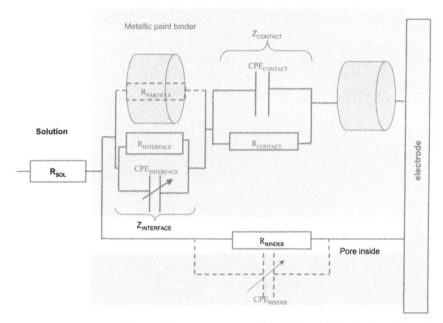

FIGURE 12-6 Impedance diagram representing dielectric behavior of wet film of metallic paint

If the impedance associated with the steel surface is negligible, the impedance of the zinc paint system will correspond to that of the zinc particles in the film. This system can be described using the uniform transmission-line equivalent circuit depicted in Figure 12-6. The total impedance of the system Z^* for a paint film of thickness L_F is presented by Eq. 12-4:

$$Z^* = R_{SOL} + L_F \frac{Z_{CONTACT} R_{BINDER}}{Z_{CONTACT} + R_{BINDER}} +$$

$$\frac{(Z_{CONTACT}^2 + R_{BINDER}^2)\cosh(L_F \sqrt{\frac{Z_{CONTACT} + R_{BINDER}}{Z_{INTERFACE}}}) + 2 Z_{CONTACT} R_{BINDER}}{(Z_{CONTACT} + R_{BINDER})\sqrt{\frac{Z_{CONTACT} + R_{BINDER}}{Z_{INTERFACE}}} \sinh(L_F \sqrt{\frac{Z_{CONTACT} + R_{BINDER}}{Z_{INTERFACE}}})}$$

(12-4)

The model assumes homogeneous percolation due to the high concentration of zinc particles. The analysis for wet coatings reveals that for progressively more degraded paint the binder impedance with pores filled with electrolyte becomes completely resistive (R_{BINDER}), and the electrochemical and mass-transport processes involving the particles ($Z_{CONTACT}$ and $Z_{INTERFACE}$) determine the measured impedance outputs, unless the porosity is so high that all conduction occurs through R_{BINDER}. EIS plots often show only two time constants. Typically, high-frequency (~10 MHz) relaxation is observed due to the contact impedance $Z_{CONTACT}$. This impedance often increases with time as the particles drift apart and the average distance between them increases. The second relaxation at ~100 kHz is related to the electrochemical reaction between the particles and the aqueous media, represented by $Z_{INTERFACE}$. As metallic-paint degradation accelerates, the contact impedance continuously increases and shifts to the lower frequencies, and $Z_{CONTACT}$ becomes a major contributor to the total system impedance. The transmission line due to the $Z_{INTERFACE}$ is barely seen at high kHz frequencies as the interfacial kinetics of anodic dissolution typically become more facile and the interfacial impedance decreases.

As was mentioned above, oxygen reduction in the Zn particles, often coupled with finite diffusion of oxygen dissolved in the electrolyte, also takes place. This cathodic process creates a third time constant and impedance Z_{REACT} and a parallel process to the contact impedance $Z_{CONTACT}$ and the interfacial impedance of anodic Zn dissolution $Z_{INTERFACE}$ (represented here as Z^* from Eq. 12-4 minus the R_{SOL}), resulting in a more complex total expression for the impedance $Z(\omega)$:

$$Z(\omega) = R_{SOL} + \frac{R_{REACT}}{1 + (j\omega R_{REACT} C_{REACT})^{\alpha REACT} + \frac{R_{REACT}}{R_{REACT} + Z^*}}$$

(12-5)

12.3. Electrorheological fluids and charged suspensions

EIS characterization of electrorheological fluids (ERF) represents an interesting subsegment of dielectric analysis of nonaqueous colloidal systems. Analogous to soot-contaminated lubricants (Chapter 10), ERFs can be described as colloidal suspensions where insulating base oil serves as a "continuous" medium suspending a high (typically a several dozen percentage) load of electrically polarizable particles. These ER-active particles are usually polymers, ceramics, metals, or composites such as cellulose, starch, titanium oxide, polyurethane, and polyaniline with sizes $a_p \sim$ 0.1–1 μm [14, 15, 16, 17, 18]. These particles with permittivities in $\varepsilon_p \sim$ 20–50 range are suspended in insulating oil with permittivity $\varepsilon_M \sim$ 2. In the presence of a high external AC or DC electric field, typically a few kV/mm, the ER particles are polarized due to drastic differences in dielectric properties of the suspended particles and the base oil. In the "activated" suspended state, ER fluids exhibit a drastic and fast increase in viscosity on the order of 100,000 times that typically occurs within 5-10 ms. This large change in viscosity essentially allows for an almost instantaneous electric-field-activated transition of ER materials between liquid and solid states.

The field-induced dipoles on the particles cause them to align and form chains or fibrillated structures that bridge the electrode gap, resulting in a large reversible viscosity change (Figure 12-7). This viscosity modification can be easily controlled by adjusting the strength of the applied electric field V. Ever since the discovery of the ER concept by Winslow in the late 1930s [19], a significant amount of research on various applications of ERFs has been implemented. ERFs are now used in industrial applications such as hydraulic valves, clutches, brakes, engine mounts, and shock absorbers. Absence of agglomeration, reversible dispersibility, low power consumption, and millisecond response to an applied field are all important factors for an ideal ERF.

The electrostatic polarization theory is commonly employed to describe ER response. The model assumes that ER fluids are dispersions of nonionic polarizable particles in a low dielectric medium and that free charges and charge-transfer electrochemical processes can be neglected. This model is based on the fact that, due to the permittivity mismatch between the particles ε_p and the continuous phase ε_M, the dipolar particles are polarized and aligned with the neighboring particles. When an electric field is superimposed on the point dipole interaction, the orientation of the dipoles in relation to the exter-

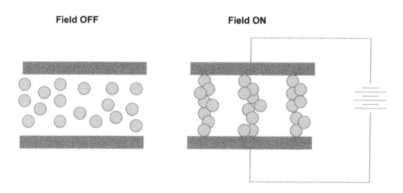

FIGURE 12-7   Electrorheological effect

nal field V becomes important. If the centers of two particles are a distance r apart and are perfectly aligned in the external field next to each other, then each particle feels a force F_{POL} from the polarization of the other particle, which is expressed as a function of the real portion of the Clausius-Mossotti factor β (Chapter 7) as:

$$F_{POL} = 12\pi a_p^2 \varepsilon_M \varepsilon_0 RE\left[\beta^2\right] V_{RMS}^2 \left(\frac{a_p}{r}\right)^4 \tag{12-6}$$

ERF dielectric response can be appropriately described by the classical Debye circuit model (Section 4-4). The model contains ~ 1 pF/cm^2 bulk base oil capacitance C_{BASE}, in parallel with Tohm range base oil resistance R_{BASE}. This combination results in a circuit with a time constant on the order of 10 seconds, typical of the impedance behavior of dielectric materials with very low ionic content. The presence of 10 to 50 percent polarizable particles results in the development of a parallel bulk-solution conduction mechanism through the particles. When compared to the ions that transport current by electrophoretic mobility, the ERF particles have larger sizes and lower mobility and are capable of becoming polarized and reoriented in the external electric field. This percolation type of conduction mechanism can be represented by a series of the particle resistance $R_{PARTICLE}$ and the contact impedance between the particles $C_{CONTACT}$ (Figure 12-8). As the ionic content is essentially absent in the ERF, no double layer "interfacial polarization" or charge-transfer effects are observed even at the mHz frequencies.

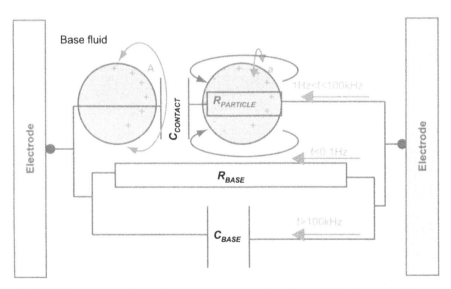

FIGURE 12-8 Impedance diagram representing dielectric behavior of electrorheological fluids

A typical ERF impedance analysis for a model system with $A = 1\text{cm}^2$ electrodes and $d = 0.1$ cm gap displays a single high-frequency Debye relaxation in the upper Hz-kHz frequency range (Figure 12-9). This critical relaxation frequency is higher for smaller (~100 nm) well-dispersed highly polarizable ($\varepsilon_p\sim50$) particles and lower for 1µm-sized less polar ($\varepsilon_p\sim20$) particles. At frequencies above the relaxation, the measured permittivity is approximately 3.0 for a 15 percent concentration of the particles. This value is in agreement with Eqs. 7-4 and 7-5 for permittivities of suspensions with nonagglomerated particles. As the phase-angle data demonstrates, at a range above the relaxation frequency conduction occurs through the base oil capacitance C_{BASE}, which has

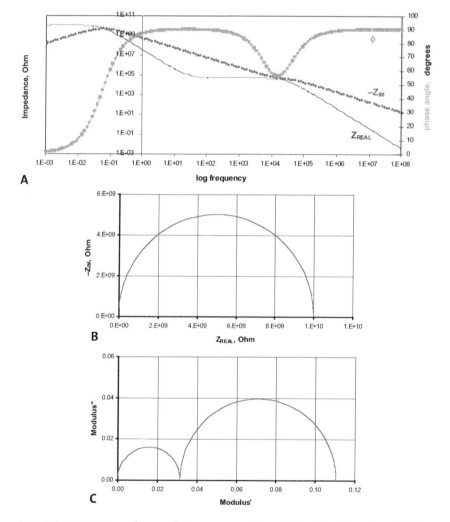

FIGURE 12-9 Impedance data representing dielectric behavior of electrorheological fluid A. Bode impedance plot; B. Nyquist complex impedance plot; C. complex modulus plot

lower associated impedance than that of the $R_{PARTICLE} - C_{CONTACT}$ series through the ERF particles. At the relaxation frequency partially resistive current starts flowing through the particles, which become polarized and start following the AC field. The phase-angle value reaches to ~ $-60°$ and at lower frequencies reverts back to $-90°$ as the capacitive particle-to-particle contact impedance $C_{CONTACT}$ dominates the circuit. The measured permittivity values below the relaxation frequency increase to ~10–20 depending on the polarized particle permittivities and concentration, reflecting a system where all particles are fully aligned. For particles with ε_p~50 and concentration ~ 50 percent, the measured low-frequency permittivity is ~15 and the circuit resistance $R_{PARTICLE}$ ~10^4–10^5 ohm. If a high-voltage gradient of several kV/mm is applied to the fluid at the AC frequency below the relaxation, the particle chains are formed, resulting in the ER effect.

In ERFs particle agglomeration may occur, in particular at elevated temperatures. At ~ 10 Hz agglomerate-related relaxation takes place, resulting in a circuit similar to the one presented in Figure 7-6B. The measured low-frequency permittivity becomes even higher (~30) in ERFs with agglomeration, according to Eq. 7-7. At mHz-range frequencies current may start flowing through the base-oil resistance R_{BASE} and the phase angle approaches $0°$. Complex modulus and phase-angle plots can often resolve all three relaxations due to the particles, their agglomerates, and the base oil better than an impedance plot (Figure 12-10).

The chemical nature of $R_{PARTICLE} - C_{CONTACT}$ parameters can be evaluated using dielectric analysis. The conduction mechanism through the particles is determined by the surface resistance of the particles and the contact capacitance between the adjacent particles, with oil playing a role of separator and measured capacitance significantly higher than C_{BASE}. Block [20] proposed several conduction mechanisms through the particles responsible for the $R_{PARTICLE}$ parameter, such as electronic conduction through the bulk of a particle, around the particle surface, or through counterion realignment in particle double layers (Schwarz model). Development of polarizable double layers and the Schwarz counterion relaxation processes around the polarized particles in an insulating nonionic medium is rather unlikely. Unlike lubricants where ionic detergents and dispersants are present, ER base fluids are essentially nonionic with resistance in Tohms and no "counterions." A conduction mechanism through the bulk of the particles appears to be possible. However, this mechanism would imply a potentially very substantial distribution in the measured $R_{PARTICLE}$ values for a large variety of ERF polarizable materials, representing polymers, metals, and ceramics. This distribution is not apparent, as most ERFs have similar $R_{PARTICLE}$ values in the 10^4–10^5 ohm range.

A surface-conduction mechanism is generally viewed as the most logical explanation for conduction through the particles. The particle surface conductance is assumed to be G_{SC} ~10^{-9} Sm, with the surface conducting area a being ~ 1 percent of the total cross-section of the particle (Section 12-2), resulting in a measured resistivity of 10^3–10^4 ohm cm and experimentally observed resistances of $R_{PARTICLE}$ ~10^4–10^5 ohm for a_p~ 0.1–1 μm. For ~ 15 to 20 percent concentration by weight, the volume fraction of the fully aligned particles should be

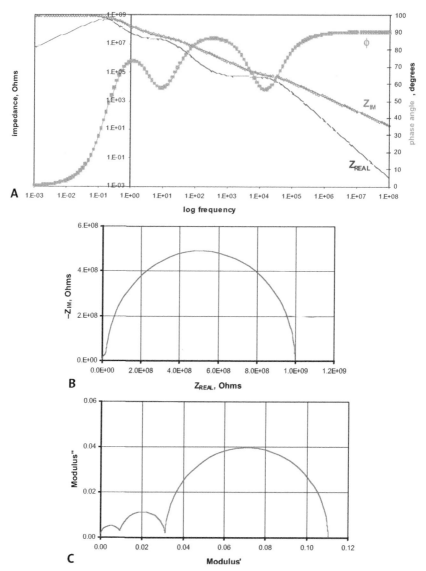

FIGURE 12-10 Impedance data representing dielectric behavior of agglomerated electrorheological fluid A. Bode impedance plot; B. Nyquist complex impedance plot; C. complex modulus plot

$\Phi\sim10$ percent. For 1 μm particles, 1cm^2 surface area of the electrodes, and 0.1 cm electrode-to-electrode separation gap, the measured resistance becomes

$$R_{PARTICLE} = \frac{1}{\Phi}\frac{a_p}{G_{SC}}\frac{d}{A} = \frac{10^{-4}\,cm}{0.1\cdot10^{-9}\,Sm}\frac{0.1\,cm}{1\,cm^2} = 10^5\,ohm.$$ For 100 nm particles under

the same experimental conditions the resistance becomes $R_{PARTICLE} =$

$\dfrac{10^{-5}\,cm}{0.1\cdot10^{-9}\,Sm\cdot1cm^2}\dfrac{0.1cm}{} = 10^4\,ohm.$ These values are in good agreement with

experimental data for a wide variety of ER materials.

The measured contact capacitance $C_{CONTACT}$ that ultimately determines the ER effect is dependent on the distance between the adjacent particles. This parameter increases for highly concentrated suspensions of well-dispersed, small, closely packed particles. These ERF modifications result in smaller particle-to-particle distances, higher effective "surface area," and higher dielectric constant of created microcapacitors, according to Eqs. 7-1 through 7-5. For a critical frequency at 10 kHz the value for the contact capacitance becomes $C_{CONTACT}\sim$ 100–1000 pF/cm^2, exceeding the base oil capacitance by a factor of 100–1000.

Agglomeration results in the particles touching each other, leading to a completely resistive conduction after enough conductive chains across the bulk solution are formed. For agglomerated clusters the corresponding values of capacitance $C_{AGG} > C_{CONTACT}$ and resistance $R_{AGG} < R_{PARTICLE}$ are experimentally observed. That may imply that the agglomerated clusters may have electronic bulk-particle conduction, one of the mechanisms initially proposed [20].

In principle two agglomeration mechanisms should be distinguished. The electrical agglomeration is induced by the kV/mm applied electric field and is a major driving force in the ER effect. The induced polarization on the particles causes their aggregation into columns aligned along the field direction. These columns are responsible for very high solid yield stress (viscosity equivalent) reaching 130 kPa at $V = 5$ kV/mm when sheared perpendicular to the columns. The current density below 4 μA /cm^2 at $V < 2$ kV/mm for a sample containing 30 percent particles was reported [21, 22]. The electrical agglomeration is reversible, and after the electric field is removed the particle chains are broken under a minimal shear and the viscosity of the fluid becomes very low. On the other hand, the chemical agglomeration of particles, such as is induced at higher temperatures, is largely detrimental to exploitation conditions of ERF-based devices. The chemical agglomeration is irreversible (that is, removal of the electric field does not break up the agglomerates), and even high shear typically is not sufficient to completely break the agglomerates. The "no field" viscosity of chemically agglomerated fluids is high. When the electric field is applied, the low values of R_{AGG} result in a dramatic increase in current and power consumption. Both rheological and electrical effects are detrimental to equipment performance, as the application of ER fluids is based on systems with very large viscosity changes between "field-on" and "field-off" conditions, while power consumption should be kept as low as possible. As recent studies in nanometer-sized polarizable particles demonstrated [21, 22], a high concentration of small, well-dispersed conductive particles is essential for transformational improvements in performance characteristics of ERF-based devices.

12.4. Impedance of metal-oxide films and alloys

A number of metal-oxide films such as ruthenium, iridium, zirconium, and tungsten oxides are known for their supercapacitive and corrosion-resistive properties. Several impedance-based methods have been developed to analyze kinetics of formation, growth, and degradation of these special films as the metal substrates are oxidized [23]. Many of these studies have been based on the Mott-Schottky equation (Eq. 5-21) for estimating the space charge layer capacitance and/or the double layer capacitance of passive films.

In a majority of cases, the impedance response of metal-oxide films can be represented by a $R_{SOL} - R_{FILM} | C_{FILM} - R_{CT} | C_{DL}$ circuit. This is a typical model based on solution resistance, film barrier impedance, and oxygen-related charge transfer at the metal-metal oxide interface. The model can be generally compared to that for PANI and other electroactive polymers presented in Chapter 9. For example, EIS monitoring of IrO_x–TiO_x surface film based on the above model demonstrated significant increase in R_{FILM} at high frequency, R_{CT} at low frequency, decreases in C_{FILM} and C_{DL}, and resulting loss of the film's supercapacitive properties with deactivation [24]. When the surface film barrier resistance R_{FILM} is very high, the metal-oxide electrodes can be represented by a simpler $R_{SOL} - CPE_{DL} | R_{CT} - C_{FILM}$ circuit. In that case with deactivation of the metal oxide film the original "supercapacitive" –90° line is slowly replaced by a partially capacitive resistive transmission-line model, with slopes different from an ideal –90°. Pseudocapacitive response of the degrading film results in development of nonuniform resistive film thickness and the appearance of a new transmission-line impedance response [25].

Complex impedance of a solid-state electrochemical cell in the study of oxygen transport through solid mixed-oxide electrolyte ($SrCo_{0.5}Fe_{0.5}O_{3-\delta}$) can again be based on the original model with additional surface reaction and gas-diffusion elements. This model includes diffusion impedance to oxygen transport through the film Z_{DIFF}, resistance to surface exchange reaction R_{SURF}, and the gas space capacitance C_{GAS}. The expression for the total system impedance is presented in Eq. 12-7 with corresponding equivalent circuit and Nyquist plots (Figure 12-11) [26]:

$$Z = R_s + (\frac{1}{R_{CT}} + j\omega C_{DL})^{-1} + (\frac{1}{R_{SURF} + Z_{DIFF}} + j\omega C_{GAS})^{-1} \qquad (12\text{-}7)$$

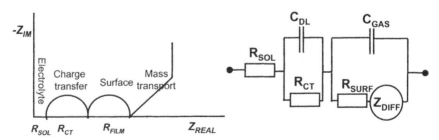

FIGURE 12-11 Equivalent-circuit and impedance plot for oxygen kinetics in solid-state electrolyte

The low-frequency impedance response can be modeled as a parallel combination of gas space capacitance C_{GAS} (Eq. 12-8) with charge transfer represented by surface exchange reaction R_{SURF} (Eq. 12-9) and finite diffusion $Z_{DIFF} = Z_O$ (Eq. 12-10):

$$C_{GAS} = \frac{16F^2 V_g C^*_{OX}}{R_G T} \tag{12-8}$$

$$R_{SURF} = \frac{V_M |(\frac{dV}{d\delta})_\delta|}{2FAD_{OX}} \tag{12-9}$$

$$Z_O = \frac{V_M |(\frac{dV}{d\delta})_\delta|}{2FA\sqrt{j\omega D}} \coth\sqrt{\frac{j\omega L_F^2}{D}} = \frac{D_{OX} R_{SURF}}{\sqrt{j\omega D}} \coth\sqrt{\frac{j\omega L_F^2}{D}} \tag{12-10}$$

Where C^*_{OX} = oxygen concentration in gas space, V_g = volume of gas space, V_M = molar volume of the sample, A = surface area of the sample, k_V = oxygen-discharge reaction rate constant, D_{OX} = oxygen diffusion coefficient, $dV/d\delta$ = thermodynamic factor (slope of coulometric titration curve at a selected sample non-stoichiometry δ), and L_F = diffusion layer or film thickness.

The growth kinetics of WO_3 was studied in hydrochloric acid solution as the tungsten-oxide barrier film was growing as a response to anodic oxidation of tungsten (Eq. 12-11), with oxygen vacancies being treated as main charge carriers across the film [23]:

$$W + H_2O \longleftrightarrow WO_2 + 4H^+ + 4e$$
$$2WO_2 + H_2O \longleftrightarrow W_2O_5 + 2H^+ + 2e \tag{12-11}$$
$$W_2O_5 + H_2O \longleftrightarrow 2WO_3 + 2H^+ + 2e$$

The analyzed circuit was presented as $R_{SOL} - R_{FILM} | C_{FILM} - R_{SURF} | L_{SURF} - C_{PC}$ combination where the film impedance and tungsten oxide surface kinetics impedance elements can be placed in series or in parallel depending on proposed physical path for the current through the structure (Figure 12-12). The Nyquist plot reveals a high-frequency semicircle representing parallel combination of the barrier film resistance R_{FILM} and capacitance $C_{FILM} = \frac{\varepsilon_{WOX}\varepsilon_0 A}{L_F}$

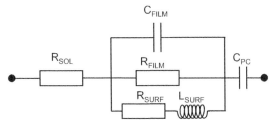

FIGURE 12-12 Equivalent circuit representing the growth kinetics of WO_3 film in acidic solution

related to the oxide film thickness L_{FILM} and its dielectric properties $\varepsilon_{WOX}\sim20$. At potentials higher than $+0.5V$ an inductive semicircle with time constant $\tau = R_{SURF}L_{SURF}$ appears in low frequencies. This impedance feature is due to formation, time, and current variations of the surface charge in the growing tungsten-oxide film. At very low frequencies a pseudo-capacitance

$$C_{PC} = \xi\frac{q}{V} = \xi\frac{zF}{V_M}\frac{\Delta L_F}{V} \sim 2mF/cm^2$$ due to passivation- induced modulation of

the film thickness emerges. $V_M\sim32.3\ cm^3/mol$ is the molar volume of WO_3 and $\xi = 0.44$ is the current efficiency of film formation (proportionality coefficient between the current density and developed charge q on the electrode) [23].

With increase in the applied potential V from 0 to 7 V, both R_{FILM} (~2 kohm) and R_{SURF} (~2 kohm) increased in a similar linear manner, while C_{PC} and C_{FILM} (~1 μF) both decreased and L_{SURF} (~1 kHcm²) steadily decreased. The approach allowed calculating the half-jump distance l (nm) for positively charged oxygen vacancies on the barrier tungsten-oxide film, the diffusion coefficient of oxygen vacancies inside the film D_{OX}, and oxide film thickness $L_F = \sqrt{D_{OX}/\omega_{CRIT}}$ from the circular relaxation frequency of the high-frequency oxide-film relaxation ω_{CRIT}:

$$\frac{l}{A}\frac{dR_{FILM}}{dV} = \frac{R_G T}{2FIV}$$

(12-12)

$$D_{OX} = \frac{IR_G T}{4e_0 FAC^*_{OX}\ V}$$

(12-13)

where I = measured current due to the flux of oxygen vacancies through the film, A = surface area, V = strength of an electric field inside WO_3 film ($\sim6*10^6$ V/cm), and $C^*_{OX} \sim 10^{20}$ cm^{-3} = concentration of oxygen vacancies in the film. Calculations lead to values of $L_F \sim 0.2nm$, diffusion coefficient $D_{OX}\sim 10^{-18}$ to 10^{-14} cm²/s, and $l \sim 0.21$ nm.

12.5. Li-ion kinetics in alkaline batteries

Impedance spectroscopy has been widely applied to performance monitoring of lithium alkaline batteries, most prominently to the study of Li^+ insertion into graphitized carbon anode during the battery charging. Despite large variability in battery chemistries, the basic battery design is fairly standard (Figure 12-13). The batteries consist of porous electrodes composed of particles of energy-storing graphite, oxide-active materials, and conductive current collector held together by a polymeric binder. They also include electrolyte (typically organic polymeric gel in the case of Li batteries) and a separator.

During the battery discharge, the current flowing within the battery is carried by the movement of Li^+ ions from the anode to the cathode and through the no-aqueous electrolyte and separator diaphragm. During charging, an external electrical power source (the charging circuit) forces the current to pass in the reverse direction; the positive terminal from the charging circuit has to be connected to the cathode of the battery, and the anode has to be connected

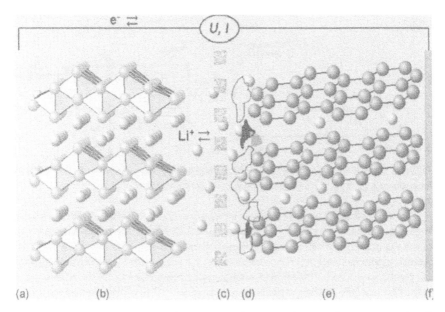

FIGURE 12-13 Schematic illustration of a typical Li-ion battery: A. aluminum cathode; B. oxide active material; C. porous separator soaked with electrolyte; D. inhomogeneous charge exchange layer; E. graphite active material; F. copper anode

to the negative terminal of the external circuit. The lithium ions then migrate from the cathode to the anode, where they become embedded in the porous electrode material in a process known as intercalation. The electrolyte is aprotic because the voltage of the battery (+4.2V) exceeds the water-stability region for both cathode and anode. The batteries are based on electronic conduction through the particles and ionic conduction through the electrolyte in cavities between the particles. The charge-transfer process occurs at the surface of each particle where the ionic and electronic conduction regions meet. This process is further complicated by the ion diffusion into the bulk of graphite particles.

The electrolyte impedance of Li batteries is typically measured between two blocking (steel) or nonblocking (Li) electrodes. The impedance response of the polymeric electrolyte (Z_{SOL}) appears as a depressed semicircle in upper kHz frequencies and changes with the electrolyte thickness. Sometimes four-electrode measurement is performed for the electrolyte characterization. This experimental setup eliminates the effect of the electrode interfaces by inserting two pseudoreference Li electrodes into the sample between two outside current electrodes (Section 8-4). The choice of conductive solvent is based on a compromise between its conductivity over a wide temperature range and its ability to develop a passivating layer on the intercalating host surface; a mixture of propylene carbonate/ethylene carbonate or a more complex derivation of this mixture is frequently employed. A ~15μm microporous separator prevents the electrodes from shortening directly.

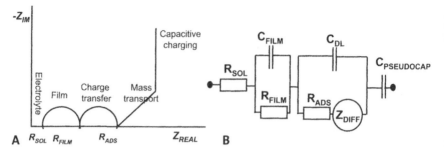

FIGURE 12-14   Li-ion kinetics in graphitized carbon anode: A. Simplified impedance plot; B. representative equivalent circuit

Li⁺-ion diffusion and kinetics in solid-state materials such as batteries are highly potential-dependent [28]. Anodes and cathodes are frequently investigated separately by EIS using a three-electrode configuration with a Li-metal counterelectrode and reference electrodes. Charging and discharging studies are performed as a function of potential vs. Li reference electrode or as a function of Li⁺ content in the graphite anode material. When the potential vs. the Li reference electrode approaches 0V, the intercalation process is complete and the highest presence of Li⁺ in the anode material is observed.

A general consensus on the exact interpretation and graphical representation of EIS data on Li⁺ intercalation in graphitized carbon has not been developed. The most general description of the process, which includes data for three frequency regions—high (100 kHz–10Hz), medium (10Hz–1Hz), and low (under 1Hz)—is offered here (Figure 12-14A). For the anode, the high-frequency solution resistance R_{SOL} is followed by a semicircle corresponding to the passage of Li⁺ ions through a passivation film $C_{FILM} | R_{FILM}$. The film impedance is fairly constant through the charging/discharging cycles and is determined by the film geometry. In principle a separate lump circuit $C_S | R_S$ (not shown in Figure 12-14) can be added to the system to represent electron transfer through the bulk particle material; sometimes two or three semicircular features can be distinguished in the high-frequency region [29]. It is followed at low frequencies by Faradaic impedance of Li⁺ ion electroadsorption at the interface represented by $C_{DL} | (R_{ADS} - Z_{DIFF})$. An explanation of the Faradaic process as being mainly limited by Li⁺ electroadsorption at the surface of graphite particles with corresponding process resistance represented by either R_{CT} or R_{ADS} has been presented [30]. Z_{DIFF} corresponds to a solid-state diffusion of Li⁺ ions in graphite particles. Finally, pseudocapacitance $C_{PSEUDOCAP}$ related to the charging of a crystalline structure by a Li⁺ insertion is observed as a –90° line at the lowest frequencies. The system can be represented by a series of: $R_{SOL} - C_{FILM} | R_{FILM} - C_S | R_S - C_{DL} | (R_{ADS} - Z_{DIFF}) - C_{PSEUDOCAP}$, with a representative equivalent-circuit diagram and Nyquist plot shown in Figure 12-14 [**31**, **32**, **33**]. When Li⁺ content is very low and the potential difference vs. Li reference electrode is large, the film impedance can be seen as a high-frequency capacitive semicircle, followed by very large Faradaic impedance.

The impedance diagram often does not allow segregating the film impedance, adsorption, and charge/mass-transfer processes, similar in fact to the case of electroactive films described in Chapter 9. The kinetic steps commonly observed in the batteries may also be represented by a "transmission line" model [34, p. 460] with membrane resistance R_{SOL} and each element composed of $C_{FILM} \mid R_{FILM} - C_{DL} \mid (R_{ADS} - Z_{DIFF})$, with diffusion having a finite absorption boundary at low frequencies. As Li$^+$ intercalation progresses and the Li$^+$ presence increases, the Faradaic impedance decreases and the diagram starts to resemble Figure 12-14B. The diffusion impedance for Li$^+$ ions follows the Warburg diffusion expression (Eq. 5-39) as:

$$Z_{DIFF} = \frac{RT}{z^2 F^2} \frac{1-j}{AC^* \sqrt{2D\omega}} = \sigma_D \omega^{-1/2} (1-j) \tag{12-14}$$

where the diffusion parameter σ_D for mass transport inside a solid film is experimentally determined as a function of Z vs. $\omega^{1/2}$:

$$\sigma_D = \frac{V_M}{zF} \frac{(dV/dx)}{A\sqrt{D}} \tag{12-15}$$

where V_M = molar volume of the carbon particles (~42 cm^3), x = molar share of diffusing lithium ions, D = diffusion coefficient, and dV/dx is a slope of a coulometric titration curve. The Li$^+$ diffusion coefficient can be determined as:

$$\sqrt{D} = V_M (\frac{dV}{dx}) \frac{1}{zFA\sigma_D} \tag{12-16}$$

For mass transport through a diffusion layer or film of thickness L_F, the real impedance Z_{REAL} becomes independent of frequency at low frequencies when $\omega < 2D/L_F^2$:

$$Z_{REAL}(\omega \to 0) = R_D = \frac{V_M}{zF} \frac{(dV/dx)}{A} \frac{L_F}{\sqrt{2D}} \tag{12-17}$$

The imaginary impedance becomes:

$$Z_{IM}(\omega \to 0) = \frac{1}{\omega C_{PSEUDOCAP}} = \frac{V_M}{zF} \frac{(dV/dx)}{A} \frac{1}{\omega L_F} \tag{12-18}$$

The entire low-frequency impedance can be represented by a combination of limiting resistance R_D and insertion pseudocapacitance $C_{PSEUDOCAP} = C_{IN} = 1/R_D\omega$. Provided that R_D can be determined from the intercept of the $-90°$ line with the Z_{REAL} axis and $C_{PSUEDOCAP}$ from the transmission-line model, Li$^+$ ion diffusion coefficients can be determined as:

$$D = \frac{L_F^2}{3R_D C_{PSEUDOCAP}} \tag{12-19}$$

LiMn$_2$O$_4$ and its derivatives have been materials of choice for lithium-alkaline battery cathodes. This selection is mainly because of the fast interfacial kinetics and higher capacity of such materials. The total impedance of cathodes is similar to that of anodes, being composed of surface film, diffusion, and charge-transfer impedances as $C_{FILM} | R_{FILM} - C_{DL} | (R_{CT} - Z_{DIFF})$. Cathode impedance changes as a function of discharge of the battery, with the Faradaic component increasing.

For a complete battery system, the impedances of all components must be added together. A resulting EIS diagram is composed of a small high-frequency electrolyte and separator impedance (Z_{SOL}), followed by the impedances of insulating layers on cathodes and anode ($C_{FILM} | R_{FILM}$); electroadsorption, charge-transfer, and solid-state diffusion impedances ($C_{DL} | (R_{CT} - Z_{DIFF})$; and formation of new capacitive crystalline structures at µHz frequencies ($C_{PSEU-DO}$). Compared to Figure 12-14B, additional semicircles may appear in the medium- and low-frequency ranges.

12.6. Impedance analysis of proton exchange mebrane fuel cells

Proton-exchange-membrane (PEM) fuel cells are being developed for applications in transportation, as well as stationary and portable power sources. They are considered as promising power sources because they offer a highly efficient and environmentally friendly solution for energy conversion. Their distinguishing features include lower temperature/pressure ranges (50–100 °C), carbon-supported platinum particle catalysts, and special proton-exchange polymer electrolyte membranes, such as Nafion®. The fuel cell consists of two porous electrodes at which the energy conversion process takes place. Molecular hydrogen is supplied to the anode, while oxygen from the air is supplied to the cathode. Hydrogen molecules give up electrons to the anode in a catalytic process, producing electric current across an external circuit. In the oxidation process, hydrogen is converted into protons or positively charged hydrogen ions (H^+). When the electrons, drawn through the external circuit by the voltage, arrive at the cathode, they react with oxygen in the second catalytic oxygen reduction process, producing negatively charged oxygen ions that react with the protons to form water (Figure 12-15).

The reliability of fuel cells is improving. However, as with any device involving complicated chemical, physical, electrical, and mechanical processes, the effects of various operational conditions on different components of the system often result in loss of performance. Much of the current research on PEM fuel cells is focused at several main objectives, such as developing higher-activity cathode reaction catalysts, reduction of catalyst poisoning by impurity gases such as carbon monoxide (CO), and solving the problems of membrane dehydration and excess water (flooding). One of the main barriers to the commercialization of proton exchange mebrane fuel cell systems, especially for automotive use, is the high cost of the platinum electrocatalysts. Although platinum is used as a catalyst for both anode and cathode reactions, developing alternative oxygen-reduction catalysts for the cathode is the more challenging task. In the past decade, there have been significant advances in reducing the platinum loading of both the anode and the cathode through the development of highly dispersed Pt nanoparticles on a carbon support, thin-film Pt electrodes, and Pt alloy nanoparticles supported on carbon.

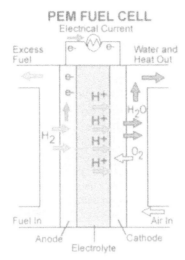

FIGURE 12-15   Schematic description of the components in a proton-exchange-membrane fuel cell

Just as is the case in many solid-state electrochemical applications, there is a significant difficulty in applying reference electrodes in the impedance and other electrochemical testing of fuel cells. The experiments typically (but not always) are conducted in a two-electrode arrangement. Conclusions about relative contributions of different segments of PEFC to the measured impedance are achieved on the basis of changes in their responses to variations in several easily modified operational parameters. These parameters include drawn current density, temperature, membrane thickness, fuel composition (concentration of hydrogen and presence of CO contamination for the anode reaction, oxygen concentration for the cathode process), and humidity level. This approach frequently allows various sections of the impedance spectra to be assigned to cathode-, anode-, and electrolyte-related contributions.

EIS analysis has been effectively utilized to identify and separate different ohmic, mass-, and charge-transfer limiting processes [**34**, p. 521]. Impedance models for polymer electrolyte fuel cells have been developed, and analytical solutions for several cases have been derived [**35**]. As a first approximation, the impedance of a typical PEM half-cell in the absence of significant mass-transport resistances can be represented by the $L_{CBL} - R_{MBR} - R_{CT} \mid CPE_{DL}$ model. The high-frequency inductive impedance of cables L_{CBL} and membrane ohmic resistance R_{MBR} components are in series with polarization resistance and double-layer capacitance (expressed as CPE_{DL}) of either cathode or anode.

The equivalent circuit for a complete PEM fuel cell can be applied initially to a "symmetrical" gas supply while varying current densities. At low current densities (< 400 mA/cm^2) the mass-transport limitations are very minor, and a simple Nyquist diagram with two relaxation times emerges. The overall impedance is dominated primarily by the cathodic process, with resulting higher cathodic overpotentials. The equivalent circuit is represented by two parallel combinations of charge-transfer resistances R_{CT} and double-layer capacitors $C_{DL} \sim 20$ μF/cm^2 for the anode and cathode processes. The double-layer capacitances are typically represented by CPE elements with the α–parameter in the 0.75–0.95 range (Figure 12-16A).

Higher-density currents result in water mass-transport limitations and the appearance of additional low-frequency impedance relaxation. This diffusion impedance due to water transport within the polymer electrolyte morphology is often represented by the finite diffusion processes affecting both the anodic (Z_{OA}) and cathodic (Z_{OC}) processes (Figure 12-16B). An increase in anodic impedance is often related to partial drying out of the anode/membrane interface.

Porous electrodes introduce additional interfaces for cathode and anode and the need for a further refinement of the equivalent circuit model according to the De Levie model (Eq. 7-66). The low-frequency impedance is often complicated by the adsorption kinetics of the oxygen-reduction process intermediates on the cathode, represented by a series combination of $R_{ADS} - L_{ADS}$. Carbon-monoxide impurities in hydrogen fuel poison the anodic catalysis process, as adsorbed CO limits platinum sites for the hydrogen reaction. The CO adsorption on the platinum catalyst results in the appearance of low-frequency inductive impedance, represented by a $R_K - L_K$ series in parallel with the anodic charge-transfer and double-layer charging impedances (Figure 12-16C). The complex impedance simulation diagram for a complete PEM fuel cell is shown in Figure 12-16D.

A detailed analysis of various contributing factors to PEM fuel-cell impedance using locally resolved EIS analysis of segmented cells was presented for pure hydrogen-oxygen and hydrogen-air gas feeds [36, 37]. The experimental design utilized a pseudoreference carbon filament electrode incorporated in the polymer electrolyte [37]. This segmentation approach allowed distinguishing among separate cathode and anode impedance contributions. Linear EIS measurement of the whole cell is limited to determining averaged impedance-related parameters, such as current density or total impedance, without providing detailed information on the individual segments' contributions to the combined cell impedance.

The anode process was described by three capacitive loops. The high-frequency relaxation corresponds to the hydrogen charge transfer and the double-layer capacitance; medium-frequency relaxation represents hydrogen adsorption and desorption (possibly complicated by CO poisoning); low-frequency relaxation is characteristic of water transport and corresponding finite diffusion limitations (Figure 12-17A). The cathode process was represented by a single capacitive feature (water diffusion Z_{OC} was found to be negligeable), corresponding to the oxygen kinetics charge transfer and double-layer capacitance, and a low-frequency inductive feature, corresponding to adsorption of oxygen-derived intermediate species (Figure 12-17B). Because the low-frequency cathode and anode features appear in the same range of frequencies for a whole cell impedance (Figure 12-17C), there is a competition between the anode and cathode processes in their contribution to the total impedance of the system, and the inductive feature is often not seen. In the combined impedance of the cell, the high-frequency feature is related to both anodic and cathodic charge transfers. This feature is often split

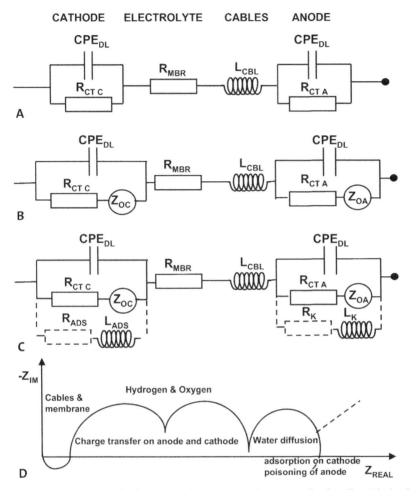

FIGURE 12-16 Equivalent circuit representing PEM fuel cell with hydrogen/oxygen supply for: A. low current densities; B. high current densities; C. with catalysis and adsorption complications; D. the Nyquist plot

into two semicircles, with the high-frequency semicircle representative of hydrogen, and medium frequency of oxygen charge transfer, which is dependent on the oxygen concentration in the supplied air. Oxygen reduction on the cathode dominates the combined charge-transfer impedance. The low-frequency feature of the plot is composed of a capacitive feature representing water finite diffusion limitations Z_{OA}, and the inductive feature representing adsorption of oxygen-derived intermediates on the cathode or CO poisoning on the anode. The high-frequency intercept with the real impedance axis is characteristic of the membrane ohmic resistance R_{MBR} (Figure 12-17D).

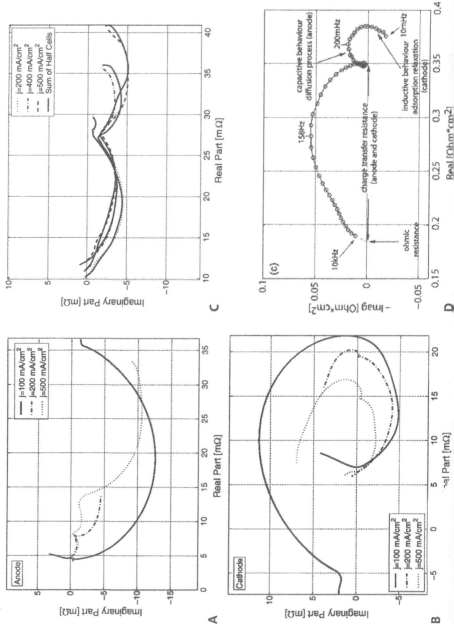

FIGURE 12-17 The impedance spectra measured in a one-dimensional H_2/O_2 PEM fuel cell studied by locally resolved EIS [36] for: A. half cell between the pseudoreference electrode and the anode; B. half cell between the pseudo-reference electrode and the cathode; C. the whole cell; D. the impedance model at current densities $j = 100–500$ mA/cm² [37] (with permission from The Electrochemical Society and the authors)

The impedance spectrum changes can be used as an effective diagnostic tool to detect changes in levels of humidity, current density, catalyst poisoning, and other operational parameters of the fuel cell. At high current density and high humidity the oxygen charge-transfer kinetics and hydrogen diffusion on the anode are rate-limiting, and only two capacitive semicircles are generally observed. At low current density and low humidity the diffusion impedance is less important. Oxygen and hydrogen charge transfer become dominant in the high- and medium-frequency range (Figure 12-18). At low frequencies, however, two inductive features related to adsorption of intermediates on the cathode and poisoning on the anode may be observed (Figure 12-18).

EIS is also capable of diagnosing several common causes of fuel-cell failure. For instance, membrane drying results in impedance increases across all frequency ranges but in particular at high frequencies. A PEM FC flooding with water from the cathode reaction causes impedance to increase only at low frequencies, with a capacitive ($\phi \to -90°$) line emerging. The case of anode catalyst poisoning reveals the inductive impedance loop at medium and low frequencies [38]. Variability of fuel used in catalytic reactions on porous electrodes also can be studied by EIS using the transmission line model [39, 40]. When, for instance, the cathodic process is investigated, two cases were considered:

- H_2/O_2 feed, where cathodic charge transfer is present and the catalyst layer can be represented by a transmission-line model combining electrode charge-transfer reaction, electrolyte resistance, and double layer capacitance (Figure 12-19A)

- H_2/N_2 feed, where cathodic charge transfer is absent and the catalyst layer can be represented by a transmission-line model combining electrolyte resistance and double layer capacitance (Figure 12-19B)

R_{MBR} represents resistance to proton conduction in the membrane. The high-frequency intercept with Z_{REAL} provides a value for R_{MBR}, which is determined by proton conductivity as $\sigma = d / AR_{MBR}$, where d = the sample thickness and A = membrane area. In the catalyst layer total $R = n*R_l$ is distributed electrolyte resistance, total R_{CT} is distributed charge-transfer resistance, total

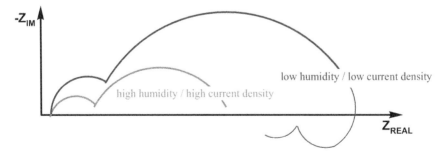

FIGURE 12-18   Nyquist plot diagram representing typical PEM fuel cell impedance as a function of humidity and current density

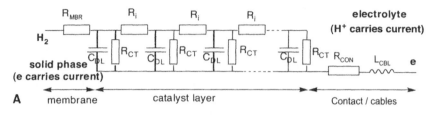

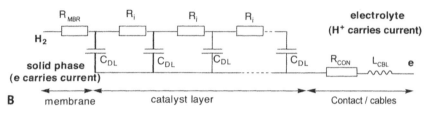

FIGURE 12-19 Equivalent circuit for fuel cell under A. H_2/O_2 feed; B. H_2/N_2 feed

C_{DL} is distributed double-layer capacitance, R_{CON} is the electronic and contact resistance of the electrolyte, and L_{CBL} is the induction due to the cables.

The analysis is based on the porous model derived by De Levie (Eq. 7-66) graphically presented in Figure 7-20C, and the expression for the impedance of H_2/O_2 system becomes:

$$Z(\omega) = \sqrt{R\frac{R_{CT}}{1 + j\omega R_{CT}C_{DL}}} \coth \sqrt{\frac{R}{\dfrac{R_{CT}}{1 + j\omega R_{CT}C_{DL}}}} \qquad (12\text{-}20)$$

The combined Nyquist plot for the H_2/O_2 system in Figure 12-20A (with R_{MBR} high-frequency contribution subtracted out) reveals the high frequency $-45°$ Warburg diffusion line followed by a charge-transfer semicircle. Additional low-frequency inductive loops may be observed sometimes as a contribution from the cathode kinetics due to the relaxation of oxygen-derived intermediate species in electrochemical-chemical-electrochemical mechanisms (Figure 12-16D). For the N_2/H_2 feed there is no Faradaic reaction on the cathode, and the charge transfer resistance $R_{CT} \to \infty$. The electrode impedance becomes:

$$Z(\omega) = \sqrt{R\frac{1}{j\omega C_{DL}}} \coth \sqrt{Rj\omega C_{DL}} \qquad (12\text{-}21)$$

At high frequencies, the combined impedance plot reveals a $-45°$ diffusion limited line as $Z_{\omega \to \infty} \sim \sqrt{R\dfrac{1}{j\omega C_{DL}}}$. At low frequencies $Z_{\omega \to 0} \sim R_{CT} + R/3 = 1/j\omega C_{DL} + R/3$. The losses in the DC polarization curve due to the proton resistance in the catalyst layer are approximately equal to $R/3$, and a capacitive

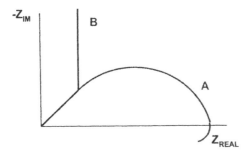

FIGURE 12-20 Impedance plot schematics for A. H_2/O_2 feed; B. H_2/N_2 feed for PEFC porous-electrode kinetics

$-90°$ line parallel to the Z_{IM} axis representing a finite diffusion with a reflecting boundary is observed (Figure12-20B).

12.7. Impedance corrosion monitoring

Corrosion is defined as spontaneous degradation of a reactive material by an aggressive environment. It develops as a result of the simultaneous occurrence of at least one anodic (metal oxidation) and one cathodic (reduction of dissolved oxygen) reaction. The corrosion process is electrochemical in nature, as both these processes are of the charge-transfer type. Hence electrochemical techniques, and in particular EIS, are suitable for estimating corrosion rate k_{CORR}:

$$k_{CORR} = \frac{\Delta W}{\Delta t} = \frac{MI_{CORR}}{zF} \qquad (12\text{-}22)$$

where ΔW = the metal weight loss, Δt – elapsed time, M = the atomic weight, z = the valence of dissolution, and I_{CORR} = the anodic metal dissolution current. The Stern-Gary equation provides the direct relationship between the steady-state corrosion current I_{CORR} and the "DC" polarization resistance R_{POL} across the interface, which in that case becomes an equivalent of Faradaic impedance Z_F [34, p. 344]:

$$R_{POL} = Z_F = \frac{1}{2.303 I_{CORR}} \frac{b_A b_C}{(b_A + b_C)} \qquad (12\text{-}23)$$

There b_A and b_C are the Tafel coefficients for the anodic and cathodic partial reactions. Values of the Tafel slopes were successfully used for the corrosion of iron in acid media [41], where both anodic and cathodic reactions take place through a two-step reaction. Usually the Tafel slopes of the steps, which are not rate-determining, cannot be easily obtained. In that case a calibration procedure, often based on weight loss data, must be developed.

Determining the polarization resistance R_{POL} is the primary task of corrosion monitoring. Provided that the Tafel coefficients can be obtained with a fair degree of accuracy, the corrosion current I_{CORR} is inversely proportional to R_{POL}, which in its turn can be determined from EIS data analysis. The advantage of EIS in corrosion monitoring is its ability to easily and conveniently cover several relaxation-time constants in one measurement and to develop a more accurate assessment of R_{POL}. The difficulty is that corrosion is a temporally distributed process, resulting in changes in the EIS diagrams with time and requiring analysis at very low frequencies where R_{POL} is determined. This situation causes stability problems in a system that is viewed from the beginning as kinetically nonstable due to the corrosion process. Often a simplistic determination of polarization resistance R_{POL} from voltammetry curves is a faster and more accurate way to understand corrosion kinetics.

When the corrosion current I_{CORR} is at steady state ($I_{CORR} = I_{AN} = -I_{CT}$), the polarization resistance is identical to the charge-transfer resistance, as $R_{POL} = R_{CT}$. R_{CT}, and therefore I_{CORR}, can be easily identified at the low-frequency range of the impedance spectrum. In the past polarization resistance was often approximated as the total low-frequency impedance, determined in practice by the difference of the measured high- and low-frequency impedances:

$$R_{POL} = |Z(j\omega)|_{\omega \to 0} - |Z(j\omega)|_{\omega \to \infty} \qquad (12\text{-}24)$$

This approach, as will be shown below, is somewhat misleading. It has been shown that AC impedance measurements over the entire frequency range can give more accurate information about impedance parameters related to corrosion-rate estimation under various conditions. The high-frequency limit for the bulk-solution conductivity region should be determined from Eq. 6-10. At very low frequencies the response may become inductive, resulting in the measured $R_{POL} < R_{CT}$ (Figure 12-21), and special precautions should be adopted to interpret the data from a simple "DC" measurement based on Eq. 12-24.

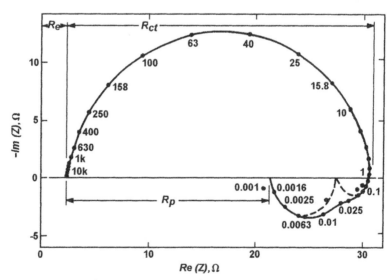

FIGURE 12-21 Impedance measurement of a corroding electrode: Holzer-type iron in deoxygenated $1M, H_2SO_4$; disk electrode rotation speed = 1600 RPM; A = 0.2cm² (figure reproduced by permission of Solartron Analytical, a part of AMETEK Advanced Measurement Technology [42])

Charge-transfer resistance has been shown to be related to the measured corrosion current I_{CORR} according to Eq. 12-23 in cases of partial diffusion control and multistep electron transfer, where the voltammetry polarization technique would fail. On the other hand, in cases of passivation kinetics the polarization method results in a more accurate estimation of I_{CORR} [42].

EIS has been applied extensively to the analysis of the corrosion mechanism of iron and other metals in aqueous solutions. To characterize a given corrosion process, it is practically advisable to obtain a full AC frequency scan of the system, including sufficient low-frequency response and small amplitude voltage perturbation with cyclic voltammetry, before acquiring the response data such as current, voltage, and polarization resistance R_{POL}. Most corrosion kinetics studies have been done on uniformly corroding surfaces where the dissolution of the metal is uniform all over the surface in contact with the electrolyte [43]. Localized corrosion and stress corrosion cracking can also be analyzed by impedance methods such as local EIS (Section 13-4).

Corroding systems can become unstable and change during a single EIS measurement, especially if the mHz frequency range is utilized. Corroding interfaces are inherently nonlinear, producing higher harmonics and introducing additional sources of error in the linear EIS measurement. In the past higher harmonic nonlinear EIS analysis has been used to determine the corrosion kinetics [44]. As with the other examples of nonlinear EIS (Section 13-3), the challenge of this analysis application to corrosion kinetic modeling originates from the difficulties of deconvolution of nonlinear contributions from several diffusion and charge transfer processes.

Analysis of EIS data allows identifying the most important stages of corrosion, such as active dissolution, active-passive transition, and passive state [45]. For instance, the active dissolution state is an example of complex multistage kinetics with intermediate adsorbed species [37]. That often results in several capacitive or inductive loops, depending on the processes' kinetic rate constants (Section 7-4). For instance, in the case of iron placed in sulfuric acid, corrosion may involve up to three adsorbed forms of oxidized iron intermediates. Depending on the solution pH and applied potential, the Nyquist diagram displays one or two inductive loops originating from the potential dependence of the surface coverage of the intermediates (Figure 12-21). When corrosion inhibitors are introduced, the inductance remains, although it becomes related to the surface coverage of the inhibiting species (typically hydrogen or hydrogen-bonded organic compounds) at lower frequencies and to the anodic iron-derived intermediates at higher frequencies (Figure 12-22A).

Coating of metal with protective insulating polymer coating (Section 12-1) or paint (Section 12-2) films is one of the most widely used corrosion-protection technologies. The isolating part of the film gives rise to a characteristic high-frequency capacitive response, represented by a parallel combination of $R_{FILM} | C_{FILM}$. For coated metals water penetration becomes an issue, as it leads to continuous progress of the charge transfer reaction, although at lower rates on heterogeneous surfaces. It also can be compared with pitting corrosion kinetics when the charge transfer process is very different in the pit vs. the rest (passive) of the surface. Faradaic impedance measurements on painted iron (Figure 12-22B) showed that the metal substrate is corroded in a similar way

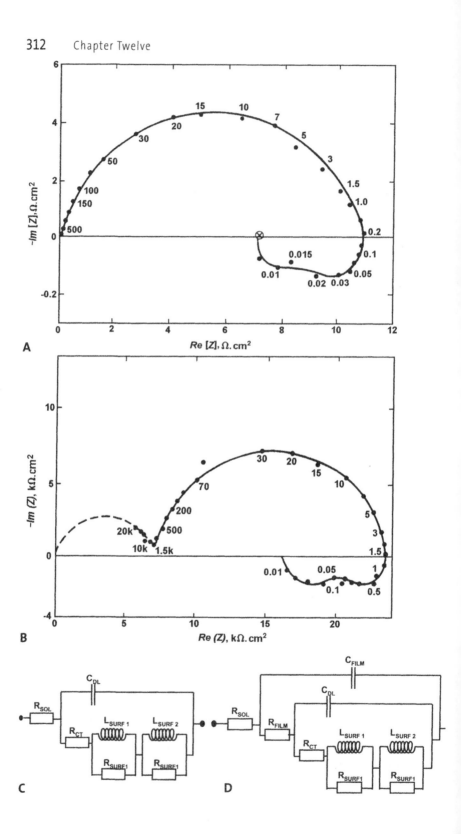

with and without the coating. However, the painted substrate develops corrosion kinetics at a substantially reduced overall area, only at the places where the paint layer develops porous flaws or delamination. Representative equivalent-circuit diagrams for these two cases are shown in Figure 12-22C and D).

The phenomenon of metal passivation, or sudden electrochemical potential-induced transition from an active dissolution state to a passive state, is responsible for the low corrosion rates observed in many metals and alloys. This condition has usually been attributed to the electrochemical formation of metal-oxide protection films or coverage of the surface by corrosion films. In either case, the metal becomes partially protected from the environment, and the corrosion current drops sharply. This condition is of enormous practical importance, as the integrity of metallic structures of great structural significance can be maintained by keeping their electrochemical potentials at some predetermined ("passivation") value.

The passive state can be represented by a system composed of solution resistance R_{SOL} and the impedances of metal-film $Z_{M/F}$, film Z_{FILM}, and film-solution $Z_{F/S}$ interfaces [34, p. 365]. The metal-film interface impedance $Z_{M/F}$ is composed of a parallel combination of R_e resistance of electron and R_{CAT} cation charge transfers and capacitance C_{SC} due to the surface charge layer. The electron transfer resistance R_e is assumed to be the smallest ($R_e \ll R_{CAT} \sim 1/\omega C_{SC}$ at low frequencies), and therefore it becomes the only detectable portion of the circuit, as $Z_{M/F} = R_e$.

In the film the electronic transport and diffusion of cation and anion vacancies are important. If the ionic diffusion-driven transport dominates, it can be represented by the semi-infinite $-45°$ Warburg diffusion line, which may also be expressed at the low-frequency finite adsorption boundary conditions for thinner films as $Z_{DIFF}(\omega) = \sigma_D (1-j)/\sqrt{\omega}$. Otherwise a simple parallel combination of film resistance and capacitance $R_{FILM} | C_{FILM}$ is observed.

For the film-solution interface there is equilibrium between the supply of ions from the solution and from the film, leading to possible kinetic variations. Experimentally observed situations for the resulting impedance $Z_{F/S}$ include a single capacitive charge-transfer semicircle with characteristic R_{CT} or its variations if different processes become important and are resolved into two "positive" capacitive-resistive semicircles (Figure 12-23D) or one positive and one with negative resistance or inductance (Figure 12-23A). When the small $R_{SOL} + R_e$ and the diffusion impedance Z_{DIFF} are added to the film-solution impedance $Z_{F/S} = R_{CT}$, typically a simple Randles response is observed. The finite Warburg diffusion in the film Z_{DIFF} dominates at low frequencies and R_{CT} due to the film-solution kinetics present at medium frequencies. At higher frequencies a separate film impedance $R_{FILM} | C_{FILM}$ semicircular feature may be observed in series with small $R_{SOL} + R_e$.

FIGURE 12-22 Impedance measurement of a corroding iron electrode A = 60cm² in 0.5M, H₂SO₄ A. aerated + 2mM propargylic alcohol, disc electrode rotation speed = 1600 rpm; B. iron coated with 40 mm epoxy paint ICI 5802022 [46] (reproduced by permission of Solartron Analytical, a part of AMETEK Advanced Measurement Technology [42]); C. representative equivalent circuits for case A at corroding potential; D. for case B

An example of this approach is a study of corrosion of stainless-steel anodes in an aggressive hydrochloric-acid environment [47]. Two capacitive semicircles were distinguished—high-frequency impedance related to the formation and growth of the corrosion film $R_{FILM} \mid C_{FILM}$, and a low-frequency feature related to the film-electrolyte charge-transfer resistance R_{CT} coupled with finite transmission boundary diffusion of ions (protons) through the film Z_{DIFF}. The circuit can be represented as $R_{SOL} - R_{FILM} \mid C_{FILM} - C_{DL} \mid (R_{CT} + Z_{DIFF})$. Over the time of the experiment the corrosion process caused a gradual increase in the film thickness L_{FILM} and an increase in R_{FILM}. However, the charge transfer and diffusion processes became more facile, with R_{CT} and diffusion impedance Z_{DIFF} decreasing, and inductive effects often developing due to the intermediates-controlled kinetic steps (Section 7-4).

An example of EIS diagram changes that are typically observed with passivation is presented in Figure 12-23. At the active state (negative potentials) the corrosion current is high, and a "reverse loop" similar to potentiostatic control shown in Figure 12-23A is seen, with measured "negative resistance" at low frequencies due to increasing coverage of the surface by adsorbed intermediates and their "relaxations." At higher positive potentials R_{POL} increases and measured corrosion current I_{CORR} decreases (Figure 12-23B). With further potential increase there is a transition from an inductive (Figure 12-23C) to a resistive-capacitive response (Figure 12-23D), with the highest corresponding polarization resistance value R_{POL} and lowest corrosion current.

Kinetics in the film can also be affected by the formation of pores and pits that according to the De Levie theory will modify the EIS response. Formation of active-passive corrosion pits on 304 stainless steel exposed to a sodium-chloride corrosive solution can be represented by the electrical equivalent circuit composed of two parallel branches [48, 49]. The first branch represents the surface area without active pits (passive layer), where only inactive pits are present, and the second branch corresponds to the impedance of the active pits. The impedance inside the inactive pits can be represented by a transmission line, according to De Levie (Eq. 7-66). For relatively high frequencies the impedance of such a circuit can be represented by the passive layer film capacitance C_{PL} in parallel with the impedance in the passive pores or inactive pits $Z_{PORE} = Z_{IP}$ (Section 7-5):

$$Z_{IP} = \sqrt{R_{IP} Z_{INTERFACIAL}} = \sqrt{\frac{R_{IP}}{j\omega C_{IP}}} \tag{12-25}$$

where R_{IP} is the electrolyte resistance per unit length of inactive pit and C_{IP} is the electrical capacity per unit length of inactive pit. The parts of the surface area including the active pits are represented by the impedance:

$$Z_{AP} = R_{EP} + \frac{R_{CT}}{1 + j\omega C_{DL} R_{CT}} \tag{12-26}$$

where R_{CT} is the charge-transfer resistance inside the active pits or cracks, C_{DL} is the double-layer capacity inside the active pits, and R_{EP} is the electrolyte resistance inside the active pits or cracks. The impedance analysis of the data

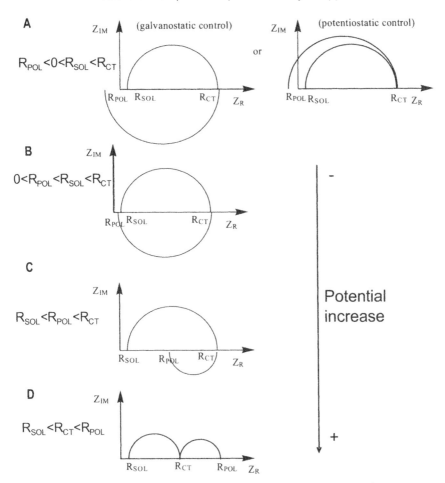

FIGURE 12-23   Impedance plot for corrosion processes at various potentials with potential for corroded systems displayed; R_{SOL} = solution resistance, R_{CT} = charge-transfer resistance, R_{POL} = polarization resistance

as a function of time of stainless-steel exposure to corrosive electrolyte allows determining the charge-transfer-controlled corrosion rate inside an active pit (Eq. 12-22). The active pitting surface area expansion is proportional to the double-layer capacity C_{DL} changes. Changes of the electrolyte resistance inside the pit R_{EP} can be also be detected.

A practical example of the characterization of the corrosion of oxides forming on zirconium and zirconium alloys in water through monitoring time-dependent changes in R_{FILM} (approximated by Z_{REAL} at 1Hz) was also presented [50]. An oscillatory cyclic pattern was observed, with a slow rise in R_{FILM} as the protective oxide film was formed and its thickness increased. The pinnacle was reached when a crack in the film appeared, followed by a rapid decrease in R_{FILM}, with the lowest value corresponding to formation of a new protective layer separating the metal surface and the degraded oxide.

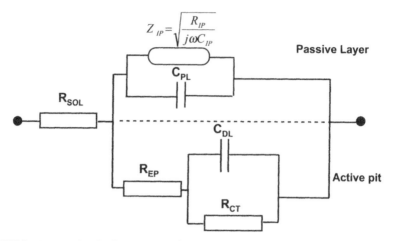

FIGURE 12-24   Equivalent circuit for pitting corrosion

References

1. S. Tourzain, *Some comments on the use of the EIS phase angle to evaluate organic coating degradation*, Electrochim. Acta, 2010, 55, pp. 6190–6194.
2. S. Krause, *Impedance methods*, Encyclopedia of Electrochemistry, A. J. Bard (Ed.), Wiley-VCH, Vol. 3, 2001.
3. A. Lasia, *Nature of the two semi-circles observed on the complex plane of porous electrodes in the presence of a concentration gradient*, J. Electroanal. Chem., 2001, 500, pp. 30–35.
4. C. Hitz, A. Lasia, *Experimental study and modeling of impedance of the her (hydrogen evolution reaction) on porous Ni electrodes*, J. Electroanal. Chem., 2001, 500, pp. 213–222.
5. M. Tomkiewicz, *Impedance of composite media*, Electrochim. Acta. 1993, 38, 14, pp. 1923–1928.
6. P. Campestrini, H. Terryn, J. Vereecken, J. H. W. de Wit, *Chromate conversion coating on aluminum alloys*, J. Electrochem. Soc., 2004, 151, 6, pp. B370–B377.
7. J.-B. Jorcin, G. Scheltjens, Y. Van Ingelgem, E. Tourwe, G. Van Assche, I. De Graeve, B. Van Mele, H. Terryn, A. Hubin, *Investigation of the self-healing properties of shape memory polyurethane coatings with the 'odd random phase multisine' electrochemical impedance spectroscopy*, Electrochim. Acta , 2010, 55, 21, pp. 6195–6203.
8. H. Marchebois, C. Savall, J. Bernard, S. Touzain, *Electrochemical behavior of zinc-rich powder coatings in artificial sea water*, Electrochim. Acta, 2004, 49, pp. 2945–2954.
9. H. Marchebois, M. Keddam, C. Savall, J. Bernard, S. Touzain, *Zinc-rich powder coatings characterization in artificial sea water EIS analysis of the galvanic action*, Electrochim. Acta, 2004, 49, pp. 1719–1729.
10. C. M. Abreu, M. Izquiedro, M. Keddam, X. R. Novoa, H. Takenouti, *Electrochemical behaviour of zinc-rich epoxy paints in 3% NaCl solution*, Electrochim. Acta, 1996),41, 15, pp. 2405–2415.
11. A Collazo, X. R. Novoa, C. Perez, B. Puga, *The corrosion protection mechanism of rust converters: an electrochemical impedance spectroscopy study*, Electrochim. Acta, 2010, 55, pp. 6156–6162.

12. C. Gabrielli, F, Huet, A. Sahar, *Dynamic analysis of charge transport in fluidized bed electrodes: impedance techniques for electroactive beds*, J. Appl. Electrochem., 1994, 24, 481–488.

13. D. G. Han, G. M. Choi, *Simulation of impedance for 2-D composite*, Electrochim. Acta, 1999, 44, pp. 4155–4161.

14. J. Wei, L. Zhao, S. Peng, J. Shi J, Z. Liu Z, W. Wen, *Wettability of urea-doped TiO2 nanoparticles and their high electrorheological effects*, J. Sol-Gel Sci. Tech., 2008, 47, 3, pp. 311–315.

15. T. Tilki, M. Yavuz, C. Karabacak, M. Cabuk, M. Ulutuerk, *Investigation of electrorheological properties of biodegradable modified cellulose/corn oil suspensions*, Carbohyd. Res., 2010, 345, 5, pp. 672–679.

16. D. P. Park, J. Y. Hwang, H. J. Choi, C. A. Kim, M.S. Jhon. *Synthesis and characterization of polysaccharide phosphates based electrorheological fluids*, Mat. Res. Innov., 2003, 7, 3, pp.161–166.

17. J. Yin, X. Zha, X. Xia, L. Xiang, Y. Qiao, *Electrorheological fluids based on nano-fibrous polyaniline*, Polymer., 2008, 49, 20, pp. 4413–4419.

18. Y. Liu, P. P. Phule, *Structure formation in novel electrorheological (ER) fluids based on ultrafine particles of electronic ceramics*, Polymer Preprints,1994, 35, 2, pp. 347–348.

19. W. M. Winslow, *Induced fibrillation of suspensions*, J. Appl. Phys., 1949, 20, pp. 1130–1140.

20. H. Block, J. P. Kelly, A. Qin, T. Watson, *Materials and mechanisms in electrorheology*, Langmuir, 1990, 6, pp. 6–14.

21. W. Wen, X. Huang, S. Yang, K. Lu, P. Sheng, *The giant electrorheological effect in suspensions of nanoparticles*, Nature. Mater., 2003, 2, pp. 727–732.

22. J. Zhang, X. Gong, C. Liu, W. Wen, P. Sheng, *Electrorheological fluid dynamics*, Phys. Rev. Lttr., 2008, 101, pp. 194503-1–194503-4.

23. M. Metikos-Hukovic, Z. Grubac, *The growth kinetics of thin anodic WO_3 films investigated by electrochemical impedance spectroscopy*, J. Electroanal. Chem., 2003, 556, pp. 167–178.

24. T. A. F. Lassali, J. F. C. Boodts, L. O. S. Bulhoes, *Faradaic impedance investigation of the deactivation mechanism of Ir-based ceramic oxides containing TiO2 and SnO2*, J. Appl. Electrochem., 2000, 30, pp. 625–634.

25. C. Bock, V. I. Birss, *Irreversible decrease of Ir oxide redox kinetics*, J. Electrochem. Soc., 1999, 146, 5, pp. 1766–1772.

26. S. Diethelm, A. Closset, J. Van Herle, K. Nisancioglu, *Determination of chemical diffusion and surface exchange coefficients of oxygen by electrochemical impedance spectroscopy*, J. Electrochem. Soc., 2002, 149, 11, pp. E424–E432.

27. I. Betova, M. Bojinov, V. Karastoyanov, P. Kinnunen, T. Saario, *Estimation of kinetic and transport parameters by quantitative evaluation of EIS and XPS data*, Electrochim. Acta, 2010, 55, pp. 6163-6173.

28. K. Dokko, M. Mohamedi, Y. Fujita, T. Itoh, M. Nishizawa, M. Umeda, I. Uchida, *Kinetic characterization of single particles of $LiCoO_2$ by AC impedance and potential step methods*, J. Electrochem. Soc., 2001, 148, 5, pp. A422-A426.

29. M.-S. Wu, P.–C. J. Chiang, J.–C. Lin, *Electrochemical investigation on capacity fading of advanced lithium-ion batteries after storing at elevated temperatures*, J. Electrochem. Soc., 2005, 152, 6, pp. A1041–A1046.

30. S. Franger, S. Bach, J. Farcy, J.-P. Pereira-Ramos, N. Baffier, *An electrochemical impedance spectroscopy study of new lithiated manganese oxides for 3V application in rechargeable Li-batteries*, Electrochim. Acta, 2003, 48, pp.891–900.

31. I. Betova, M. Bojnov, P. Kinnunen, K. Makela, T. Saario, *Conduction mechanism in oxide films on ferrous alloys studied by impedance spectroscopy in symmetrical and asymmetrical configurations*, J. Electroanal. Chem., 2004, 572, 2, pp. 211–223.

32. Y.-C. Chung, H.-J. Sohn, *Electrochemical impedance analysis for lithium ion intercalation into graphitized carbons*, J. Electrochem. Soc., 2000, 147, 1, pp. 50–58.

33. M. Holzapfel, A. Mertinent, F. Alloin, B. Le Gorrec, R. Yazami, C. Montella, *First lithiation and charge/discharge cycles of graphite materials, investigated by electrochemical impedance spectroscopy*, J. Electroanal. Chem., 2003, 46, pp. 41–50.

34. E. Barsukov, J. R. MacDonald, *Impedance Spectroscopy*, John Wiley & Sons, Hoboken, New Jersey, 2005.

35. M. Eikerling, A. A. Kornyshev, *Electrochemical impedance of the cathode catalyst layer in polymer electrolyte fuel cells*, J. Electroanal. Chem., 1999, 475, 2, pp. 107–123

36. I. A. Schneider, H. Kuhn, A. Wokaun, G. G. Scherer, *Fast locally resolved electrochemical impedance spectroscopy in polymer electrolyte fuel cells*, J. Electrochem. Soc.. 2005, 152, 10, pp. A2092–A2103.

37. H. Kuhn, B. Andreaus, A. Wokaun, G.G. Scherer, *Electrochemical impedance spectroscopy applied to polymer electrolyte fuel cells with a pseudo reference electrode arrangement*, Electrochim. Acta, 2006, 51, 8-9, pp. 1622–1628

38. J.-M. La Canut, R. A. Abouatallah, D. A. Harrington, *Detection of membrane drying, fuel cell flooding, and anode catalyst poisoning on PEMFC stacks by electrochemical impedance spectroscopy*, J. Electrochem. Soc., 2006, 153, 5, pp. A857–A864.

39. M. Eikerling, A. A. Kornyshev, *Electrochemical impedance of the cathode catalyst layer in PEM fuel cells*, J. Electroanal. Chem.,1999, 475, pp. 107–115.

40. R. Makharia, M. F. Mathias, D. R. Baker, *Measurement of catalyst layer electrolyte resistance in PEFCs using electrochemical impedance spectroscopy*, J. Electrochem. Soc., 2005, 152, 5, pp. A970–A977.

41. I. Epelboin, C. Gabrielli, M. Keddam and H. Takenouti, *A model of the anodic behaviour of iron in sulphuric acid medium*, Electrochim. Acta, 1975, 20, pp 913–916

42. C. Gabrielli, *Identification of electrochemical processes by frequency response analysis*, Solartron Analytical Technical Report 004/83, 1998, pp. 1–119.

43. I. Epelboin, C. Gabrielli, M. Keddam and H. Takenouti, *Proc. of the ASTM Symposium Progress in Electrochemical Corrosion Testing*, F. Mansfeld and U. Bertocci (Eds.), 1979, pp 150–160.

44. J.-P. Diard, B. Le Gorrec, C. Montella, *Corrosion rate measurements by non-linear electrochemical impedance spectroscopy. Comments on the paper by K. Darowicki, Corros. Sci., 1995, 37*; Corros. Sci.,1998, 40, pp. 495–508.

45. P. Agarwal, M. E. Orazem, L. H. Garcia-Rubio, *Application of the Kramers-Kronig relations in electrochemical impedance spectroscopy*, Electrochemical Impedance—Analysis and Applications, J. R. Scully, D. C. Silverman, M. W. Kendig (Eds.), ASTM 04-011880-27, Philadelphia, 1993, pp. 115–139.

46. l. Beaunier, I. Epelboin, J. C. Lestrade, H. Takenouti, *Electrochemical and scanning microscope study of pained Fe*, Surf Technol., 1976, 4, pp. 237–254.

47. M. C. Li, C. L. Zeng, S. Z. Luo, J. N. Shen, H. C. Lin, . N. Cao, *Electrochemical corrosion characteristics of type 316 stainless steel in simulated anode environment for PEMFC*, Electrochim. Acta., 2003, 48, pp. 1735–1741.

48. S. Krakowiak, K. Darowicki, P. Slepski, *Impedance of metastable pitting corrosion*, J. Electroanal. Chem., 2005, 575, 1, pp. 33–38.

49. I. Epelboin, M. Keddam and H. Takenouti, *Use of impedance measurements for the determination of the instant rate of metal corrosion*, J. Appl. Electrochem., 1972, 2, pp. 71–79.

50. J. Schefold, D. Lincot, A. Ambard, O. Kerrec, *The cyclic nature of corrosion of Zr and Zr-Sn in high-temperature water (663K). A long-term in situ impedance spectroscopic study*, J. Electrochem. Soc., 2003, 150, 10, pp. B451–B461.

CHAPTER 13

EIS Modifications

13.1. AC voltammetry

AC voltammetry is an extension of classical linear sweep electrochemical techniques such as cyclic voltammetry (CV) [1]. Analogous to CV, in AC voltammetry the mean DC potential V is imposed potentiostatically at arbitrary values that usually are different from the equilibrium value. A DC ramp with a comparatively slow sweep rate v and an AC signal $V_{AC} = V_A \sin \omega t$ with an amplitude V_A are superimposed and applied to a working electrode, and the response AC current and its phase angle as a function of V are registered. The function of the DC potential is to enforce and change the surface concentrations of the electroactive species at the electrode surface. The surface concentrations depend on the perturbation rates and magnitudes of the input DC and AC signals, and periodicity and on the type of analyzed system.

A simplification of the theoretical treatment can be achieved when the perturbation AC frequency is sufficiently high to separate the AC and DC contributions of the response signal. Often V is varied linearly on a long time scale (low frequency), while the higher-frequency AC perturbation V_{AC} (typically 10 Hz to 100 kHz) is used. The mean surface concentrations of the electroactive species are exactly the same as they would be in a linear sweep experiment without a superimposed AC potential. Due to the different time scales of the long-term diffusion caused by the DC potential V and the rapid diffusional fluctuations caused by the AC potential of an amplitude V_A, the surface concentration set up by the DC potential has the same effect as a bulk concentration on the AC process. As a result of this feature, AC voltammetry allows to interrogate process dynamics on different time scales, to explore the kinetics and thermodynamics of different processes, or to selectively target specific process dynamics, such as parallel reactions. Its main strength lies in the quantitative characterization of electrode processes, and it can also be used for analytical purposes.

The magnitude of Faradaic impedance Z_F for a simple reversible system is determined by both diffusion and charge-transfer resistance (Section 5-6) as:

$$Z_F = \sqrt{\left(R_{CT} + \frac{\sigma_D}{\sqrt{\omega}}\right)^2 + \left(\frac{\sigma_D}{\sqrt{\omega}}\right)^2} \tag{13-1}$$

A reversible redox process involving oxidized species with a bulk-solution concentration $C_{OX}{}^*$ determined by the amplitude of the AC current output I_{AC}, can be derived [1, p. 389], where V_0 is a standard potential:

$$Z_F = \frac{V_A}{I_{AC}} = \frac{4R_GT}{z^2F^2AC_{OX}^*\sqrt{\omega D}}\cosh^2(\frac{zF}{2R_GT}(V - V_0)) \tag{13-2}$$

At the peak maximum, the applied DC-potential V equals the standard potential V_0. Since for this condition $V = V_0$ and $cosh(0) = 1$, the peak current becomes:

$$I_p = \frac{z^2F^2AC_{OX}^*\sqrt{\omega D}}{4R_GT}V_A \tag{13-3}$$

From Eq. 13-3 it can be seen that the peak current is proportional to z^2, $\omega^{1/2}$, and C_0^*. This concentration dependence allows analytical electrochemistry applications of AC voltammetry.

The peak current is also proportional to the amplitude of the AC potential V_A. This is, however, limited to small amplitudes—that is, to quasilinear conditions. In principle, AC voltammetry can be carried out at considerably larger V_A values in order to increase the sensitivity and signal-to-noise ratio of the method. Although under these conditions the peak current is not proportional to the AC amplitude any longer, it could be demonstrated that the area under the peak is proportional to the perturbation amplitude and the analyte concentration. Since the rules for semi-infinite diffusion apply, a phase angle of $-45°$ for the impedance Z_F vs. frequency ω dependence can be observed for any reversible electrode reaction.

In contrast to classical cyclic voltammograms, AC-cyclic voltammograms have a clear baseline, which is advantageous for quantitative measurements. By extending AC linear sweep voltammetry by a reverse scan, AC cyclic voltammetry is obtained. If the surface concentrations of the electroactive species are the same at the same potential for forward and reverse scans, the peaks for forward and reverse scans are expected to be identical. If the DC process is not fully reversible, the surface concentrations of the electroactive species are different at a given DC potential for forward and reverse scans— that is, for quasi-reversible systems a displacement of the peaks for forward and reverse scan can be observed. This displacement can be used to derive kinetic parameters of the electrode reaction. For instance, the derivation [1, p. 393] (or Eq. 5-33) for sluggish one-step heterogeneous quasireversible and/or irreversible kinetics with a rate constant k_{heter}:

$$R_{CT} = \frac{R_GT}{z^2F^2Ak_{heter}C_{OX}^*}\left(\sqrt{\frac{D_{RED}}{D_{OX}}}\right)^\alpha \frac{1+e^{\frac{-zF(V-V_0)}{R_GT}}}{e^{\frac{-\alpha zF(V-V_0)}{R_GT}}} \tag{13-4}$$

As the AC frequency ω decreases, the diffusion limitations dominate the response and the peak current is proportional to $\omega^{1/2}$. At very low frequencies R_{CT} is small compared to the diffusion contribution, which can be broadly defined as $\sigma_D/\omega^{1/2}$, and the system looks reversible, with Eq. 13-2 becomes valid. With increase in the AC frequency, the charge-transfer resistance becomes predominant and the peak current I_p becomes independent of the frequency. At higher frequencies $R_{CT} \gg \sigma_D/\omega^{1/2}$ and $Z_F \rightarrow R_{CT}$, and the amplitude of the alternating current I_{AC} and the peak value of the current I_p can be determined as:

$$I_{AC} = \frac{V_A}{R_{CT}} = \frac{z^2 F^2 A k_{heter} C^*_{OX}}{R_G T} \left(\sqrt{\frac{D_{OX}}{D_{RED}}}\right)^{\alpha} \frac{e^{\frac{-\alpha z F (V - V_0)}{R_G T}}}{1 + e^{\frac{-z F (V - V_0)}{R_G T}}} V_A \qquad (13\text{-}5)$$

$$I_p = \frac{z^2 F^2 A k_{heter} C^*_{OX}}{R_G T} \left(\sqrt{\frac{D_{OX}}{D_{RED}}}\right)^{\alpha} \alpha^{\alpha} (1 - \alpha)^{1-\alpha} V_A \qquad (13\text{-}6)$$

AC voltammetry allows precise, rapid kinetic characterization measurements at solid electrodes. The challenge with AC voltammetry lies in the interpretation of the current response, which is often related to nonlinearity of the electron-transfer process and a large capacitive double-layer contribution in experiments when voltage waveform changes rapidly. This nonlinearity is minimized in both AC voltammetry and classical EIS by minimizing the amplitude of the AC voltage perturbation V_A, and large capacitive contribution is suppressed by using Fourier-transformed processing of the signal.

13.2. Potentiodynamic and Fourier-transform impedance spectroscopy

A typical EIS experiment often requires several minutes to an hour to record a full impedance spectrum and generally does not provide an adequate framework for studying time-dependent systems. This limitation is particularly valid for studies of Faradaic electrochemical reactions that affect the lowest frequency range of the impedance spectrum and therefore require longer sampling time. Information about heterogeneous kinetics can be obtained by studying various voltage-dependent features of the impedance equivalent-circuit parameters. As has been shown already in several examples, the DC response of the system plays an essential role in dictating the detailed equivalent-circuit representations found in such cases. These effects become particularly important for Faradaic systems, where significant DC currents are generated in the voltammetry scans.

For several decades cyclic voltammetry (CV) has been the technique of choice in investigating charge-transfer kinetic processes. Simultaneous determination of both EIS and CV parameters is very important, as exclusively CV-based analysis often misses important electrochemical aspects of the system, such as the double-layer capacitance, details of the double-layer structure, monolayer adsorption through functional dependence of C_{DL} on electrochemical potential, and electrochemical activity of nanostructural deposits through charge-transfer-resistance dependence on potential. However, it is often difficult to quantitatively compare potentiodynamic CV data for voltage-dependent currents with potentiostatic AC impedance results. In recent years, a number of time-resolved AC impedance techniques have been reported that can effectively address the aforementioned disadvantages of stationary-state EIS. These methods allow combining EIS with CV in potentiodynamic EIS (PDEIS) [2, 3, 4] and Fourier-transform EIS (FTEIS) [5] for characterization of electrochemical responses of interfacial processes.

"Multisine" and "wave mixing" methods have been introduced into commercial equipment. These methods involve the application of an input signal comprised of a number of sinusoids, resulting in "white noise." A number of different frequencies may be incorporated into a single signal, reducing the time needed to complete an impedance measurement in some cases from hours to minutes. The time-to-frequency transformation is based on natural laws and may be implemented by Laplace or Fourier transformations, and Fourier transforms can be obtained in real time during PDEIS measurement [4]. In addition, a voltammetric measurement can be discretely converted to a frequency domain representation for a number of input signals [6]. The simplest time-to-frequency domain conversion is the Fourier transformation of coulostatically induced potential transients $V(t)$ recorded following application of charge q. The Fourier transform of continuous-time function $V(t)$ on a finite time interval $[0, t]$ is:

$$V(f) = \int_0^T V(t)\exp(-j2\pi ft)dt \tag{13-7}$$

When $V(t)$ is sampled at discrete evenly spaced time intervals Δt, $V(f)$ can be approximated by:

$$V(f) = F\{V(t)\} = \Delta t \sum_i^{N-1} V_i \exp(-j2\pi ft_i) \tag{13-8}$$

where $t_i = i\Delta t$, $V_i = V(i\Delta t)$, $i = 0, 1, 2..., N$ for a frequency band $[0, f_0]$ with $f_0 = 1/2\Delta t$. This transformation approach considers the response as discrete data points rather than a continuous function. The rate at which the response is sampled and the duration over which it is sampled determine the extent of the frequency domain that is resolvable. For example, if a transient response is sampled at 1000Hz ($\Delta t = 0.001$) for 0.5 s, the lower limit of resolution is $2Hz = 1/0.5\sec$ and the upper limit is $500Hz = 1/(2*0.001\sec)$. The electrochemical impedance at each frequency can be evaluated as:

$$Z(j\omega) = K_F \frac{1}{I_G} F\{V(t)\} \tag{13-9}$$

where K_F is a constant depending upon the implementation of the Fourier transform and I_G is a generic term used to represent the signal responsible for the transient. For the case of coulometric perturbation I_G is the amplitude of the current pulse q/t; for coulostatic perturbation with voltage source V and capacitor C it becomes equal to VC. Implementation of Fourier transform must be done carefully to avoid mathematical errors (aliasing) with data free of excessive noise to ensure correct mathematical transformation of the signal [6].

FTEIS is an experimentally convenient technique for such potentiodynamic impedance measurements. In this approach, a series of Nyquist (or Bode) spectra is obtained in parallel with the recording of CV data. DC voltage-dependent electrode equivalent-circuit models can be developed through CNLLS analysis of the impedance spectra. However, in FTEIS postexperiment processing is often cumbersome and inconvenient, and FTEIS limits the amplitude of the excitation signal to under 10 mV (for aqueous solutions) because of the nonlinearity of the electrochemical system.

PDEIS also employs virtual instruments for probing the interfaces and analysis of the potential-dependent electrochemical response, which offers some advantages over traditional FT techniques. In PDEIS consecutive probing with a stream of wavelets in a kHz-high Hz frequency range in addition to integral probing and analysis is offered. The method reveals in a single potential scan the real-time multidimensional (3D) spectra that characterize complex impedance as a function of both AC frequency and electrode potential, along with potentiodynamic (or cyclic) voltammogrammetry. After the scan, a built-in equivalent electric circuit analyzer decomposes the AC part of the interface response into constituents related to different interfacial processes and structures. Similarly to CV, PDEIS is suitable for both stationary and nonstationary systems and provides information on the time-frequency-potential variance of the electrochemical response in each case. The limitation of PDEIS compared to standard EIS is fundamentally related to a smaller number of acquired data points allowing for accelerated data collection, resulting in a narrower frequency range and less frequent sampling. Typically 20 to 30 frequencies are processed per each potential step (staircase DC potential ramp), which is sufficient to determine the essential features of the Nyquist plot. In aqueous systems the DC steps are in ~1-5 mV range, and amplitude of the AC signal is ~10mV for Hz - kHz frequency range [3].

13.3. Nonlinear higher-harmonics impedance analysis

In a linear EIS analysis a small AC perturbation amplitude V_A is typically chosen so that the inherently nonlinear electrochemical system can be approximated by a linear system. In this approximation the measured impedance is independent of the amplitude of the perturbation $V_{AC}(t) = V + V_A \sin(\omega t)$, and there is no higher-harmonic response. Additionally, higher-order harmonics are normally filtered out during EIS measurements as a part of noise reduction.

However, in the specific case of EIS analysis of highly resistive materials, AC voltage amplitudes on the order of $V_A = 1$ volt are typically being applied to ensure the passing of sufficient current through a sample and achieve better signal-to-noise ratio. These high AC voltage amplitudes produce significant nonlinearities and harmonics other than "fundamental" or "first harmonic" in the current response, that can be utilized in nonlinear EIS (NLEIS) as an important pattern-recognition factor.

It is useful to consider the second (2ω) and higher ($n\omega$, $n>2$) harmonics for the characterization of electrochemical systems. The bulk electrolyte solution obeys Ohm's law and therefore can be represented by a small resistive component $R_{SOL} \sim 1$–10 ohm, which can be considered linear. However, in tissues and highly resistive samples, such perturbation voltages may lead to changes in viscosity or number of ions per volume, and the high-frequency bulk-solution response of the system may become nonlinear. Nonlinearity also occurs when significant capacitance C_{BULK} in nonpolar resistive samples is present (for instance, in organic colloids). That would also be the case for very high electric fields. According to the Debye-Huckel theory, every charged particle or ion is surrounded by a symmetrical ionic atmosphere. In the case of fluids placed in an external high electric field, the movements of ions may occur so fast that a retarding ionic atmosphere does not have time to form, and the fluids show higher nonlinear conductivity (Wien effect [7]). For the electrode process it has been shown that there is a voltage amplitude limit for linear behavior at about 20–100 mV, with corresponding frequency-dependent currents of ~ 10–100 mA/cm^2 in MHz range and around 5 μA/cm^2 at mHz frequency range. Frequency harmonic analysis shows that nonlinearity appears at voltage AC amplitudes as low as $V_A = 20$ mV.

Traditional "linear" EIS has pattern-recognition problems, in particular in the low-frequency range, where multiple kinetic models based on charge transfer, diffusion, and adsorption can be proposed and often are equally successful in providing a good fit with experimental impedance data. The advantage of techniques based on the nonlinearity of electrochemical systems is that the second and higher harmonic signals are relatively free of "nearly linear" elements contributing to the total impedance response, such as aqueous-solution resistance R_{SOL} and double-layer charging currents. Since double-layer capacitance C_{DL} behaves in much more linear fashion than Faradaic processes, the charging currents of the double layer can mainly be found at the base excitation frequency ω. In addition to this mostly linear double-layer charging effect, the bulk-solution resistance R_{SOL} is also largely filtered out, as the solution conductivity is essentially linear, especially in more conductive "supported" media. Higher harmonics therefore are of great interest, because they arise almost exclusively from a variety of strongly nonlinear mainly heterogeneous kinetic processes such as electrochemical charge transfer, diffusion, and adsorption [8]. NLEIS allows studying the charge- and mass-transport processes at the interface relatively free from interference. Hence, the determination of heterogeneous and coupled kinetic parameters is frequently easier at higher harmonics. By correlating these components, one can in principle obtain unique and reliable information about electrode reaction mechanisms not available from linear EIS analysis and often resolve the interpretation ambiguities of linear impedance data. Higher harmonics typically are detected by frequency-response analyzers, but also lock-in amplifiers can be tuned to detect a multiple of excitation frequencies (Section 8-2), and the fast Fourier transform multiple-sine method (Section 8-3) can also be employed.

Another advantage of nonlinear impedance analysis is that measurement of several harmonics may facilitate extraction of kinetic parameters at a single DC "offset" potential V [8] not available from small-amplitude fundamental-frequency impedance measurement. NLEIS can be used to calculate all the harmonics of the current response to a sinusoidal potential perturbation $V_{AC}(t) = V + V_A sin(\omega t)$ and derive the nonlinear impedance. Results from a simulation study can be compared with experimental NLEIS data, leading to more accurate quantification and modeling of the impedance data and better interpretation of the electrochemical kinetic processes [8, 9, 10, 11, 12, 13].

In the previous chapters only the linearized form of the current-potential relation has been used. The other terms of the Taylor series expansion of current vs. potential were neglected. In reality current-potential relationships show a considerable amount of curvature (Figure 8-14). The Taylor series expansion of the Butler-Volmer equation predicts that:

$$\Delta I = \left(\frac{dI}{dV} \right)_V \Delta V + \frac{1}{2} \left(\frac{d^2 I}{dV^2} \right)_V \Delta V^2 + ... \tag{13-10}$$

When a pure sinusoidal voltage V_{AC} is applied to an electrochemical cell, the waveform of the resulting current is very often distorted due to this nonlinear current-potential relationship unless the excitation voltage amplitude V_A is sufficiently small. The response signal can be described as:

$$I = I_{DC} + I_0 sin(\omega t + \phi) + I_1 sin(2\omega t + \phi_1) + I_2 sin(3\omega t + \phi_2) + .. + Noise \tag{13-11}$$

where I_{DC} is the DC component of the current and $I_1, I_2, I_3...$ are harmonic distortion components. Precise measurements of these current responses at each harmonic can be used (Eq. 13-11) to evaluate unknown terms in Eq. 13-10. For example, Nakata [14] developed a diagnostic criterion for detection of nonlinearities in a simple circuit represented by a parallel combination of voltage-dependent capacitor C and conductor G subjected to a perturbation by a sinusoidal potential $V_{AC}(t) = V + V_A sin(\omega t)$. The "real" and "imaginary" currents through the conductor I_{REAL} and capacitor I_{IM} can be expressed as:

$$I_{REAL}(t) = V_{AC}(t)G(V_{AC}) = V_{AC}(t)[G_0 + G_1 V_{AC} + G_2 V_{AC}^2 + G_3 V_{AC}^3] =$$

$$[V + V_A \cos \omega t][G_0 + G_1(V + V_A \cos \omega t) + G_2(V + V_A \cos \omega t)^2 + G_3(V + V_A \cos \omega t)^3] =$$

$$= G_0 V + G_1[V^2 + 0.5V_A^2] + G_2 V[V^2 + 1.5V_A^2] + G_3[V^4 + 3V^2 V_A^2 + 0.375V_A^4] +$$

$$+ \cos \omega t[G_0 V_A + 2G_1 VV_A + 3G_2 V_A\{V^2 + 0.25V_A^2\} + G_3 VV_A\{4V^2 + 3V_A^2\}] +$$

$$+ 0.5V_A^2 \cos 2\omega t[G_1 + 3G_2 V + 6G_3 V_A(V^2 + V_A)] +$$

$$+ V_A^3 \cos 3\omega t[G_2 + G_3 V] +$$

$$+ 0.125V_A^4 G_3 \cos 4\omega t =$$

$$= I_{REALDC} + I_{REAL1} \cos \omega t + I_{REAL2} \cos 2\omega t + I_{REAL3} \cos 3\omega t + I_{REAL4} \cos 4\omega t$$

$$\tag{13-12}$$

$$I_{IM}(t) = \frac{dq}{dt} = \frac{dq}{dV}\frac{dV}{dt} = C(V_{AC})\frac{dV_{AC}}{dt} = [C_0 + C_1 V_{AC} + C_2 V_{AC}^2 + C_3 V_{AC}^3]\frac{d(V_0 + V_A \cos \omega t)}{dt} =$$

$$= -\omega V_A \sin \omega t[C_0 + C_1(V + V_A \cos \omega t) + C_2(V + V_A \cos \omega t)^2 + C_3(V + V_A \cos \omega t)^3] =$$

$$= -\omega V_A \sin \omega t[C_0 + C_1 V + C_2 V^2 + C_3 V^3 + 0.25V_A^2\{C_2 + 3C_3 V\}] -$$

$$-0.5\omega V_A \sin 2\omega t[C_1 V_A + 2C_2 V V_A + 3C_3 V_A V^3 + 0.5C_3 V_A^3] -$$

$$-0.25\omega V_A^3 \sin 3\omega t[C_2 + 3C_3 V] -$$

$$-0.125\omega V_A^4 C_3 \sin 4\omega t =$$

$$= -I_{IM1} \sin \omega t - I_{IM2} \sin 2\omega t - I_{IM3} \sin 3\omega t - I_{IM4} \sin 4\omega t$$

(13-13)

The total current is:

$$I(t) = I_{REAL}(t) + I_{IM}(t) \tag{13-14}$$

From the separate lines of Eqs. 13-12 and 13-13 real and imaginary components of impedance at first (fundamental), second, third, and fourth harmonics can be calculated from the known voltage signal parameters and measured frequency-dependent current. The values for the characteristic total capacitance $C(V_{AC})$ and conductance $G(V_{AC})$ of the circuit can be computed. Comparison of the experimental and calculated frequency-dependent data for each harmonic serves as a diagnostic criterion that the system can indeed be represented by a simple parallel $G \,|\, C$ combination. Poor fit between the experimental and the calculated frequency-dependent impedance or current functions implies that a more complicated kinetic mechanism is responsible for the measured impedance characteristics.

Harmonic analysis has been used widely for determining corrosion rates [15,16, 17], adsorption impedance [8, 18], effects of diffusion [19, 20], and other complex multistage kinetic processes [21]. Applications of nonlinear EIS often include introducing perturbation current, measuring current and voltage, Fourier transformation of time-dependent input and output signals, determining real and imaginary components of the output signals for the higher harmonics, conversion back into the frequency domain, and extraction of harmonics coefficients. The problem with NLEIS lies in its extensive use of complex mathematical simulations that require dedicated software packages (such as Mathematica®) utilizing the numerical nonlinear fitting function. The process often attempts to evaluate several possible realistic equivalent-circuit models that are entered into the calculation. Preliminary parameters describing nonlinear diffusion, adsorption, and charge-transfer impedance elements should be introduced into the calculation initially, while solution resistance and double-layer capacitance are assumed to be constant [18]. The following minimization routine is performed on a series of random starting values for each of the kinetic parameters, and the fit values of the models are used to develop the best fit parameters for each of the models at each experimental potential.

13.4. Local EIS

In conventional EIS experiments the electrode response to a perturbation signal corresponds to a measurement averaged across the whole electrode surface area. However, electrochemical systems show nonuniform current and
potential distributions, resulting in CPE behavior. Such distributions can be
studied by the local EIS method, which employs in situ probing of local current density distribution in the vicinity of the working electrode surface. Local
EIS (LEIS) relies on the fact that AC current density in the solution very near
the working electrode is proportional to the local impedance properties of the
electrode [22]. The AC current spreads in the solution as a function of the distance from the electrode surface, and as a consequence the LEIS results depend on the distance between the probe and the surface. That allows for spatially resolved LEIS measurements of the surface topography and kinetics at
the electrochemical interface.

In order to determine the current densities normal to the surface, the AC
potential drop is measured between the planes parallel to the electrode surface, employing a dual electrode scanning microprobe (Figure 13-1). The magnitude of the local impedance for the applied magnitude of the AC potential
$V_A(\omega)$ between the working and reference electrodes can be calculated from
the potential drop differential $\Delta V(\omega)_{PROBE}$ between two microprobes, conductivity of the electrolyte σ, and distance l between the two microelectrodes normal to the surface, using the following relationship:

$$Z(\omega)_{LOCAL} = \frac{V_A(\omega)l}{\Delta V(\omega)_{PROBE}\,\sigma} \tag{13-15}$$

The assumption made in the derivation of Eq.13-15 is that the current density at the tip of the two-electrode probe and the current density at the electrode surface are the same, which means that all the current measured at the
two-electrode microprobe is assumed to be flowing normally to the surface of
the working electrode. In reality, however, the AC currents spread in the solution as a function of the distance from the working electrode surface. The resolution of LEIS depends on the distance of the two-electrode probe from the
working electrode surface, which is sometimes being affected by 3D heteroge-
neities and local CPE behavior. The spatial resolution of LEIS is also limited by
the size of the two-electrode probe, since it is not possible to resolve features
smaller than the probe. A reduction in size of the probe is also accompanied by
an increase in its impedance, which makes it more difficult to measure the current density accurately.

A series of recent papers by Orazem and coworkers [23, 24, 25, 26, 27, 28]
demonstrated the applicability of the LEIS method to detailed investigations
of electrochemical kinetics and mass-transport limitations of the reactions.
These papers demonstrated the effects of the distribution of current associated
with electrode 3D and 2D geometries, resulting in complex impedance behavior with an associated CPE element at high frequencies due to the radial distribution of the electrochemical potential and resulting heterogeneity in the

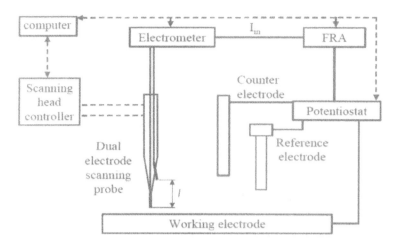

FIGURE 13-1 Schematic of an experimental setup for LEIS [22] (Uniscan instruments) (with permission from the author)

bulk solution local impedance, and complex impedance behavior at low frequencies associated with well-known intermediate adsorption kinetics (Section 7-4). The spatial current distribution due to electrode geometry effects are reflected in the local ohmic impedance, which has complex distributed behavior at both high and low frequencies.

LEIS have been applied, among other uses, to broad investigations of electrochemical kinetics [23, 24, 25, 26, 27], corrosion and passivation of metal surfaces [28], and studies of defects in organic coatings [29]. Recently this method was employed to assess the influence of electrode geometry and surface roughness, current and potential distributions along the electrode surface, and adsorption effects on high- and low-frequency CPE behavior observed in many impedance measurements [23, 24, 25, 26, 27].

13.5. Scanning photo-induced impedance microscopy (SPIM)

Another impedance-based imaging technique for laterally resolved characterization of thin films or electrochemical systems is scanning photo-induced impedance microscopy (SPIM) [22]. It is based on photocurrent measurements at field-effect structures. In their simplest arrangement, field-effect structures consist of a semiconductor substrate with a thin insulator and a gate electrode. This gate electrode can be a metal film, resulting in a metal-insulator semiconductor or electrolyte-insulator semiconductor, where the electrolyte is in direct contact with the insulator, and a reference electrode is required to fulfill the function of the gate electrode (Figure 13-2).

Different regions of the semiconductor in a field-effect structure can be addressed with light. A DC voltage is applied between the semiconductor substrate and the gate electrode. Due to the presence of an insulator, this does

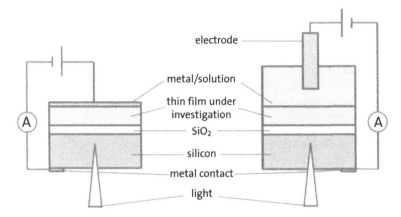

FIGURE 13-2 Experimental arrangements for SPIM: A. gas phase measurements; B. measurements in electrolyte solutions [22] (with permission from the author)

not produce a DC current but controls the space charge region in the semiconductor at the interface between the semiconductor and the insulator. Electron hole pairs created during illumination of the semiconductor separate in the field of the space charge region if it is biased towards inversion. The resulting current causes electrons and holes to collect on both sides of the insulator, effectively charging the capacitor. When the illumination is turned off, the capacitor discharges and a current flows in the opposite direction. A modulated or chopped light beam focused on one region of the semiconductor substrate results in a local AC current. When the space charge region is biased towards accumulation, no photocurrent flows.

If the system under investigation—for example a thin film—is deposited on a semiconductor insulator structure, the photocurrent decreases. The difference in photocurrent is a function of the complex impedance of the added film layer. Local differences in impedance can be detected in the scanning mode. An additional variation of the modulation frequency of the light results in AC impedance spectra with lateral resolution. In the case of SPIM, a local AC voltage is created due to the photoeffect within the semiconductor by a focused, modulated light beam, leading to a localized AC current. In contrast to other localized impedance techniques such as LEIS, no probe or microelectrode needs to be positioned above the sample. It is, however, required that the system under investigation be deposited on a field-effect structure.

References

1. A. J. Bard, L. R. Faulkner, *Electrochemical Methods, Fundamentals and Applications*, John Wiley & Sons, New York, 2001.
2. K. Darowicki, P. Slepski, *Dynamic electrochemical impedance spectroscopy of the first order electrode reaction*, J. Electroanal. Chem., 2003, 547, pp. 1–8.

3. G. A. Ragoisha, A. S. Bondarenko, *Potentiodynamic electrochemical impedance spectroscopy*, Electrochim. Acta., 2005, 50, pp. 1553–1563.

4. G. A. Ragoisha, A. S. Bondarenko, N. P. Osipovich, E. A. Streltsov, *Potentiodynamic electrochemical impedance spectroscopy: lead underpotential deposition on tellurium*, J. Electroanal. Chem., 2004i 565, pp. 227–234.

5. C. M. Pettit, P. C. Goonetilleke, C. M. Sulyma, D. Roy, *Combining Impedance spectroscopy with cyclic voltammetry: measurement and analysis of inetic parameters for Faradaic and non-faradaic reactions on thin-film gold*, Anal. Chem., 2006, 78, pp. 3723–3729.

6. N. Birbilis, B. N. Padgett, R. G. Buchheit, *Limitations in microelectrochemical capillary cell testing and transformation of electrochemical transients for acquisition on microcell impedance data*, Electrochim. Acta, 2005, 50, pp. 3536–3544.

7. S. Grimnes, O.G. Martinsen, *Bioimpedance and Bioelectricity Basics*, Academic Press, 2000.

8. D. A. Harrington, *Theory of electrochemical impedance of surface reactions: second-harmonic and large-amplitude response*, Can. J. Chem., 1997, 75, pp. 1508–1517.

9. S. L. Carlson, M. E. Orazem, O. D. Crisalle, L. Garcia-Rubio, *On the error structure of impedance measurements. Simulation of FRA instrumentation*, J. Electrochem. Soc., 2003, 150, 10, pp. E477–E490.

10. S. L. Carlson, M. E. Orazem, O. D. Crisalle, L. Garcia-Rubio, *On the error structure of impedance measurements. Simulation of PSD instrumentation*, J. Electrochem. Soc., 2003, 150, 10, pp. E491–E500.

11. S. L. Carlson, M. E. Orazem, O. D. Crisalle, L. Garcia-Rubio, *On the error structure of impedance measurements. Series expansions*, J. Electrochem. Soc., 2003, 150, 10, pp. E501–E511.

12. J. E. Garland, C. M. Pettit, D. Roy, *Analysis of experimental constraints and variables for time resolved detection of Fourier transform electrochemical impedance spectra*, Electrochim. Acta, 2004, 49, pp. 2623-2635.

13. J.-S. Yoo, S. -M. Park, *An electrochemical impedance measurement technique employing Fourier transform*, Anal. Chem., 2000, 72, pp. 2035–2041.

14. S. Nakata, N. Kido, M. Hayashi, M. Hara, H. Sasabe, T. Sugawara, T. Matsuda, *Chemisorption of proteins and their thiol derivatives onto gold surface: characterization based on electrochemical nonlinearity*, Biophys. Chem.,1996, 62, pp. 63–72.

15. B. Bozzini, I. Sgura, *A non-linear AC spectrometry study of the electrodeposition of Cu from acidic sulphate solutions in the presence of PEG*, J. Appl. Electrochem., 2006, 36, pp. 983–989.

16. K. Darowicki, J. Majewska, *The effect of a polyharmonic structure of the perturbation signal on the results of harmonic analysis of the current of a first-order electrode reaction*, Electrochim. Acta, 1998 44, pp. 483–490.

17. J.-P. Diard, B. Le Gorrec, C. Montella, *Corrosion rate measurements by non-linear electrochemical impedance spectroscopy. Comments on the paper by K. Darowicki, Corros. Sci., 1995, 37, 913*, Corrosion Science, 1998, 40, pp. 495–508.

18. V. F. Lvovich, M. F. Smiechowski, *Non-linear impedance analysis of industrial lubricants*, Electrochim. Acta, 2008, 53, pp. 7375–7385.

19. V. V. Elkin, V. Y. Mishuk, V. N. Alekseev, B. M. Grafov, *Polarization diagram of the electrochemical impedance of second order: a theory that accounts for the diffusion of reactants*, Russ. J. Electrochem., 2001, 37, 4, pp. 399–408.

20. V. V. Elkin, V. Y. Mishuk, M. A. Abaturov, V. N. Alekseev, B. M. Grafov, *Second-order impedance spectroscopy of an electroreduction reaction with allowance for the Frumkin correction*, Russ. J. Electrochem., 2003, 39, 4, pp. 350–364.

21. J.R. Wilson, D.T. Schwartz, S.B. Adler, *Nonlinear electrochemical impedance spectroscopy for solid oxide fuel cell cathode materials*, Electrochim. Acta, 2006, 51, pp.1389–1402.

22. S. Krause, *Impedance Methods*, Encyclopedia of Electrochemistry, A. J. Bard (Ed.), Wiley-VCH, Vol. 3, 2001.

23. V. Mie-Wen Huang, V. Vivier, M. E. Orazem, N. Pebere, B. Tribollet, *The apparent constant-phase-element behavior of an ideally polarized blocking electrode*, J. Electrochem. Soc.,2007), 154, 2, pp. C81–C88.

24. V. Mie-Wen Huang, V. Vivier, M. E. Orazem, N. Pebere, B. Tribollet, *The global and local impedance response of a blocking disk electrode with local constant-phase-element behavior*, J. Electrochem. Soc., 2007, 154, 2, pp. C89–C98.

25. V. Mie-Wen Huang, V. Vivier, M. E. Orazem, N. Pebere, B. Tribollet, *The apparent constant-phase-element behavior of a disk electrode with Faradaic reactions*, J. Electrochem. Soc., 2007, 154, 2, pp. C99-C107.

26. S. –L. Wu, M. E. Orazem, B. Tribollet, V. Vivier, *Impedance of a disc electrode with reactions involving an adsorbed intermediate: local and global analysis*, J. Electrochem. Soc., 2009, 156, 1, pp. C28–C38.

27. J. –B. Jorcin, M. E. Orazem, N. Pebere, B. Tribollet, *CPE analysis by local electrochemical impedance spectroscopy*, Electrochim. Acta, 2006, 51, pp. 1473–1479.

28. I. Frateur, V. Mei-Wen Huang, M. E. Orazem, N. Pebere, B. Tribollet, V. Vivier, *Local electrochemical impedance spectroscopy: Considerations about the cell geometry*, Electrochim. Acta, 2008, 53, pp. 7386–7395.

29. M.W. Wittmann, R.B. Leggat, S.R. Taylor, *The detection and mapping of defects in organic coatings using local electrochemical impedance spectroscopy*, J. Electrochem. Soc., 1999, 146, 11, pp. 4071–4075.

CHAPTER 14

Conclusions and Perspectives of EIS

Since the original concepts of impedance spectroscopy were articulated by Oliver Heaviside more than 125 years ago, the impedance method has continued to evolve along with modern science and technology, supporting and fueling an ever-expanding number of contemporary industrial applications and research areas. During the most recent decades, advances in EIS technique have been closely linked to those of computer science and electrical engineering, enabling the great progress achieved in the development of commercial EIS instrumentation. That enabled further advancements in the method—both in expanding it to an even broader universe of emerging fields of use and in understanding the method's underlying principles, best experimental practices, and data interpretation in a multitude of practical applications.

EIS changed the ways electrochemists interpret the electrode-solution interface. With impedance analysis, a complete description of an electrochemical system can be achieved, as the data contains all necessary electrochemical information. The technique offers the most powerful analysis of the status of electrodes, monitors, and probes in many different processes that occur during electrochemical experiments, such as adsorption, charge and mass transport, and homogeneous reactions. EIS offers huge experimental efficiency and results that can be interpreted in terms of linear systems theory, modeled as equivalent circuits, and checked for discrepancies with the Kramers-Kronig transformations.

EIS is playing an absolutely critical role in solving the great challenges of the 21st century. Among other fields, advances in healthcare therapies and biomedical devices, performance optimization of new renewable, sustainable sources of energy such as batteries and fuel cells, monitoring mechanical stability of structures as large as bridges and refineries and as small as interdigitated implantable sensors—all rely on impedance-based devices. Intelligent use of the principles of EIS becomes essential to further socioeconomic advances.

The future of the impedance method lies in its applicability to studies of an almost unlimited number of experimental processes and applied systems. These EIS applications directly address many industrial issues of worldwide importance, such as corrosion and anodic behavior of metals and composite materials, states of electrodes during charging/discharging cycles of batteries and fuel cells, surface characterization of polymer-modified structures, and biomedical implants and sensors. EIS remains flexible, operationally simple, inexpensive, and compatible with nearly all possible operational environments method for in vivo diagnostic monitoring and laboratory analysis. EIS applicability to fast, real-time, in-situ measurements and utilization of multiple hyphenated techniques such as combined spectroelectrochemical/gravimetric/impedance measurement during electrochemical experiments allow for continuous development of impedance-based technologies and products. The incredible flexibility of the method and its ability to be combined with other techniques for in-situ and ex-situ analysis, fueled by continuous advances in scientific instrumentation, present a significant opportunity to resolve the remaining interpretation challenges of electrochemical impedance spectroscopy analysis. Developing a better understanding and reducing the effects of the most significant EIS limitations, such as data-interpretation ambiguities and pattern-recognition problems, will result in further expansion of EIS application to studies of physical, chemical, mechanical, and electrical properties of experimental systems.

Abbreviations and Symbols

a, a_p	ion (particle) size (radius)	[cm]
a_x, a_y, a_z	particles axial dimensions, Eq. 7-18	[cm]
A	surface area of working electrode	[cm^2]
A_{MBR}	surface area of membrane	[cm^2]
A_R	Arrhenius constant, Eq. 5-11	[unitless]
A, B, G	parameters from multistep kinetics analysis, Eqs. 7-59–7-61	[1/ohm, 1/H, ohm H]
A^*, B^*, C^*, D^*, E^*	parameters from cellular electrokinetics, Eq. 11-9	[unitless]
b, b_A, b_C	Tafel coefficients for anodic and cathodic processes, Eq. 7-52	[V]
c_K	cell constant, Eq. 8-9	[1/cm]
$C \, (C_1, C_2, C_N \ldots)$	capacitance (of materials or circuit segments)	[F]
C_{ADS}	adsorption capacitance, Eq. 5-37	[F]
$C_{ADS\,IHP}$	adsorption capacitance for inner Helmholtz layer	[F]
$C_{ADS\,OHP}$	adsorption capacitance for outer Helmholtz layer	[F]
C_{AGG}	capacitance of particles' agglomerates	[F]
C_{BASE}	base fluid material capacitance	[F]
C_{BINDER}	organic binder capacitance, Chapter 12	[F]
C_{BULK}	bulk material capacitance	[F]
C_C	complex capacitance, Eq. 5-8	[F]
$C_C{}'$	real part of complex capacitance, Eq. 5-8	[F]
$C_0 = C_{LF}$	complex capacitance at low frequency, Eq. 5-8	[F]
$C_{CONTACT}$	contact capacitance particle-to-particle, Chapter 12	[F]
C_D	limiting diffusion capacitance	[F]
C_{DD}	second diffusion layer capacitance, Chapter 5	[F]
C_{DL}	double layer capacitance	[F]
$C_{DIFFUSE}$	diffuse layer capacitance, Eqs. 5-15	[F]
C_{EFF}	effective equivalent capacitance, Eqs. 3-4 – 3-6	[F]
C_{FILM}	film (adsorbed) capacitance, Chapter 12	[F, F/cm^2]
$C_{FREE\,ADD}$	free additives capacitance, Chapter 10	[F]
C_{GAS}	gas space capacitance, Eq. 12-7	[F]

C_{GEOM}	geometrical (bulk) capacitance, Chapter 6	[F]
$C_{HELMHOLTZ}$	Helmholtz layer capacitance, Eqs. 5-15	[F]
C_{IDT}	capacitance of interdigitated electrode, Eqs. 8-11	[F]
$C_{INF} = C_{HF}$	complex capacitance at high frequency, Eq. 5-8	[F]
C_{IHP}	inner Helmholtz layer capacitance	[F]
C_{INPUT}	capacitance of impedance measuring device	[F]
$C_{INTERFACE}$	interfacial capacitance	[F]
C_{IP}	inactive pits capacitance, Chapter 12	[F]
C_{LF}	complex capacitance at low frequency, Eq. 5-8	[F]
C_{LOAD}	parasitic capacitive load, Chapter 8	[F]
C_M	media capacitance, Chapter 11	[F]
C_{MBR}	membrane capacitance, Eq. 7-10, Chapter 11	[F]
$C_{NETWORKS}$	dipolar networks capacitance, Chapter 10	[F]
C_{OHM}	cable insulator capacitance, Eq. 5-1	[F]
C_{OHP}	outer Helmholtz layer capacitance	[F]
C_{PL}	passive layer capacitance, Chapter 12	[F]
$C_{PSEUDOCAP}$	pseudocapacitance	[F]
C_R	contact capacitance of RE and CE, Eq. 8-21	[F]
C_{REACT}	reaction capacitance	[F]
C_{RC}	coupling capacitance between RE and CE, Eqs. 8-15–8-17	[F]
C_{WC}	coupling capacitance between WE and CE, Eqs. 8-15–8-17	[F]
C_{WR}	coupling capacitance between WE and RE, Eqs. 8-15–8-17	[F]
C_S	capacitance of electron transfer through bulk particle	[F]
C_{SOOT}	soot dispersed particles capacitance, Chapter 10	[F]
$C_{SOOT\ AGG}$	soot agglomerated particles capacitance, Chapter 10	[F]
C_{SC}	surface charge capacitance, Eq. 5-21	[F]
C_{SH}	shunt capacitance, Chapter 11	[F]
$\Delta C = C_{LF} - C_{HF}$	capacitance increment	[F]
C^*, C^*_i	bulk solution concentration of species type i	[mol/cm³]
$C^*_{OX,}\ C^*_{RED}$	bulk solution concentration of oxidant, reductant, Eq. 5-22	[mol/cm³]
$C_{OX,}\ C_{RED}$	surface concentration of oxidant, reductant, Eq. 5-22	[mol/cm³]
C_{HEAT}	heat capacitance, Eq. 7-44	[J/(kg K)]
d	distance between WE and CE, sample thickness	[cm]
D_D	dipole moment, Eq. 7-20	[unitless]
D, D_i	diffusion coefficient of species type i	[cm²/sec]
$D_{OX,} D_{RED}$	diffusion coefficient of oxidant, reductant, Eq. 5-25	[cm²/sec]
$e_0 = 1.6\ 10^{-19}$	elementary charge	[C]
E_A	activation energy, Eq. 5-11	[J/mol]

f	frequency	[Hz]
f_C	critical relaxation frequency	[Hz]
$f_{C\,HF}, f_{C\,LF}$	critical relaxation for high (low) frequency process	[Hz]
f_D	critical relaxation frequency for diffusion process	[Hz]
f_{HI}, f_{LO}	cut-off for high (low) frequency process, Eqs. 6-7–6-9	[Hz]
f_{MW}	Maxwell-Wagner frequency, Eq. 7-36	[Hz]
$F = 96500$	Faradaic constant	[C/mol]
$F[\ ]$	Fourier function	[unitless]
F_{BROWN}	Brownian force, Eq. 7-48	[N]
F_{DEP}	dielectrophoretic force	[N]
F_{DRAG}	drag force, Eq. 7-30	[N]
F_{EL}	electrical force	[N]
F_G	gravitational force	[N]
F_{POL}	polarization force, Eq. 12-6	[N]
$g = 9.8$	gravitational acceleration constant, Eq. 7-39	[cm/sec^2]
G_{AC}	AC conductance, Eq. 5-8	[Sm, 1/ohm]
G_{MBR}	membrane conductance, Eq. 7-10	[Sm]
G_{OHM}	cable conductance, Eq. 5-1	[Sm, 1/ohm]
G_{SC}	particle surface conductance, Chaper 12	[Sm]
ΔG_{ADS}	adsorption Gibbs energy, Eq. 11-14	[J]
h	height (elevation)above the electrode plane, Eq. 7-38	[cm]
H	levitation height above the electrode plane, Eq. 7-39	[cm]
i_0	exchange current, Eq. 5-27	[A]
I	current	[A]
I_A	AC current amplitude	[A]
I_{CORR}	corrosion current , Eq. 12-22	[A]
I_F	Faradaic current, Eq. 7-57	[A]
I_{OUT}	output current	[A]
$I_{PARASITIC}$	parasitic current through a sample, Eq. 8-20	[A]
I_{SAMPLE}	current through a sample, Eq. 8-18	[A]
i	species type	[unitless]
$j = \sqrt{-1}$	complex number	[unitless]
J_i	flux of charged species i, Eq. 1-18	[mol/sec]
J_0	Bessel function, Eq. 8-14	[unitless]
k_0	general kinetic rate constant, Eqs. 5-25, 5-26	[1/sec, cm/sec]
k_1, k_2	general kinetic rate constants, Eq. 7-20	[1/sec]
k_{heter}, k_{CORR}	heterogeneous (corrosion) kinetic rate constant, Eq. 5-28	[cm/sec]
k_f, k_b	kinetic rate constant for forward, backward reaction	[cm/sec]
K_1, K_2	reaction rates, Eq. 7-51	[mol/(cm^{2}sec)]
$K_B = 1.38 \cdot 10^{-23}$	Boltzmann constant	[J/K]
K_F	Fourier transformation constant, Eq. 13-9	[unitless]

$K(k)$	elliptic integral, defines geometry for IDT electrodes, Eq. 8-12	[unitless]
l	length of pores, Eq. 7-65; electrodes Eq. 8-10	[cm]
l_D	diffusion length, Eq. 7-11	[cm]
L_D	diffusion layer thickness	[cm]
L_H	Helmholtz layer thickness	[cm]
L_F	film thickness	[cm]
L	inductance	[H]
L_K	kinetic reaction inductance	[H]
L_{LOAD}	parasitic inductive load, Chapter 8	[H]
$L_{OHM,}\,L_{CBL}$	cable ohmic inductance, Eq. 5-1	[H]
$L_{SORP,}\,L_{ADS}$	(ad)sorption process inductance	[H]
L_{SURF}	surface process inductance	[H]
M	molecular weight, Eq. 1-21	[g]
$M^* = 1/\varepsilon*$	complex modulus	[unitless]
$M_{REAL}(M'),M_{IM}(M'')$	real and imaginary modulus, Eqs. 2-5–2-6	[unitless]
n	number of (pores), Eq. 7-65	[unitless]
$N = 0, 1, 2..$	integer, number of IDT electrodes, Eq. 8-10	[unitless]
$N_A = 6.02 \cdot 10^{23}$	Avogadro's number	[1/mol]
N_C	concentration of cells, Eq. 11-3 – 11-4	[cells/cm^3]
N_D	"doping density", Eq. 5-21	[1/cm^3]
P	material polarization density, Eq. 1-12	[C/m^2]
$q = ze$	particle charge, Eq. 7-23, Chapter 11	[C]
q_{ADS}	adsorption charge density, Eq. 11-11	[C/cm^2]
q_{DL}	double layer charge density, Eq. 7-30	[C/cm^2]
q_0	counterion (surface) charge density, Eq. 7-15, 11-4	[1/cm^2]
q_{SURF}	surface charge density, Eq. 7-29	[C/cm^2]
Q	CPE constant, Eq. 3-1	[sec$^\alpha$/ohm]
$Q(CPE)_{ADS}$	CPE constant representing adsorption dispersion	[sec$^\alpha$/ohm]
$Q(CPE)_{BULK}$	CPE constant representing bulk relaxation dispersion	[sec$^\alpha$/ohm]
$Q(CPE)_{DL}$	CPE constant representing the double layer dispersion	[sec$^\alpha$/ohm]
$Q(CPE)_{FILM}$	CPE constant representing the adsorbed film dispersion	[sec$^\alpha$/ohm]
$Q(CPE)_{INTERFACE}$	CPE-interfacial capacitance dispersion	[sec$^\alpha$/ohm]
$Q(CPE)_L$	CPE constant representing pseudocapactive film charging	[sec$^\alpha$/ohm]
$Q(CPE)_{PORE}$	capacitance dispersion in pores, Chapter 7	[sec$^\alpha$/ohm]
r	radius of: electrode, Eq. 5-56; pores, Eq. 7-70	[cm]
$R\,(R_1,\,R_2,\,R_N...)$	resistance (of materials or circuit segments)	[ohm]
R_{ABS}	absorption resistance	[ohm]
R_{ADS}	adsorption resistance, Eq. 5-36	[ohm]
R_{AGG}	resistance of particles agglomerates, Chapter 7	[ohm]

R_{BASE}	base fluid material resistance, Chapter 10	[ohm]
R_{BINDER}	organic binder resistance, Chapter 12	[ohm]
R_{BULK}	bulk material resistance	[ohm]
$R_{CONTACT}$	contact particle-to-particle resistance, Chapter 12	[ohm]
R_{CP}	resistance of cytoplasm (interior fluid), Chapter 11	[ohm]
R_{CT}	charge transfer resistance	[ohm]
R_D	limiting diffusion resistance	[ohm]
R_{DC}	resistance at DC	[ohm]
R_e, R_{CAT}, R_{AN}	resistance of electron (cation, anion) transfer, Chapter 12	[ohm]
R_{EN}	encapsulation (proteins adsorption) resistance, Chapter 11	[ohm]
R_{EP}	easy path resistance, Chapter 7	[ohm]
R_{FILM}	film (adsorbed) resistance, Chapter 12	[ohm]
$R_{FREE\ ADD}$	free additives resistance, Chapter 10	[ohm]
R_{INPUT}	resistance of impedance measuring device	[ohm]
$R_{INTERFACE}, R_{INT}$	interfacial resistance	[ohm]
R_{IP}	inactive pits resistance, Chapter 12	[ohm]
R_K	kinetic reaction resistance	[ohm]
R_{LOAD}	parasitic resistive load, Chapter 8	[ohm]
R_M	resistance of current measuring circuit, Eq. 8-27	[ohm]
R_{MBR}	membrane ohmic resistance, Chapter 12	[ohm]
$R_{NETWORKS}$	dipolar networks resistance	[ohm]
R_{OHM}	ohmic (cable) resistance, Eq. 5-1	[ohm]
$R_{PARTICLE}$	particle resistance, Chapter 12	[ohm]
R_{POL}	polarization resistance, Eq. 12-23	[ohm]
R_{PORE}	resistance in pores, Eq. 7-65	[ohm]
R_{REACT}	reaction resistance	[ohm]
R_R	resistance of reference electrode, Eq. 8-15	[ohm]
R_S	resistance of electron transfer through bulk particle	[ohm]
R_{SOL}	solution resistance	[ohm]
R_{SOOT}	soot dispersed particles resistance	[ohm]
$R_{SOOT\ AGG}$	soot agglomerated particles resistance	[ohm]
R_W	infinite diffusion resistance	[ohm]
R_{WO}	high frequency resistance for diffusion impedance, Eq. 5-59	[ohm]
R_0	finite diffusion resistance	[ohm]
R_{1_EST}	estimated solution resistance, Eq. 2-3	[ohm]
R_W, R_R, R_C	resistances of WE, RE and CE, Eq. 8-15–8-17, 8-21	[ohm]
$R_C (R_1, R_2)$	cells radius, Chapter 11	[cm]
$R_G = 8.31$	gas constant	[J/(mol K)]

s	width of space between two adjacent IDT electrodes, Eq. 8-12	[cm]
$t, \Delta t$	time, time period	[sec]
T_t	wave period, Eqs. 8-6 – 8-7	[sec]
$T, \Delta T$	temperature, temperature change	[K, °C]
T_0	equilibrium glass transition temperature, Eq. 5-12	[K, °C]
u_i	solution electrophoretic mobility of species type i	[cm/(sec V)]
u_{EP_JC}	electrophoretic mobility induced charges, Eq. 7-27	[cm/sec]
u_0	double layer mobility of species type i	[cm/(sec V)]
v_i	velocity of species type i	[cm/sec]
v_{DEP}	velocity of species under the influence of DEP force	[cm/sec]
$v_{ELECTROTHERMAL}$	velocity of species under the influence of electrothermal force	[cm/sec]
V	voltage, applied electrochemical potential	[V]
V_A	AC voltage amplitude	[V]
V_{EQ}	equilibrium electrode potential, Eq. 5-22	[V]
V_{FB}	flat band potential, Eq. 5-21	[V]
V_g	gas space volume	[cm³]
V_{IN}	input voltage, Chapter 8	[V]
V_0^{IN}	input voltage amplitude, Eq. 8-2	[V]
V_M	molar volume	[mL/mol, cm³]
V_{MIN}	minimal required potential, Eq. 7-49	[V]
$V_{MIN\,CT\,(DIFF)}$	potential of lowest charge transfer (diffusion) resistance	[V]
$V_0 = V_{HIGH} - V_{LOW}$	measured voltage between two reference points, Eq. 8-20	[V]
V_{OUT}	output voltage, Chapter 8	[V]
V_P	particle volume, Eq. 7-19	[cm³]
V_{PEAK}	AC peak voltage amplitude	[V]
$V_{PP} = 2V_{PEAK}$	AC voltage peak-to-peak amplitude	[V]
V_{PSD}	phase sensitive detector voltage, Eq. 8-3	[V]
V_R	voltage drop across contact impedance of RE, Eq. 8-21	[V]
V^{REF}	reference voltage, Eq. 8-1	[V]
V_0^{REF}	reference voltage amplitude, Eq. 8-1	[V]
$V_{RMS} = V_{PEAK}/\sqrt{2}$	root-mean-square voltage amplitude	[V]
V_{SAMPLE}	measured voltage through a sample, Eq. 8-18	[V]
V_t	electric field strength in tangential direction, Eq. 7-23	[V/cm]
V_0	standard redox potential, Eq. 5-30	[V]
ΔV	voltage difference	[V]
w	width of IDT electrode finger, Eq. 8-12	[cm]
W	power per volume, Eq. 7-42	[W/cm³]

ΔW	metal weight loss, Eq. 12-22	[g]
X	spatial coordinate	[cm]
x	molar share of diffusing ions, independent variable	[unitless]
$y = 2.6$	association parameter, Eq. 1-21	[unitless]
z, z_i	charge of species type i (number of electrons)	[unitless]
Z	impedance	[ohm]
Z^*	complex impedance	[ohm]
$Z_A, |Z|$	impedance amplitude (module)	[ohm]
Z_{ADS}	adsorption impedance	[ohm]
Z_{ARC}	impedance of CPE|R element, Eq. 5-29	[ohm]
Z_C	impedance of capacitive circuit	[ohm]
Z_{CELL}	combined impedance of biological cell, Chapter 11	[ohm]
$Z_{CONTACT}$	contact particle-to-particle impedance, Chapter 12	[ohm]
Z_{CPE}	impedance of CPE element, Eq. 3-1	[ohm]
Z_{CP}, Z_{DC}, Z_{LC}	impedance of cytoplasm for "dead" and "live" cells, Chapter 11	[ohm]
Z_{DIFF}	general diffusion impedance	[ohm]
$Z_{DIFF\,ADS}$	diffusion impedance hindering adsorption, Chapter 10	[ohm]
Z_{EQ}	equivalent impedance	[ohm]
Z_f	general impedance at frequency f, Eqs. 6-10–6-11	[ohm]
Z_F	Faradaic impedance, Eq. 5-45	[ohm]
Z_{FILM}	film impedance	[ohm]
Z_{IM}	imaginary impedance	[ohm]
$Z_{INTERFACE}$	interfacial impedance	[ohm]
Z_M	impedance of media	[ohm]
Z_{MBR}	membrane impedance, Chapter 11	[ohm]
$Z_{MEASURED}$	measured impedance, Eq. 8-18	[ohm]
$Z_{M/F}, Z_{F/S}$	impedance of metal-film, film-solution interface, Chaper 12	[ohm]
Z_P	impedance of particles	[ohm]
Z_R	impedance of resistive circuit, contact of RE, Eq. 8-21–8-22	[ohm]
Z_{REAL}	real impedance	[ohm]
Z_{REDOX}, Z_{REACT}	impedance of redox reaction	[ohm]
Z_S	Schwarz (counterion) diffusion impedance	[ohm]
Z_{SOL}	solution impedance	[ohm]
Z_{SORP}	sorption impedance	[ohm]
Z_W	Warburg (infinite) diffusion impedance	[ohm]
Z_0, Z_{0C}, Z_{0A}	finite diffusion impedance (on cathode, anode)	[ohm]
$Z_{0\_REFLECTING}$	finite diffusion impedance, reflecting boundary, Eq. 5-47	[ohm]
$Z_{0\_TRANSMITTING}$	finite diffusion impedance, transmitting boundary, Eq. 5-47	[ohm]

α_A	particles shape factor, Eq. 7-6	[unitless]
$\alpha, \beta = 0 \le 1$	shape parameters of relaxation peak or CPE, Eq. 1-16	[unitless]
$\alpha, \beta = 0 \le 1$	Tafel reaction order (transfer) coefficients, Eq. 5-22	[unitless]
$\alpha, \beta, \gamma, \delta$	potentiostat and analyzed system constants, Eq. 8-27	[unitless]
$[\beta] = 0 \le 1$	Clausius-Mossotti factor, Eq. 7-1	[unitless]
$[\beta_R] = 0 \le 1$	real portion of Clausius-Mossotti factor, Eq. 7-1, 7-34	[unitless]
χ	share of voltage drop across the double layer, Eq. 7-28	[unitless]
χ^2	least square fit parameter, Eq. 8-28	[unitless]
δ	distance between counterions, wall thickness, Eq. 7-12	[cm]
$(1-\delta)$	fractional material volume change, Eq. 5-13	[unitless]
δ_A	particles agglomeration factor, Eq. 7-6	[unitless]
$\varepsilon^*, \varepsilon_M{}^*, \varepsilon_P{}^*$	complex permittivity of material (media, particles)	[unitless]
ε	relative permittivity of material (dielectric constant)	[unitless]
ε'	real permittivity	[unitless]
ε''	imaginary permittivity of material (loss factor)	[unitless]
$\varepsilon_{CP}, \varepsilon_{DC}, \varepsilon_{LC}$	permittivity of cytoplasm, "dead" and "live" cells, Chapter 11	[unitless]
ε^*_{CP}	equivalent permittivity of cellular cytoplasm, Chapter 11	[unitless]
$\varepsilon_\infty, \varepsilon_{HF}$	high frequency complex permittivity of material	[unitless]
ε_{LF}	low frequency complex permittivity of material	[unitless]
ε_M	relative permittivity of media (dielectric constant)	[unitless]
ε_{MBR}	relative permittivity of cellular membrane, Chapter 11	[unitless]
ε_P	relative permittivity of particles (dielectric constant)	[unitless]
$\Delta\varepsilon = \varepsilon_{LF} - \varepsilon_{HF}$	permittivity increment	[unitless]
$\Delta\varepsilon_\alpha, \Delta\varepsilon_\beta$	permittivity increment, α-and β-relaxations, Eq. 7-10, 7-15	[unitless]
$\Delta\varepsilon_{RX}$	permittivity increment, chemical reaction, Eq. 7-20	[unitless]
$\varepsilon_0 = 8.85 \cdot 10^{-14}$	constant permittivity of a vacuum	[F/cm]
ϕ	phase angle (shift)	[degrees, °]
ϕ_D	potential at Helmholtz plane, Eq. 5-18	[V]
ϕ_{WALL}	potential at the electrode (wall) plane, Eq. 7-29	[V]
$\Delta\phi$	electrochemical potential difference	[V]
$\Phi = 0 \le 1$	fraction of particles, Eq. 7-4	[unitless]
γ_{ADS}	adsorption coefficient, Eq. 11-13	[unitless]

Γ	amount of adsorbed species, Eqs. 5-37 – 5-38	[mol/cm^2]
Γ_{MAX}	maximum amount of adsorbed species, Eqs. 7-54	[mol/cm^2]
η	solvent (material) viscosity	[Pa sec, cP, kg/(cm sec)]
κ	thermal conductivity, Eq. 7-43	[J/(cm sec K)]
λ_{DEBYE}	Debye length	[cm]
λ_P	penetration depth, Eq. 7-70	[cm]
ν	flow rate, Eq. 1-19	[cm^3/sec]
ν_{EO}, ν_{EP}	electroosmotic, electrophoretic velocity, Eq. 7-31 – 7-32	[cm/sec]
$\nu_{EP\_ION\,(PARTICLE)}$	electrophoretic velocity of ions (particles), Eq. 7-23 – 7-24	[cm/sec]
ν_{EP_IC}	electrophoretic velocity for induced charges, Eq. 7-27	[cm/sec]
$\pi = 3.14$	geometrical constant	[unitless]
ϖ	geometry factor, Eq. 7-20	[unitless]
$\theta = 0 \leq 1$	surface coverage, Eq. 7-52	[unitless]
$\theta_{EQ} = 0 \leq 1$	equilibrium surface coverage, Eq. 7-55	[unitless]
ρ	resistivity of material	[ohm cm]
ρ_D	material density, Eq. 5-14	[g/cm^3]
σ	conductivity of material	[Sm/cm]
$\sigma_{CP}(\sigma_C), \sigma_{DC}, \sigma_{LC}$	conductivity of cytoplasm, "dead" and "live" cells, Chapter 11	[Sm/cm]
σ^*_{CP}	equivalent conductivity of cellular cytoplasm, Chapter 11	[unitless]
σ_D	Warburg coefficient, Eq. 5-40	[ohm/sec$^{0.5}$]
σ_M, σ_{SOL}	conductivity of media (solution)	[Sm/cm]
σ_{MBR}	conductivity of cellular membrane, Chapter 11	[unitless]
σ_P	conductivity of particles	[Sm/cm]
σ_{PB}	bulk conductivity of particles, Eq. 7-37	[Sm/cm]
σ_{PS}	surface conductivity of particles, Eq. 7-37	[Sm/cm]
τ	time constant	[sec]
$\tau_C = 1/\omega_C$	critical relaxation time	[sec]
$\tau_1, \tau_2, \tau_\alpha, \tau_{RX}$	time constant for processes 1, 2, α, reaction, etc	[sec]
τ_{HF}, τ_{LF}	time constant for high (low) frequency process	[sec]
$\omega = 2\pi f$	radial frequency	[radian]
$\omega_C = 2\pi f_C$	radial critical relaxation frequency	[radian]
$\omega_D = 2\pi f_D$	radial critical relaxation frequency for diffusion process	[radian]
ω_K	critical relaxation frequency for homogeneous reaction	[radian]
$\omega_{MAX} = 2\pi f_{MAX}$	radial frequency where Z_{IM} is maximum value, Eq. 3-2	[radian]
$\omega_{RC} = 2\pi f_{RC}$	radial critical relaxation frequency for charge transfer	[radian]
ω_{ROT}	rotation rate of rotating disc electrode, Eq. 5-53	[1/sec]
ξ	current efficiency of film formation, Eq. 12-12	[unitless]

ψ_{BIAS}	bias error, Eq. 8-26	[unitless]
ψ_{FIT}	fitting error, Eq. 8-26	[unitless]
ψ_{STOCH}	stochastic error, Eq. 8-26	[unitless]
ζ	zeta-potential	[V]

Index

Printed and bound by CPI Group (UK) Ltd, Croydon, CR0 4YY
09/11/2021

03091045-0001